# Atoms, Bytes and Genes

"Atom", "byte" and "gene" are metonymies for techno-scientific developments of the 20th century: nuclear power, computing and genetic engineering. Resistance continues to challenge these developments in public opinion. This book traces the debates over atoms, bytes and genes which raised controversy with consequences, and argues that public opinion is a major factor of the development of modern techno-science. The level and scope of public controversy is an index of resistance, examined here with a "pain analogy" which shows that just as pain affects movement, resistance affects techno-scientific mobilization: it signals that something is wrong, and this requires attention, elaboration and a response to the challenge. This analysis shows how different fields of enquiry deal with the resistance of mentalities in the face of industrial, scientific and political activities inspired by projected technological futures.

**Martin W. Bauer** is Professor of Social Psychology at London School of Economics and Political Sciences and the Editor of *Public Understanding of Science*.

# Routledge Advances in Sociology

*For a full list of titles in this series please visit www.routledge.com.*

93 **Heritage in the Digital Era**
Cinematic Tourism and the
Activist Cause
*Rodanthi Tzanelli*

94 **Generation, Discourse, and
Social Change**
*Karen R. Foster*

95 **Sustainable Practices**
Social Theory and Climate Change
*Elizabeth Shove and
Nicola Spurling*

96 **The Transformative Capacity
of New Technologies**
A Theory of Sociotechnical Change
*Ulrich Dolata*

97 **Consuming Families**
Buying, Making, Producing
Family Life in the 21st Century
*Jo Lindsay and JaneMaree Maher*

98 **Migrant Marginality**
A Transnational Perspective
*Edited by Philip Kretsedemas,
Jorge Capetillo-Ponce and
Glenn Jacobs*

99 **Changing Gay Male Identities**
*Andrew Cooper*

100 **Perspectives on Genetic
Discrimination**
*Thomas Lemke*

101 **Social Sustainability**
A Multilevel Approach to
Social Inclusion
*Edited by Veronica Dujon, Jesse
Dillard, and Eileen M. Brennan*

102 **Capitalism**
A Companion to Marx's
Economy Critique
*Johan Fornäs*

103 **Understanding European
Movements**
New Social Movements,
Global Justice Struggles,
Anti-Austerity Protest
*Edited by Cristina Flesher
Fominaya and Laurence Cox*

104 **Applying Ibn Khaldūn**
The Recovery of a Lost Tradition
in Sociology
*Syed Farid Alatas*

105 **Children in Crisis**
Ethnographic Studies in
International Contexts
*Edited by Manata Hashemi and
Martín Sánchez-Jankowski*

106 **The Digital Divide**
The Internet and Social Inequality
in International Perspective
*Edited by Massimo Ragnedda
and Glenn W. Muschert*

107 **Emotion and Social Structures**
The Affective Foundations of
Social Order
*Christian von Scheve*

108 **Social Capital and Its
Institutional Contingency**
A Study of the United States,
China and Taiwan
*Edited by Nan Lin, Yang-chih Fu
and Chih-jou Jay Chen*

109 **The Longings and Limits of Global Citizenship Education**
The Moral Pedagogy of Schooling in a Cosmopolitan Age
*Jeffrey S. Dill*

110 **Irish Insanity 1800–2000**
*Damien Brennan*

111 **Cities of Culture**
A Global Perspective
*Deborah Stevenson*

112 **Racism, Governance, and Public Policy**
Beyond Human Rights
*Katy Sian, Ian Law and S. Sayyid*

113 **Understanding Aging and Diversity**
Theories and Concepts
*Patricia Kolb*

114 **Hybrid Media Culture**
Sensing Place in a World of Flows
*Edited by Simon Lindgren*

115 **Centers and Peripheries in Knowledge Production**
*Leandro Rodriguez Medina*

116 **Revisiting Institutionalism in Sociology**
Putting the "Institution" Back in Institutional Analysis
*Seth Abrutyn*

117 **National Policy-Making**
Domestication of Global Trends
*Pertti Alasuutari and Ali Qadir*

118 **The Meanings of Europe**
Changes and Exchanges of a Contested Concept
*Edited by Claudia Wiesner and Meike Schmidt-Gleim*

119 **Between Islam and the American Dream**
An Immigrant Muslim Community in Post-9/11 America
*Yuting Wang*

120 **Call Centers and the Global Division of Labor**
A Political Economy of Post-Industrial Employment and Union Organizing
*Andrew J.R. Stevens*

121 **Academic Capitalism**
Universities in the Global Struggle for Excellence
*Richard Münch*

122 **Deconstructing Flexicurity and Developing Alternative Approaches**
Towards New Concepts and Approaches for Employment and Social Policy
*Edited by Maarten Keune and Amparo Serrano*

123 **From Corporate to Social Media**
Critical Perspectives on Corporate Social Responsibility in Media and Communication Industries
*Marisol Sandoval*

124 **Vision and Society**
Towards a Sociology and Anthropology from Art
*John Clammer*

125 **The Rise of Critical Animal Studies**
From the Margins to the Centre
*Nik Taylor and Richard Twine*

126 **Atoms, Bytes and Genes**
Public Resistance and Techno-Scientific Responses
*Martin W. Bauer*

# Atoms, Bytes and Genes
Public Resistance and
Techno-Scientific Responses

Martin W. Bauer

NEW YORK AND LONDON

First published 2015
by Routledge
711 Third Avenue, New York, NY 10017

and by Routledge
2 Park Square, Milton Park, Abingdon, Oxfordshire OX14 4RN

First issued in paperback 2016

*Routledge is an imprint of the Taylor & Francis Group,
an informa business*

© 2015 Taylor & Francis

The right of Martin W. Bauer to be identified as author of this work has been asserted in accordance with sections 77 and 78 of the Copyright, Designs and Patents Act 1988.

All rights reserved. No part of this book may be reprinted or reproduced or utilised in any form or by any electronic, mechanical, or other means, now known or hereafter invented, including photocopying and recording, or in any information storage or retrieval system, without permission in writing from the publishers.

**Trademark Notice:** Product or corporate names may be trademarks or registered trademarks, and are used only for identification and explanation without intent to infringe.

*Library of Congress Cataloging-in-Publication Data*
A catalog record has been requested for this book.

ISBN 13: 978-0-415-79353-7 (pbk)
ISBN 13: 978-0-415-95803-5 (hbk)

Typeset in Sabon
by IBT Global.

To Sandra and Ana, the two loves of
my life who resist my machinations

# Contents

|   |   |   |
|---|---|---|
| | *List of Figures* | xi |
| | *List of Tables* | xiii |
| | *List of Excursions* | xv |
| | *Foreword* | xvii |
| | *Acknowledgments* | xxi |
| 1 | Introduction: Movement Redirected by Resistance | 1 |
| 2 | Mobilising a Different Future | 9 |
| 3 | The Atom: Bombs and Power | 30 |
| 4 | Environment, Safety and Sustainability | 69 |
| 5 | Ten Propositions on Learning from Resistance | 90 |
| 6 | The "Bytes" of Mainframes, PC and Social Media | 114 |
| 7 | Public Opinion and Its Discontents | 148 |
| 8 | Genes, Biotechnology and Genomics | 175 |
| 9 | Some Further Observations on Resistance | 212 |
| | *Appendix 1: Notes on Social Movement and Social Influence* | 233 |
| | *Appendix 2: Chronologies of Atoms, Bytes and Genes* | 245 |
| | *References* | 251 |
| | *Index* | 283 |

# Figures

| | | |
|---|---|---|
| 2.1 | The spiral of directing projects by steering and regulating. | 15 |
| 3.1 | Atomic bombs and nuclear power stations in the world, 1945–2011. | 31 |
| 3.2 | The number of operational nuclear power stations and the electricity generated in percent of the national total in 30 countries. | 34 |
| 3.3 | The index of nuclear news in the UK Press 1946-2012. | 43 |
| 3.4 | European and US sentiment on nuclear power with fitting trends. | 46 |
| 3.5 | The Chernobyl and Fukushima effect on public opinion in comparison. | 48 |
| 3.6 | Declining projections of nuclear electricity. | 57 |
| 3.7 | The degree of realization of nuclear projects in additional percentage to plan. | 59 |
| 4.1 | Environmental news references in the UK press 1946-2010. | 83 |
| 4.2 | US opinion polls on whether people are aware (yes or no) of the 'greenhouse effect' and 'global warming' between 1986 and 2008. | 84 |
| 4.3 | Peak of the 3rd cycles of environmental news between 1988 and 1992, the percentage of respondents for whom the issue is important, and percentage of respondents reporting five or more behaviours supporting sustainability in Britain. | 85 |
| 6.1 | Numbers of computer in use worldwide. | 116 |
| 6.2 | News intensity and slant on 'computers' in the British press, 1946–2004. | 119 |
| 6.3 | Computer means job killer, deskilling, stressor or more interesting work in UK and Germany. | 132 |
| 6.4 | Households with Internet access in the US and the UK, UK press coverage of computing and Internet, and NASDAQ composite stock index 1990–2007. | 137 |
| 7.1 | German "Technikfeindlichkeit" (compared to USA). | 153 |

| | | |
|---|---|---|
| 7.2 | Knowledge, interest and attitudes in EU-12 countries. | 157 |
| 7.3 | Science news in the UK press. | 160 |
| 7.4 | Relationship between knowledge of science and scientific "ideology." | 163 |
| 8.1 | News coverage of biotechnology in Europe, US and Japan. | 186 |
| 8.2 | Number of references in the British national press to biotechnology, 1980–2006 and to agricultural biotechnology as percentage of the total coverage. | 188 |
| 8.3 | Framing of biotechnology before and after the watershed years. | 193 |
| 8.4 | Evaluation of different applications of biotechnology grouped into two clusters: RED and GREEN. | 194 |
| 8.5 | The dynamics of European optimism over biotechnology from 1991 until 2005. | 196 |
| 8.6 | Diffusion of GM soya among the three most important growers and globally. | 203 |
| 9.1 | Resistance as counter-force to change and its factors. | 217 |
| 9.2 | Analogical reasoning from pain to resistance. | 223 |
| 9.3 | The characteristic time pattern of the pain experience. | 229 |
| 9.4 | The functional perspective of resistance. | 230 |

# Tables

| | | |
|---|---|---|
| 1.1 | Innovating the Analysis of Resistance Through Controversies | 6 |
| 2.1 | The Different Operating Principles of the Three Arenas | 18 |
| 5.1 | A Typology of Collective Learning from Resistance | 106 |
| 7.1 | Periods, Problems and Propositions on the Public | 156 |
| 9.1 | Levels of Analysis and the Functional Analogy between Them | 224 |
| A2.1 | Chronology of the Dual-Path of the Atom | 245 |
| A2.2 | Key Phases of Anti-Nuclear Protest: Civil and Military | 246 |
| A2.3 | Major Nuclear Power Incidents and Accidents | 246 |
| A2.4 | The Computer's Long Past and Short History | 246 |
| A2.5 | Dates in the History of Genetic Engineering and Biotechnology | 248 |

# Excursions

| | | |
|---|---|---:|
| 2.1 | Techno-scientific meta-narratives of historical direction (telos) | 23 |
| 3.1 | Once bad luck, twice suspicious, three times a system failure | 49 |
| 4.1 | The "Waldsterben" and the Shetland spill | 74 |
| 4.2 | The risks of professional activism | 78 |
| 4.3 | Consumers and product safety | 87 |
| 6.1 | A critical mass of one "Unabomber" | 130 |
| 6.2 | Computer addiction and phobias | 141 |
| 6.3 | Altering the decision making criteria | 143 |
| 6.4 | Resistance and the split in the project team | 143 |
| 7.1 | German "Technikfeindlickeit" and the uses of opinion polls | 153 |
| 8.1 | Naming the new development | 176 |
| 8.2 | Monsanto's Life Science and the North Atlantic GeneRush | 179 |
| 8.3 | Key public debates on genetic engineering | 184 |
| 8.4 | Biotechnology democracy in Switzerland | 197 |
| 8.5 | The Brazilian soya miracle 1996 to 2005 | 203 |
| 9.1 | Folk wisdom about pain and change | 222 |

# Foreword

In his timely and subtly argued book, Martin Bauer develops a number of considerations that should make us less pessimistic about our ability to halt the juggernauts of projects and programs of which only some are set up for our own good. What is resistance and how would one make one's resistance effectual? Of course, one must bear in mind that just as one person's freedom fighter is another's terrorist, so one person's honestly felt and passionately pursued resistance might be thought by another as a stubborn refusal to face the facts, a mindless attempt to hang on to the old and rusty ways.

The core of this book is the profoundly significant insight, that what needs to be explained in the public reception of the goods and programs of techno-science is not resistance movements as such, but the many cases where no resistance appeared. How has it come about that some of the eminently resistible techno-science projects have not encountered resistance? This is a many layered enigma. Resistance to electricity has long since disappeared (my grandmother shared Thurber's grandma's worry that the stuff might leak out of a socket without a plug). Ought not there to be a notable absence of resistance to the establishment of the means for generating this useful commodity?

By treating the development of societies as human associations in which the means of life are the products of techno-science, Martin Bauer asks the social psychological question: how does it come about that people become part of techno-civilisation? In my own case, I went into my first career as a chemical engineer, drawn by the magic and romance of the vast plants and subtle chemistry needed to produce chemicals on an industrial scale—O the wonder of the Harber process, of the Bessemer converter, of the production of nylon! H. G. Wells' masterly evocation of the romance of great industrial undertakings, *Men Who Would be Gods*, captures this mood wonderfully well. I think there must a great many paths to, and forms of resistance to the progress of techno-science projects, and perhaps each project brings forth its own band of resisters, though I suppose we may find ourselves recognising some old friends from rent-a-crowd dramas. Much resistance is turned aside—farmers do use artificial fertilisers and they will surely grow GM crops. Vaccination is almost universally unresisted, and so is blood transfusion. Cars stream across the landscape. Is the accommodation to and assimilation of techno-science best seen as building a new kind of community rather than just a new version of traditional human associations? These questions are particularly worth asking when we reflect on the truth of the adage—if someone has invented it, it will be used.

Looking at the case of nuclear engineering, whether for military or peaceful purposes, we can see two powerful streams of public opinion converging and diverging from a common shallow understanding of the physics involved—one was ban the bomb and reactors are dangerous while other was hurrah for the balance of terror and benefits of cheap power. Isn't it curious that in the debate about inviting French and Chinese companies to build the next generation of nuclear power stations in the UK, Chernobyl and Fukushima have scarcely been mentioned!?

Techno-science must be seen against the background of a new romanticism. The special contemporary concepts of sustainability and security only make sense in the light of an idea of Nature promoted by the huge coverage of the lives of plants and animals as they are lived, more or less in the absence of any human beings except David Attenborough and his cameraman. The interplay between nature and material production is odd—why sustainability? It cannot be in the interests of bats, moths or whales. It surely highlights the possibility of a means for the production of what is needed for human life in such a way that the sources of the required materials are not exhausted. But why not burn up all the fossil fuel, warm the earth and move on to electric cars? Why make the matter of clean water and healthy food a governmental responsibility—because that is surely what recruiting techno-science to food production, and air and water quality has done. There are deep moral issues here in the stripping away from the concept of 'person' its essential moral core.

These brief reflections show how important Martin Bauer's project is. If techno-science is interpreted as a social movement, its growth and the resistances to that growth, and the means it uses to overcome them by seeking a weaker point in the wall of traditional ways of life that surround it, then it must be studied not as a matter of economics nor as a matter of the growth of science, but as social, through and through. So there arises the key question: how does techno-science learn from the resistances it encounters how to overcome them, by-pass or ignore them and proceed on its triumphant way to the sort of world H. G. Wells and I once dreamed of? There may be public resistance to fracking, nuclear power (though pretty muted; why did it fade away?) but we will extract the last molecule of hydrocarbons from the earth, and that you may be sure of. The internet and the social networks might have proven to have been a political force but they have dissolved into exchanges of trivia or into arenas for libel, defamation and the courts.

A test case for the idea that techno-science is both a moving force and a site for resistance is genetic engineering—with the story of Dr Frankenstein and his monster right to hand. Offers to clone a favourite dog are lately in the media. The flurry of destruction of experimental plots of GM crops has died away. Fear of Frankenstein's monster has slowly given way to support for a kind of scientific Oxfam, the moral imperative to remove the threat of starvation. Here is a complex web of discourses, social influences, personal ambitions and irrationalities juxtaposed against the powerful rationalities of science that will doubtless yield some profound insights.

We must surely welcome Martin Bauer's project and hope that it will serve to initiate an expanding series of studies within this framework.

Rom Harré<br>
Linacre College, Oxford

# Acknowledgments

This book was delayed—the publisher's deadline reads October 2007—by the realities of current academic life, even at the LSE. Being Head of Department and the ever growing grind of 'curriculum management' is no longer compatible with mental projects such as book writing. But finally, it is done and fully updated, and I am delighted to present it to the readers.

Any book has help, and I am grateful to the colleagues at LSE Social Psychology and to my students on *LSE PS439 Science, Technology and Resistance*, who over many years appreciated this argument critically and incredulously. This argument also went into seminar papers presented in various places as far as Beijing (China), Buenos Aires (Argentina), Porto Alegre, Campinas and Rio de Janeiro (Brazil), Cornell (US), Jyvaskyla (Finland), Bielefeld (Germany), Udine and Rome (Italy), Athens (Greece), York, Manchester and Cambridge (UK).

I am infinitely indebted to my London 'commillitants' on the trails of the public understanding of science. Steve Miller, Simon Lock and in particular Jane Gregory kept going the London Inter-collegiate PUS seminars as a forum of discussion, and Jane patiently read versions of this manuscript over the years starting with the proposal to the publishers. Dorothy Nelkin and Sheila Jasanoff encouraged me to develop the argument many years ago. Dorothy called whenever she was in London, which remains a fond memory of this Grand Dame of science studies.

With John Durant, now at MIT, and my colleague and friend George Gaskell (LSE) I enjoyed researching the biotechnology controversies of the 1990s. The European project 'Biotechnology and the Public' (1994–2002) with colleagues in France, Sweden, Norway, Austria, Italy, Greece, Germany, Switzerland, Poland, Bulgaria, Japan and US, names are too many to list here, became a key testing ground for my argument. And George periodically reminded me: and where is the book?

Thanks belong to Sue Howard, whose editing kept me away from major barbarities in my use of English language; also to two Routledge editors who mobilised patience for this project, first Ben Holtzman and later Max Novick in New York. We never managed to meet in person, but exchanged only friendly e-mails.

# 1 Introduction
## Movement Redirected by Resistance

In this book, the words 'atom', 'byte' and 'gene' stand as metonyms for three global techno-scientific developments of the second part of the 20th century, namely nuclear power, computing and genetic engineering. Atoms, bytes and genes ground these developments. Atoms are the units of matter the understanding of which defines modern physics; bytes are a series of 8 bits and as such the operational unit of modern computing, and genes are fragments of a chromosome that regulate the development of life forms. Resistance stands for the challenges these developments came to face in public opinion over the years. And 'techno-science' stands for the efforts at the frontiers of uncertainty and knowledge where science and engineering are not easily distinguished. No nuclear physics without large machines, no brain nor climate science without large and small computers, no modern genetics without sequencing robots. What is technology, where is the science? Modern science is a cyborg, a man-machine-practice that incorporated machines to develop them further.

This was a difficult book to write, because all the debates are old, and still ongoing, indeed fast moving on all fronts. There is as yet little opportunity for historical distance and quiet stock-taking, not even for the atom. Writing about atoms, bytes and genes is a constant struggle against unexpected events (e.g. Fukushima, new research, new mobilisation), new issues (terrorism, enhancement, fracking) and streams of specialised social science literature. The scope of this book is impossible. The only way out of a dilemma between saying everything and nothing is *abstraction and selection*. I put forward a particular view of resistance and how it impacted these developments with a reading of the literature to develop and illustrate the framework. Space is limited and so is the patience of the reader.

But, this text needed to be written, close to an obsession, several times delayed by distractive academic duties. But encouragement from friends and foes suggested that its time has come. For many years I taught a post-graduate course at LSE with an annual group of 10–15 students, PS439 Science, Technology and Resistance, which kept this flame burning. But, obsessions inevitably hark back to biography. Being born in Switzerland in the late 1950s made me a witness to secular changes: the modernist common sense which equated science, technology and Progress, writ large, gave way to a new taken-for-granted mentality where this equation has boundary conditions: 'it depends' comes very naturally these days. As a boy I was glued to the television, which my parents rented to assist the moon excursions in the late 1960s, Neil Armstrong's 'that's one small step for [a] man, one giant leap for mankind' resonated with youthful dreams of being an astronaut. The anti-nuclear mobilisations of the 1970s alerted me as a bystander to issues of ecology and politics, while at university the computer agitated my sensibilities. As a student of psychology and economic history, I studied the arrival of digital machinery at work; in the early 1980s word processing was coming on stream, the 'PC revolution' was declared, and Orwell's premonitions of '1984' were looming large. My first paid job was to facilitate office automation at NCR in pursuit of the paperless office. However, neither did paper go away nor did computer acceptance go smoothly; this became a personal wake-up call to think again.

I consider tracing these long-term changes in common sense an important call for a social psychology that understands itself as the history of the present (Gergen, 1983). Much of this new common sense can be traced to the old 'enemies of progress' (Sieferle, 1984; 1986). Public opinion over techno-science started to shift in the 1970s, first in industrialised countries in Europe and North America, but the controversies gradually went global. Atoms, bytes and genes have remained controversial, but not equally so, and this is a puzzle.

Four threads of ideas are woven into these pages: the reassertion of public opinion over science and technology, a positive reading of resistance, the perennial struggles of the sciences with the recalcitrance of the world, and the prospects for a general social science 'resistology':

- Public opinion has reasserted itself as a factor in the development of techno-science. I will trace these historical changes through the debates over atoms, bytes and genes, all of which have at times become publicly controversial with interesting consequences.
- The level and scope of public controversy is an index of public resistance. The functionality of this resistance is elaborated on an analogy with pain. This framework allows us to see what pain does for movement, resistance does for social mobilisation: it signals that something is wrong and this demands urgent attention, elaboration and response. Pain prompts movement.
- The social sciences have innovated their dealing with resistance. This book shows many new terms and concepts that were invented for a perennial problem: how to grasp, analyse and handle the 'recalcitrance' of public opinion and common sense. I offer a panorama of how 'resistance' has changed while staying the same. The constant is 'resistance', while the discourse moves on.
- I will offer sketches on a potential new field of enquiry: a general 'resistology' across the social sciences. Such a field of enquiry might analyse how we deal with the ubiquitous recalcitrance of mentalities in the face of industrial, scientific and political projections of the future.

This book is also the story of a looming discontent with my own discipline. Social Psychology is deeply invested with visions of a technologically dominated society. Gunter Anders was a visionary when he called for a 'social psychology of things' back in the 1950s, because only the 'thing' orientation would be able to liberate social psychologists from the procrustean presumptions of adapting humans to the technological fait accompli (Anders, 2002). The 'new man' of Strategic Bomber Command (see Chapter 3), without emotions and social commitments, is a dysfunctional model of human conduct, which nevertheless has, as Charles Taylor (2007) has elaborated, a long history in Western mentality: it is an adaptation to the demands of an iron cage. This bodily and socially shielded individual cannot be the model human of psychology in the 21$^{st}$ century.

## A POSITIVE STORY LINE FOR RESISTANCE

In a world dominated by claims to scientific and technological revolutions, public resistance gets bad press and is generally looked upon as a lost cause. My book fundamentally challenges this view, both conceptually and empirically. The main thesis of the book could be stated as follows:

> It is not the 'resistance' that needs an explanation but its absence; and if there is resistance, how we rise to the challenge, is the key question.

The absence of resistance has costs. Opportunity costs arise from not being able to capitalise on resistance to reach a sustainable future. We must also ask the counter-factual question: had the USSR seen public resistance to nuclear power, the Chernobyl disaster (April, 1986) might have been averted; and maybe the same can be said of Fukushima in Japan (March, 2011). The analysis of resistance in this book offers an explicitly functional view grounded in a social psychological framework and in sociological imagination. During my undergraduate studies, I came across a then already dated paper that caught my attention. Psychologist PR Lawrence put it like this:

> When resistance does appear, it should not be thought of as something to be overcome. Instead, it can best be thought of as . . . a signal that something is going wrong. . . . . Signs of *resistance are useful in the same way that pain* is useful for the body. (Lawrence, 1954, 56; my emphasis)

This simile of resistance as pain captured my imagination. I combed through the writings of Lawrence, but found nothing more on what must have been a one-time rhetorical throw-away, though the paper had become a Harvard Business Review Classic in 1969. This metaphor deserved closer attention, but how to go about it? To elaborate a metaphor into an analogy, one needs a framework—tertium comparationis—in the light of which the source domain (pain) and the target domain (resistance) become more similar than they would otherwise appear. Two ideas were at hand to elaborate the pain-resistance analogy: the concept of 'self-monitoring' and the logic of functional comparison. In the early 1980s, I studied in Bern, where a social psychology was elaborated that was equally distant from behaviourism and cognitivism, and that fused concepts of concrete activity (Vygotzky, Rubinstein), pragmatism (James, Dewey) and animal behaviour (Tinbergen, Lorenz) into a social psychology of action, of action that was intentionally and socially constrained (see Harré and Cranach, 1982). We students were inspired by the quest for the 'right way to conduct psychology' (so the title of a Festschrift) but wrote essays on humbler matters.

First, the notion of *self-monitoring* designates a functional similarity between consciousness, emotion and pain in the guidance of activities. All these modalities of the spontaneous 'I' and the social 'me' evaluate what is important for my (our) life, yielding intentions of what to do and what to leave. In short, conscious experiences of cognition, emotion and pain are no mere embellishment of brain processes (i.e. epiphenomenalism). To the contrary, they are real by their consequences. Like any worm, humans need 'nervous' tissue and brains to move, but in addition, humans command self-monitoring and conscious experiences for good or bad.

Second, the notion of *functional analogy* designates the observation that two processes have similar functions demonstrated by comparison: the same outcomes are achieved through very different processes and structures. Hence, pain and resistance are comparable in their functions of self-monitoring, while being structurally very different. People do not usually resist new technology *because* they are in pain, though that might be a special case. A functional analogy is illustrated by the wings of an Airbus380 and the wings of a swan. What is structurally quite different, formed metal on the Airbus, stacked feathers on the swan, is functionally similar: wings and wind allow heavy bodies to fly through the sky. Varying Sartre's famous 'l'enfer c'est l'autre' (hell is other people), I am inclined to claim that *others are the pain of our project*. The key question in this context is not 'why others are a pain in the neck', there might be many good and bad reasons, but the key question is how one rises to the challenge they present.

The sociological imagination has caught up with the topic of resistance in the new millennium. Peter Sloterdijk (2005) offers a long essay on what he calls the end of the 'age of dis-inhibition and of nautical ecstasy' with a new take on resistance. Unashamedly grand narrative, he reviews the period from 1492 to 1943 (from Columbus reaching the West Indies to the battle of Stalingrad, and the gold standard of Bretton Woods). This 'longue durée' is characterised by a mentality of projecting a distant future and obsessing with postponed gratification justified by utopian hopes. The essay is a rhapsodic variation on the theme of nautical risk taking at a dangerous sea, charging into the unknown with a feeling of superiority and entitlement. The age of exploration and expansion brings hardship and requires endurance, which are only bearable under a collective psychosis that imagines what is to come: the images of El Dorado keep the sailors at the ropes. These projects of expansion and exploration have no regard for anything except their own projections. Any resistance is dismissed as pointless in the light of a necessary future. This age of 'dis-inhibition' and 'auto-persuasion' destroys and maims in the name of *Progress*. The spirit of these men of action, liberated from all moral constraints, tread a fine line between madness, crime and innovation. A 'philosophy of captains' enacts the 'nautical ecstasy' and mobilises the 'mass psychosis of utopian expectations' to immunise themselves against any awareness of side effects and negative consequences. 'They sail into the world like bullets into battle . . . projecting themselves in full flight . . . living a projectile existentialism' (In Heideggerian language 'sich entwerfend im Wurf'; projecting oneself in projection; my paraphrasings). In such rhapsodic terms, Sloderdijk characterises a mentality which finally comes to an end; resistance regains its function in public affairs and the 'age of inhibition' returns after a long period of suspension.

Others before have observed that Prometheus is shackled again, and claims to modesty and restraint proliferate with widely enhanced technological capabilities (Solomon, 1984). The ethos of pursuing a blind course of action on the basis of imagined prospects is tamed by negative feedback: collateral damage and unexpected consequences come more immediately in sight. This gives way to at least a thought of inhibition: not everything that is possible also needs to be done. (However, compare Blumenberg's soliloquium of 1997: '*But in the end we will do it, because we need to proof ourselves*'; my free translation). If the sociological imagination captures the zeitgeist, then resistance might indeed get better press than being the enemy of progress. Resistance as the scapegoat is based on strange assumptions, namely that the new is necessary and always superior than the old. Under these assumptions, resistance explains the rejection of the necessary and better in derogative terms of conservatism, stupidity, ignorance, luddism, false consciousness or risk aversion. But resistance is logically and empirically a corollary of exerting choice. If I have choice I must resist any imposition, and resistance demands choice where it is denied. And where there is choice, resistance is required to hold off the rejected options, be it the old technology that is replaced, or the alternative new technologies deemed to be worse (see Edgerton, 2008, 9ff).

Table 1.1 lists examples of conceptual innovations on resistance in each wave of techno-science. Beneath a flurry of new terminology, the scandal of public resistance persists. The fundamental problem of resistance to change persists, though new terms and concepts replace old ones. Social sciences in a conservative fold might look sceptically upon social engineering and are more likely to see resistance in a favourable light. The discourse of reaction (e.g. Hirschman, 1991) celebrates resistance by pointing out the perversity, futility and jeopardy of ill-conceived 'reform' efforts. On the other hand, the progressive modes

*Table 1.1* Innovating the Analysis of Resistance through Controversies

| Technology | Conceptual innovations on 'resistance' |
|---|---|
| Nuclear power | • Risk perception (in contrast to expert assessment)<br>• Risk society (the limits of insurance systems)<br>• Moral panics over mass media coverage of science; media bashing and media blaming<br>• Science illiteracy, Public misunderstanding of science<br>• Nuclear phobia, nuclear anxiety<br>• NIMBY (not in my back yard) and NIABY (not in anybody's backyard)<br>• Science and technology studies<br>• Technology 'stigmatised' by accident<br>• Colonisation of the life world |
| Information technology | • Computerisation as pre-emptive mobilisation<br>• the fringes of IT normality: addictions, phobias, inter-nots, refusenic, non-users, nonliners<br>• User-centred system design (an engineering precursor of 'upstream engagement')<br>• Late-adopters and laggards in diffusion research |
| Genetic engineering | • Bioethics to pre-empt resistance<br>• Up-stream engagement to avoid resistance<br>• Various formats of public participation to assimilate and accommodate resistance |
| Nanotechnology Synthetic biology | • Science fictions as diegestic prototyping<br>• The optimal timing of public engagement to avoid stirring the resistance |

link resistance to the counter-forces of history, to the dark forces of reaction and unreason, to a false consciousness out of kilter with the course of history. Unnecessary nuisance and a futile impediment on the inevitable path of progress is to be done away with. However, such revolutionary voices became quieter in the late 20th century, though it might have moved from the political to the corporate world, where the quest for 'permanent revolution' survives in new vigour in change management and process reengineering. It appears that the spirit of Leon Trotsky and his communards moved from the departments of history, philosophy and politics to that of business management.[1]

There is no way of framing resistance without making assumptions. By elucidating my assumptions, the readers can judge for themselves.

- New technology is disruptive of established practice, thus a scandal and a provocation to the routine practices of life. Artefacts create a form of life, and changing these artefacts is disruptive. There is also a

growing recognition that artefacts play an important role in making people do things in a certain way: they are an influence on others. Therefore we must expect resistance to be the normal response to this attempted influence, and we must wonder about its absence.
- Resistance is thus not only a function of disruptions, but also of the manner techno-science mobilises with blanket framing. Symbol and images filter any challenge, buffer the provocation, and mask the scandal; all questions go away. The watchdog no longer barks because it is incapacitated.
- I retain a didactic tone in several chapters, because these notes originate in the teaching of a post-graduate course. The reader might be patient with the teacher's voice in this text. I hope it is not overbearing.

## A BRIEF GUIDE FOR THE READER OF THIS BOOK

There are several ways of reading this book, and reading it linearly from Chapter 1 through to Chapter 9 might not be the best. There are two lines of reading, a theoretical and an empirical one; only Chapter 7 on public opinion bridges both conceptual and empirical contents. This suggests a non-linear reading, moving in and out of the chapter sequence as they are arranged.

### The Theoretical Plot: Chapters 5→ 1→ 9→ Appendix 1, and Back to 5 and 1

If you are a reader interested in theory you might start with Chapter 5 on the framework of Resistance and the propositions that can be derived from it; you might then move to Chapter 1 on Mobilisation and from there to Chapter 9 on resistance, extending these into the didactic Appendix 1: Notes on Social Movement and Social Influence. From there you might well go back again to Chapter 5 and 1 once more, and then close the circle with the more empirical chapters.

### The Empirical Plot: Chapters 3→ 6→ 8→ 7, Diving In and Out of Appendix 2

If you are an empirically interested reader, there might be several routes. The historically interested will follow the chronology and start with Chapter 3 on nuclear power, moving on to Chapter 4 on the dawning of ecological awareness, and from there to Chapter 6 on computers and information technology, leading finally to Chapter 8 on genetic engineering, and wrap it up with Chapter 7 on the waking of public opinion. You will realise that with each chapter you will arrive at the present, as they chart parallel developments, the beginnings of which are somewhat staggered.

If you are a young reader you start with Chapter 6 on computers and information technology, as this is what captures the imagination in the age of iPad, iPhone, and iRaq. If curious about how earlier controversies developed, you might then move to Chapter 3 on nuclear power; and from there to 8 on genetic engineering, to end up finally reading Chapter 7 on public opinion and how this emerged among these different controversies.

But in the end, as in all walks of life, there is no one best way: feel free to make up your own.

## NOTES

1. Incidentally, my first PhD student, who worked on 'practices of managing resistance in the business world' landed a straight-laced job with McKinsey Consulting in the early 2000s, having been an ardent Trotskyite student activist some years earlier. Whenever I see the 'Socialist Workers Party' trying to recruit new members during Freshers' Week at LSE, I suggest they move from the LSE Old Building to the New Academic Building, where the Management School is likely to offer more resonance for their revolutionary worldview.

# 2  Mobilising a Different Future

In January, 2013, the 'Human Brain Project' (HBP), led by a team at EPF Lausanne and University of Heidelberg got the green light: it had been allocated 1.19 billion Euros over 10 years.[1] This was big science; new style EU science funding. It transpired earlier that the core group of successful scientists had had available close to 5 million Euros for preparing this bid through media work; organising a web presence, events and meetings; and anticipating controversy before any research work had even started. The vision of modelling the human brain with computers is controversial. But considering the funding pot, seed-funding was about 0.4% of the overall budget, which on 1.19 billion Euros makes a sizable start. What is happening here? How shall we make sense of this increasingly normal event in scientific culture? This episode of early 2013 illustrates what this chapter conceptualises: how researchers, materials and symbols align into technoscientific projects and face up to challenges.

Modern techno-scientific developments increasingly shape up as collective actions that mobilise public support to change society. In everyday language we say: social movements are about values and identities. Nuclear power led us into the atomic age, where electricity would be 'too cheap to meter'. Information technology revolutionises the way we make war and peace, communicate, work, memorise and think of ourselves and others. Biotechnology changes our 'nature' and the moral foundations of society. I suspect, the Human Brain Project will lead to similar claims making, and already with us are nanotechnology and synthetic biology reengineering matter and designing life forms that do not exist outside the lab. Science and technology seen in this light are attempts to raise a world from a vision.

This theoretical chapter will elaborate the idea that techno-scientific developments can be fittingly understood as social movements recruiting support for social values and social identities. To develop this idea I rely on research on social movements and on the social psychology of group formation and social influence.[2] (See Appendix 1 for more detailed exposition of the intellectual resources). I shall do this with three questions in mind while exploring the following thesis:

Techno-science developments are quasi-social movements with an activist core and public followers. Should this be the case, we must ask:

- How are scientists and support aligned into techno-scientific projects?
- Can these processes be typified as community building, assimilation and accommodation of dissent?
- What is the role of the public opinion, if any?

The research on social movements suggests four variables for the formation of collective action: *grievances, future projects, public attention and context*. These variables define any form of activism. Activism has no contractual obligation to full-fill. Activism cannot sanction participants with hiring and firing, they are mostly volunteers; activism is assessed on mobilisation and demobilisation of participants into actions in response to grievance; envisioning the future and gaining public attention are resources to deal with these grievances in any particular context.

Social movements are fluid *networks of people* who *identify with* and *refer* to streams of opinions and beliefs about some grievance; they *express preferences* for social change (see Ahlemeyer, 1989; McCarthy and Zald, 1987, 20; Tilly, 1978, 9). This common reference might be a 'techno-science', such as nuclear power, genetic engineering or computing, that addresses scandals such as energy shortage, cancer, or human reproductive insufficiency. Social movements operate with minimal *formal co-ordination*. Social movement organisations (SMOs) comprise few cadres, staff and workers, like a small-and-medium-sized enterprise, but they mobilise lots of voluntary contributions. SMOs fall between spontaneous events like fads and fashions, riots, crowds or milieus that have no membership, and political parties, clubs or corporations with a defined and paying membership. SMOs work on a potential constituency that is never fully mobilised (Useem and Zald, 1987, 275). The purpose of SMOs is to maximise mobilisation for the issue; the available *action repertoire* for this purpose is a key feature of SMOs.

To look at techno-science as a quasi-social movement brings advantages. We avoid the *fallacy of false dilemmas*. First, it avoids separating the technical from the political, things are not either technical or political; the technical is political and the political is technical. Secondly, in social movements the material and the symbolic are mixed, movements clearly need both. Human actions are bodies guided by symbols. Finally, captured in the notion of *'inter-objectivity'*, objectification, i.e. the separation of the objective and subjective in research, is the result and not the starting point (see Latour, 1996). Society is full of objects that should fall under the constitution of society in the same way legal rules do. The moral framework by which we live includes both laws and objects. However, many fail or refuse to recognise this by treating the 'social' different from the 'technical' in order to keep the technical among experts and out of politics. The movement idea avoids having to decide whether techno-science is political or not, is material or symbolic, is objective or subjective, the origin or the result. The social movement idea steers clear of extremes by sitting on the fence. The non-political, objective, material, a-priori of these dilemmas comes under the term 'technological determinism' and the political, subjective symbolic process view is deemed 'social construction' (see discussions in Brey, 2005; Klein and Kleinmann, 2002; Pinch 1998, Winner, 1977). But the path of techno-science is neither carved in stone, as determinism wants to have it, nor entirely open to arbitrary will power as in social construction. Techno-science is rather social mobilisation by many means and through multiple constraints.

The movement concept also avoids separating means from ends in human activity. Techno-sciences are tied to aspirations for a 'better world' (see Boesch, 2005), to values and worldviews (Blumenberg, 1981 [1959]), and to a quest for social identity: what society do we want to be (see Winner, 1986)?

The purpose of this chapter is to rehearse some of the elements of this analysis. The rest of the book will focus on empirical difficulties in the social mobilisation for nuclear power, genetic engineering, and computing and information technology during the second part of the 20$^{th}$ century and into the new millennium.

## HOW FAR IS TECHNO-SCIENCE A SOCIAL MOVEMENT?

One might argue that science and technology are part of the establishment in modern society, and not social movements. Researchers working in industry, government or universities are members of reputable *national and international academies* like the Royal Society of London, or of *learned societies* of physics, chemistry, astronomy, biology or many other technical pursuits. Furthermore there are the *traditional forums for showcasing* the sciences such as the American Society for the Advancement of Science (AAAS, which owns the journal *Science*), the British Science Association (BSA), and many others, which define the institutional landscape of science. These organisations have *formal membership*, often by invitation only or by subscription, renewed by fee payments. They elect committees and participate in conferences that are like tribal gatherings.

From a historical perspective, these structures arose from social movement, but once established they are no longer social movements. What started informally during the enlightenment in learned circles of debate and experimentation, became establishment. Through historical cycles of activism, science became a taken-for-granted institution in modern society by about the 1920s (Knight, 2006). Thus past social movements liberated knowledge production from the shackles of myth and traditional religion, at least that is how the popular narrative goes.

However, social movement research also predicts that established institutions resort to social mobilisation when their influence in the polity is threatened. Useem and Zald (1987, 275ff) showed how nuclear power, in the 1960s still an integral part of energy and defence policy, in the 1970s mobilised a grass-roots nuclear movement to counter eroding influence over policy. Nuclear power was no longer in the driving seat of society. In the US, policy changes on safety, pollution, licensing and export brought loss of business and a de facto moratorium. Rising environmental concerns and concerns over safety and liability among consumers changed the public preferences. A series of disasters from the 1960s to the 1980s (Thalidomide, Three Mile Island [TMI], Chernobyl, Challenger, Bophal, Seveso, Basel) undermined the post-war common sense of progress. Risk regulations and increased private patronage led to a new context for conducting science (see Mirovski and Sent, 2005; Krimsky, 2003; Etzkowitz and Leydesdorff, 2000) with implications for the way the techno-science is done. What we have seen since the 1980s is a return of the *movement character* of science, of mobilising the public, but with a new repertoire of actions. This leads to a historical thesis:

> Social mobilisation for techno-science is a historical practice with novel means.

It seems unusual to talk of science as a social movement, as collective action analysis mostly deals with societal outsiders and political opposition movements; how can something as established as science and technology be a social movement? However, it is difficult to agree what science and technology 'really are'. They can be conceptualised very differently, and the concept of 'social movement' can indeed be usefully and gainfully applied to modern techno-science. But a modicum of doubt remains; techno-science is hardly a prototypical social movement, which leads me to talk more cautiously of a *quasi-social movement*.

## ACTIVISM AND PROJECT FORMATION

Grievances give rise to visions of a better future; they draw attention to the gap between the present and a possible future which motivates the quest for innovations. Not every idea, image, or symbol resonates in any context. Activist organisations diversify and apply their competences to new frustrations, providing career paths for activists moving between projects, and between movements and counter-movements. Partisans change allegiance, and professionals change sides. Conversions across controversies are not unheard of in techno-science: the nuclear pioneer becomes an anti-nuclear activist. Some geneticists support eugenics, others are anti-genetic modification; the computer science pioneer warns of hackers, surveillance and false promises of artificial intelligence. The activity of individuals will vary by degree of involvement, convictions and dissent, and the cultural milieu defines what counts as a scandal.

## Forming an Activist Core

Social psychology investigates the development of *joint intentionality in small groups*. In joint intentionality, actors assist each other in a common pursuit; they know of the importance of 'us' against 'them', and that 'we' all want the same: it is neither me, nor you, but us (Tuomela, 1995). Such an understanding forms an activist core. Joint actions contrast with parallel actions that are circumstantial. People open umbrellas when the rain starts, not because they have agreed to open umbrellas. Joint action arises with mutual awareness and identification with a common cause. Thus, in a performance, dancers might open umbrellas on a pre-arranged musical signal.

Group dynamics idealises teams building as the outcome of a process with phases (see Tuckman and Jensen, 1977). Initially, people might join others scandalised by a grievance, on a pretext or out of curiosity, in search for opportunities, with vague expectations and little sense of purpose except maybe dissatisfaction or even outrage. Deliberations follow about what needs to be done and how this can be achieved. This will stir conflict over values, aims and objectives, because previously held views differ; and there might be an attempt to take over and to dominate: others will resist this. The *storming phase* leads to a temporary equilibrium and the setting of a modus operandi. Through *normalising* the common purpose is set, and the project is formalised in declarations, mission statements, statutes and practice, complete with management structure, planning, procedures, risk assessment, milestones and deliverables. The project is defined and people can take a stance, defend and challenge it. Some such projects attract funding, others fail at the start. Once resources are in place and the mission clear, the group enacts their vision and seeks allies for the *performance*. Depending on the context of other projects, friends or foes, the project will grow, diversify, be taken over or fade away, known as *mourning*. Some networks continue to resurface in new formations, but the projects will change.

This group process of forming, storming, norming, performing and mourning are schematic, but they highlight the mobile nature of projects and their life cycle. Any normative edifice is a temporary balance of opinions, defended and challenged from the outside by opponents and from the inside by newcomers. A movement industry mobilises resources for an entire sector of projects that work at the new frontiers of science and technology, otherwise called nanotechnology or synthetic biology. This social mobilisation profits from a division between visionaries, who through style and content raise attention, and more moderate actors who can enjoy the attention but distance themselves from 'extremism'. One thinks of scientists who publicly reject visions of 'nanobots' or 'grey goo' but enjoy the public attention in the shadow of these 'crazy' symbols. Such movements comprise swarms of *projects* which imagine different futures, anticipate different not-yets, future perfect or futurum exactum in different lights. *Movements are triplets of subject, object and project.* Actors (subjects) identify by reference to issues (objects) and futures (projects). Issues imply actors with a design, and designs link actors and issues to a common future (see Bauer and Gaskell, 1989, 2002 and 2008).

Consider for a moment the Human Genome Project (see Chapter 8), formed by visionary geneticists who dreamt of a map of the human genome. The activist core formed through the 1980s, disagreeing over sequencing versus mapping. Then HUGO, the SMO was formed to normalise these expectations, to channel resources inclusive of a 3% clause for research into ethical, legal, social and economic consequences (ELSE), and to give the research public visibility. Biology had become the new lead science, and timing was good: the physics community with its large Hadron Supercollider project was competing for public resources. HUGO secured resources in 1990/91, with US and UK research teams playing a central role, Japanese and Italian teams a more peripheral one. This umbrella of many projects worked through internal and public controversies, over mapping or sequencing, private or public ownership, and over potential or actual social implications. Controversies caught attention and public imagination, and brought public profile. By 2000, three years ahead of schedule, the project was declared complete in a media event in the White House that involved heads of state. Follow-up projects take different directions but define the new industry of genomics research. The vision of 'synthetic biology' arises in a novel bid for large scale co-ordination.

## The Mobilisation of Resources: Assimilation and Accommodation

In this particular perspective, the 'movement' is a Machiavellian plot at several fronts, with adversaries and allies and few if any constants, but lots of risks and uncertainties. The projects either succeed in attracting allies as it stands, or must make concessions to accommodate new allies. The attribute 'machiavellian' suggests that nothing is sacred or taboo, even morality and ethics dispensable as needed (see Latour, 1988). The only constant is the push to 'move on', seeking a sustainable path to have a future.

Because of competition, every movement faces the problem of *attracting public attention*. They face the polity and public opinion as potential resources and hurdles. Public opinion comprises informal conversations and formal mass media attention. SMOs are the bandwagons from which to preach to the constituency and to persuade bystanders to come on board. Social movements face a dilemma between legitimacy in numbers and legitimacy of modus operandi; often activities designed to invite bystanders can alienate the core constituency. The professionalism of public relations can be alien to the partisans and appear as empty talk. But to seek political influence, numbers beyond partisan membership are important.

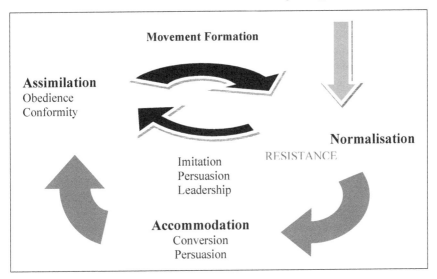

*Figure 2.1* The spiral of directing projects by steering and regulating. *Source*: Modified from Sammut and Bauer, 2011.

Scientific projects nowadays compete at various frontiers. They compete for public sponsors and private venture capital and stock market profile. They compete for attention among policy makers to recognise the movement sector as policy priority, named and framed as a future that deserves budgetary attention. Nanotechnology has succeeded in early 21$^{st}$ century, where nuclear power, information technology and biotechnology did before. Support is mobilised by naming and framing the issues among interested publics such as scientific peers, by offering them new opportunities. Attention is however also necessary in the wider community, to recruit political support and youthful enthusiasm into careers at university, research labs and high-tech industry.

The movement responds to these challenges with a characteristic mix of a) community building and public deliberation of the common future, b) lobbying the polity and venture capital to gain access to resources, c) accommodating the dissent of significant challengers and d) by strategic adaptation to anticipated concerns.

*Movements that make a future* engage these activities in a spiralling process with two feedback loops as schematised in Figure 2.1. The formation of projects has three significant moments: normalisation, assimilation and accommodation. Each moment corresponds to a modality of social influence: forming the community and *normalising* the joint-intentionality; extending the community by *assimilating* newcomers and dissenters with conformity pressure, persuasion and identification; using authoritative and prestigious leadership (primary loop); and extending the community in the face of resistance by *accommodating* the dissent with processes of self-persuasion and conversion (secondary loop).

Assimilation and accommodation are processes of conflict resolution when future perspectives diverge. They are non-violent but also non-deliberative, because they assume a hierarchy of power a > b. These processes of social influence lie mid-field between violence and power-free deliberation, and they constitute modalities of *collective learning*. Assimilation controls deviance and brings the few in line with the many and under the prestigious authority of science. This regulates and maintains temporary stability of direction. Through accommodation the movement brings the many in line with the dissident few; the joint intentionality is reconsidered. Here the voice of the few successfully turns the helm of the ship. Social movement thus includes the interplay of *regulating (primary loop) and steering (secondary loop) without a fixed telos*. Telos, the vision of the better future, is itself a resource of mobilisation (Dierkes, Hoffmann and Marz, 1996). Notions such as progress, the 'scientific solution' and the historical account of the birth of science from ignorance in the exit 'from Mythos to Logos' are meta-narratives that potentially mobilise support and disenfranchise opponents. The interplay of accommodation (steering) and assimilation (regulating) is an activity cycle. When regulating cannot absorb the disruption, steering comes into play and initiates a new cycle of regulation on a new track. The secondary loop is triggered by resistance, when authority and persuasion are no longer sufficient to contain dissent. In these situations the project needs adaptation in order to be sustainable.

Assimilation and accommodation are thus nested processes. A precondition of successful dissent is organised activity that persuades through consistency in public image and internal identity. This organisation implies control of deviance, skills and discipline. Thus conformity and compliance are necessary for effective dissent; to challenge the conformity of the many requires the conformity to the few. The hardened non-conformist is an unsuitable activist; the hardened activist is a dubious non-conformist. This logic bears risks. It seems that only an *ethos of resistance* offers a way out of this paradox of control, because resistance casts a shadow of doubt on influence per se. For the ethos of resistance, the power of assimilation and accommodation is as ubiquitous as it is dubious.

## PUBLIC ATTENTION SEEKING

Movement organisations must keep their core community. However, this core community might not be a sufficient, not least because the ambition of techno-sciences is universal. Thus, the movement cannot content itself with preaching to the converted. SMOs are on a mission to the wider public as potential partisans not least because the polity responds to this wider public. The wider public is a large stream of opinions to be mobilised, but this might be a risky undertaking, because it is not fully controllable.

## Movement Milieus and Representations

Over time the techno-movement might be able to cultivate a *'movement milieu'* (Rucht and Neidhardt, 2002, 19), a public attentive to the comments and issues of the activists. This milieu defines the recruitment basis for the movement which is able to provide resources if needed, and it is well networked into specialist news, meetings, workshops, festivals, conventions and educational settings.

In the cultivation of a milieu, the movement has to consider *existing milieus* with divergent concerns. These social milieus are the residuals of historical conflicts and social structure, but also form in a novel manner around sociological and economic changes such as lifestyles based on levels of education and consumption. High-tech societies offer opportunities for novel combinations of access to wealth and access to culture, in what is called the 'experience society'. And the shift from an industrial to a knowledge economy fosters 'post-material values' that sensitise for quality of life issues (see chapters 4 and 7).

We need to imagine these social milieus without the boundaries of exclusive 'ghettos'; members move across milieus, but these milieus participate to different degrees in the arenas of public opinion. Arenas of public opinion, which comprise perception, mass media and polity, reach into these milieus to different degrees. While perception of nuclear power or genetic engineering might be fairly widespread, given attention, access to media coverage is unequal and representation in policy is not guaranteed. Arenas have potential to reach or fail to reach milieus. The mobilising techno-movement needs to take into account the degree of participation of these milieus in the arenas. Milieus might have their favourite arena; some are politically represented, other not; most likely, new milieus do not have their concerns translated into policy. The movement might consider milieus as *territories*, some already 'conquered', others unexplored. We can remain flexible on what constitutes a territory, a nation, a geographic region or a social category (Bar-Tal, 2000). But we might sensibly say, the nuclear movement has conquered India, and it is exploring Iran.

As the techno-movement mobilises, the recruitment milieu might respond welcoming or resisting the approach for a variety of reasons. The outcome of this gradient of resistance will be the *social representation* of techno-science in the milieu (Bauer and Gaskell, 1999 and 2008).[3] This encounter of movement and milieus is often controversial and the milieu's common sense response is observable as social representation, civic epistemology (see Jasanoff, 2005) or ethno-epistemic assemblage (Michael and Brown, 2005).

## Arenas have their Eigen-logic

The community of the techno-scientific project is extended in two directions, into the polity to influence political decisions and into public opinion. Public opinion is important because the government is no longer the only patron, stock markets have entered the frame, and they consider public opinion. Public opinion is also important to recruit researchers and to put pressure on the polity; politicians consider public opinion during the electoral cycle.

The polity, mass media and public perceptions have their own logic, which guarantees a certain autonomy vis-à-vis the techno-scientific project. The movement needs to negotiate attention at all frontiers. Table 2.1 lists the elements of this Eigen-logic that distinguish the modus operandi, the achievements and overall contributions of the arenas. Every arena is an ecological environment which can offer the movement a niche for topics which are selected or ignored according to the Eigen-logic of operation, achievement and functions.

The *regulation of technology* in policy and law making is slow, often tied to the electoral cycle and changes of government. Activities have good memory and a strong concern for consistency across domains of

*Table 2.1* The Different Operating Principles of the Three Arenas

| Public arenas | Public perception | Mass media | Regulation & policy |
|---|---|---|---|
| Modus operandi | Medium cycles Some consistency Memory | Daily, weekly cycles News values Issue cycles Imperative of novelty Little consistency Hardly any memory Framing | Election cycles Bias against novelty Consistency Long memory |
| Achievements | Attention, Opinion, Attitude, Stereotype, Schemata, Awareness, Skills | Attention Diffusion Propagation Propaganda Advertising Education and training Agenda setting | Regulatory regimes Legislation Competitiveness Being represented |
| Functions | Being able to act everyday; Conversations with others | Information; Entertainment; Linking societal domains; | Blame voidance; Responsiveness to public; resourcing new technology; |

*Source*: Modified from Bauer & Gaskell, 2002.

regulation. Regulations on nuclear power should be consistent with those of genetic engineering. There is an inherent bias against new legislation; the working assumption is that existing rules cover novel developments, unless otherwise shown. The new is assimilated into existing regimes, and there is an operational resistance to change of regulatory regime. Hood et al. (1999) identified three policy regimes. 'Responsive government' reflects public perceptions and media opinion, 'client politics' reflects vested organised interests and 'minimum feasible response' tries to correct market failures or tort-law processes, in effect organises the avoidance of blame. A regime is a temporarily stable modus operandi. As the movement expands, the policy regime may switch between these types of management.

There is a growing literature which looks at mass media and social mobilisation. The mass media bear opportunity and risks for active minorities who offer a different future (see Kielbowicz and Scherer 1986). The mass media are a place where public opinion becomes visible and effective. Various actors therefore compete for media attention. Three aims are crucial: to position oneself and gain visibility, to evaluate and to frame the issue. Thus media attention, evaluation and framing are key targets for public actors to orient public opinion and to wield influence (Schaefer, 2008).

## Platforms for Dramatisation of the Issues

The community is built, maintained and extended on rhetorical contributions. SMOs persuade the core movement to stay on track, and invite the wider public to join the bandwagon. They do this by mobilising the *'three musketeers' of rhetoric*, i.e. by developing evidential arguments, by emotive appeal and by building authority, prestige and credibility. What classical rhetoric identified as the three types of argument—*logos, pathos and ethos*—still typify modern public controversies (Gross, 1990; Bauer, 2002). These rhetorical platforms have a history of tension over the *hygiene of public discourse:* what type of arguments should prevail? According to Plato's polemic, the sophists championed pathos, the emotive appeal, at the expense of truth. Aristotle admonished the construction of good arguments (logos) without denying the importance of both pathos and ethos. The Romans privileged ethos and style, the virtuous man speaking well was trusted by the community. The modern neo-positivists wanted to abolish rhetoric, focus exclusively on the constative function of language (logos) and dispense of pathos and ethos altogether; they ignored that we act in and by speaking (see Barthes, 1994; Meyer, 2008). The modern public sphere harks back to this rhetorical triplet of platforms. *Public claims making* is evaluated simultaneously with reference to objective reality and answering to truth (constative), with reference to social reality and answering to rightness (normative), and with reference to subjective reality and answering to sincerity (expressive) in order to secure reasonable outcomes. Reduction to any one platform leads to pathology of communication and mistrust among speakers (Habermas, 2001).

The techno-scientific project needs all rhetorical horses, but struggles with notions of hygiene in its core constituency. *Sound science* (logos) is preferred by a core of activists; for many this is the only admissible argument. Prestige and credibility (ethos), acting and being seen to act virtuously, is the concern of SMO. The *'social responsibility'* extends to 'movement social responsibility' (MSR) from its corporate equivalent CSR. The biotechnology movement leads the way on ethos, lobbying for bioethical self-regulation. Passing committee by avoiding harm and securing benefits for animals and humans becomes an ethos marker of a techno-scientific project. Finally, *public appeal* mobilises emotive images and language (pathos). The SMO's toolbox for attracting public attention includes stunts, creation and use of iconic images, for example of, DNA. Parading suffering and grieving people bring home problems that need to be solved; all rhetorical topoi with perennial public appeal. Art patronage is part of this repertoire. A Hollywood relations office may be the aspiration of every respectable research institution. Techno-science mobilises the art world, and atoms, bytes or genes become music, plastic and drama. The *aesthetisation of techno-science* brings artists into the laboratory, and a new kind of agitation-propaganda raises the attention of a fascinated wider community.

## Seeking Public Resonance

Communication can focus collective attention to an issue of relevance, be that to recognise a threat or a source of pleasure and entertainment. With the mass media the space-time constraints of speaker and listener are broken—mass media can be consumed whenever and wherever—but this is paid for with a lowering of the probability of successful communication (see Luhmann, 1990). How do the mass media compensate for this loss of control? They professionalise the production in genres that maximise resonance; this becomes a business model of trading public attention, measured by audience figures, with advertising space.

The mass media arena brings together speakers, topics and audiences under a logic of dramatic considerations of what makes a 'good story'. Lists of *news values* have been complied to operationalise this problem. To make it, the techno-science has to be novel, controversial, exciting, produce images, be locally relevant with a personalised human touch and command expert sources (Hansen, 1994). Stories are scandalised into moral conflicts between right and wrong. Issues of causality become issues of blame and responsibility (Neidhardt, 1993). Little is known of whether news values have changed over the past century (see Bauer and Bucchi, 2007). Mass media focus attention and reduce complexity to the effect that a course of action becomes obvious.

Other *operational routines* structure content making. Most media operate on a tight time schedule. There is editorial closure in newspapers the night before; for evening TV programs in the afternoon; magazines have a weekly or monthly cycle. Editorial closure is an anxious moment; there is always a risk of missing the big story and leaving a scoop to others. The hunt for the great story has a herding effect: every editor chases the same story. Time pressure is increasing and employment more precarious; journalists become dependent on PR contacts and press releases, and produce 'churnalism'. The balance of power between journalists and their sources is shifting in favour of the latter (see Goepfert, 2007; Brumfield, 2009). The weakness of the media is the strength of the SMO with a slick PR operation.

The *agenda setting* idea has accumulated much evidence that mass media do well at telling us what to think about, though not necessarily how to think about it. Mass media modulate attention. With a lag of one or two weeks, issue rankings of mass media and public perceptions converge. Daily news co-ordinates a curious 'inter-spirituality' (Tarde, 1901) with political significance: a call for action. People will notice what the media pay attention to, though contingent on their *need for orientation* arising from relevance and previous knowledge. On what affects people's lives directly, like crime and costs of living, media agenda setting is weak. However, on issues remote from everyday experience like pollution, national unity, energy policy, techno-science, most conversations take their cue from the mass media (McCombs, 2004, 54ff). However, media also receive the agenda. Public perceptions drive issues which the media, at the pulse of opinion, pick up; and there are policy agendas. The inter-dependency between the arenas makes for nine cause-effect constellations within and between mass media, conversations and public policy (see Bonfadelli, 2004, 241). Techno-science movements seek visibility at reduced cost by understanding the media's Eigen-logic. Buying attention by advertising is too expensive; here techno-movements learn fast from other social movements.

## Framing: Evaluation and Positioning

Mass media coverage evaluates issues on a positive-negative and ambivalent dimension. Public opinion is positioned on issues, expressing public attitudes. In controversies over water fluoridation, nuclear energy and global warming it was observed that mass media not only raised attention, but invariably turned public sentiment more negative. Media attention can induce a conservative bias; attention is already the news (see Mazur, 1990 on the *quantity of coverage*). However, this effect was not replicated over biotechnology across Europe in the 1990s. What looked on the surface like a Mazur effect: massive increases in press attention at the same time public attitudes turned negative (see Chapter 8), did not hold when controlling for differential media exposure (Gutteling, 2005).

*Priming* shows how evaluation flows from positive-negative media coverage to public perception. The issues that are currently most attended upload the criteria against which the actor will be evaluated, irrespective of what else they might be doing (Iyengar and Kinder, 1987). The news agenda primes public judgment by setting the criteria. Once nuclear power is top news, actors are evaluated on whether they support it or not; with economy crisis news, the economy is the criterion. What is called 'priming' or 'attribute agenda setting' converges on 'framing' (see Scheufele, 2000; Entman, 1993). Mass media draw attention, evaluate the issues and set the terms on which this evaluation is achieved. It shows the power of single issue movements, often a SMO, to dictate the terms of the debate with a successful campaign. Framing has thus become a core concept of social movement research. It stands for the politics of signification, the work on meaning and the contention over constructing the actuality of what is the case and what needs to be done.

*Framing* is an analytic and strategic question for a movement actor. Framing determines how we talk and think about atoms, bytes and genes (see Benford and Snow, 2000). Nuclear power can be scientific progress, modernisation, a guarantee of independence, a devil's bargain and an economically ruinous energy option. Biotechnology is progress or a threat to human dignity and the environment. Information technology is a revolution, a new Gnosis or an instrument of surveillance. Frames reduce complexity and thus make a difference. As a motor schema, they offer up diagnosis of cause-effect, moral responsibility and suggest particular courses of action, often crystalised in core images and metaphors (see Gamson and Modigiani, 1989). The nuclear mushroom cloud has a sublime appeal, both terrifying and fascinating, and used in the context of nuclear power it signifies the risk of proliferation. A nuclear accident matters differently depending on whether it is framed as a random event or the result of design flaws. Interested actors sponsor frames that offer up advantages to mobilise activists, to garner bystanders and to disarm adversaries. Frames are contested because only some will win on its terms.

One issue *creates opportunities and risks for other issues*. Take the environmental issue of 'global warming'. It brings a sense of urgency; polluters are morally responsible; we are called to self-restraint and to reform of our normal way of life. Its symbolism is rich with images of dying animals, collapsing icebergs, deserted lands, polluting chimneys etc. These images resonate widely and point the finger. To present nuclear energy as 'clean' and part of the solution captures an opportunity to prime its contribution in a positive light.

Frames resonate depending on their credibility, and fidelity in the *toolkit of culture*. Framing is often a shot in the dark of a SMO; some framing attempts are successful where others fail. The frame that comes to dominate is often nobody's design, but the balance of competing ideological commitments of the movement (see Polletta and Kai Ho, 2006, 197). Balancing acts lead to frame extension, frame merger, frame bridging and amplification, and alignment into master narratives of 'environment', 'progress' or 'human rights' which capture many different concerns. Resonance often means sacrificing the purity of ideals and commitments. In order to win the debate over abortion, some activists dropped the issue of choice in favour of compassion ('help, don't punish': Marx Ferree, 2003).

*Excursion 2.1* Techno-scientific Meta-narratives of Historical Direction (Telos)

*Techno-scientific movements have at their disposition a number of 'grand narratives' to frame the public debates. These residuals of previous mobilisation efforts continue to resonate in many contexts.*

*The notion of technology as human destiny arises from comparing humans with other animal species. This shows that tool making is part of the simian-human transition. Monkeys might carry and use stones to crack a nut or a stick to poke for termites, and teach other monkeys, but they rarely sharpen flints for the spear nor potter like humans do. Artefacts and their accumulation are a human sociality (see Boesch, 1996; Latour, 1996). Tool making compensates for the lack of genetic programming of behaviour patterns (Gehlen, 1980) and mitigates the existential anxiety that comes with bi-pedalism: with eyes higher than the savannah grass humans extend the horizon and apprehend many more possible dangers (Blumenberg, 1990). Thus, at the human-simian transition, technology is a natural destiny.*

*Technological revolutions abound in public discourse. The revolutionary trope puts any movement effort into a lineage of past struggles for liberation against the forces of darkness and oppression. The classical stories often are triplets: the Neolithic transition ends nomadic life and enables civilization with large settlement, scriptures, and domesticated animals and crops (10000–5000 BC). The agricultural revolution produces beyond subsistence and feeds an exploding population. Finally from 1750 the industrial revolution brings an explosion of productive forces and potential plenty for everyone (e.g. Festinger, 1983). Others offer finer grained phases (e.g. Ribeiro, 1972), but the trope of 'three' is old and millennial. It includes Hitler's 'Third Reich', Joachim Fiore's ages of the Father, Son and Holy Spirit, Marx's Feudalism, Capitalism, and Communism, and the Positivists' move from a Theocratic, via a Metaphysical to a Scientific Age as the climax of history.*

*Ideas of long waves of the world economy are credited to the Russian economist Kondratjev (see Van Dujin, 1983). Empirical observations cover the past 200 years and identify long cycles of ups and downs of around 50 years. A key feature of each cycle is the new technology that gives rise to a new industrial lead sector. Old techno-industrial structures collapse in the down-swing, and the inventions of the depression build up the production and attitudes of the up-swing (Trebilcock, 2002; DeGreen, 1988; Mensch, 1975). In this story, the world economy moved from wood and textile to iron and railways, to cars and chemicals, to atom, airplanes and computers, and the latest set: genetic engineering and nanotechnology. The existence of these waves is controversial (Maddison, 1991). But interest in the idea is countercyclical to the world economy; expect a revival after the global crisis of 2008/09.*

*Then there is the* linear model of diffusion *(see Rogers, 1983; Valente and Rogers, 1995) The narrative suggests a glorious division of labour between scientists who discover, engineers who innovate and turn lofty ideas into prototypes, products and property, and the social scientists who market the stuff by flogging it to consumers. The model incorporates the myth of a clear distinction between science and engineering, pure and applied sciences (see Krige, 2005).*

*These meta-narratives bring confidence and a sense of inevitability, which Blumenberg (1997) captures with his famous aphorism 'Everything about futurology: Surely, we will not do everything of which we are capable ... But in the end we will, because we need to prove ourselves.' (my free translation)*

## Issue Cycles

Observations on environmental news showed cycles. Downs (1972) suggested that news comes in ebbs-and-flows with five phases: In the initial *pre-problem stage* grievances receive little public attention. A dramatic event such as an accident triggers an *alarmed discovery* of this grievance. Now the wider public takes notice and develops a pressing sense of 'something needs to be done'. Most hope for a technical fix so that no fundamental societal change is needed. During a third phase, the costs of *solutions* become clearer and the public become aware that the grievance was beneficial for some. A solution with winners and losers poses a difficult public choice. For some the solution is costly, for others the problem is already the threat. Whether a decision is made or avoided, the public attention span is limited and likely to go into *gradual decline*; news space is taken over by other issues. Finally, *post-problem* the grievance moves into an unstable limbo. An SMO might have come into being that will monitor the issue and keep the memory, so it can flare up sporadically.

Downs (1972, 41ff) speculated on what made an issue likely to fade: only a minority is affected; the status-quo and inaction benefits the majority; the grievance has no news value but requires continuous drama. Mass media run the risk of boring their audience and therefore move to other topics. Furthermore, political issues compete with the entertainment needs. Down's pessimism about fading environmental news was not confirmed; the issue persisted, but through several cycles of active dramatisation (see Chapter 4). It is also likely that issues differ in scope. Broad issues like the environment, nuclear power, computing and biotechnology over time encompass smaller topics like water pollution, acid rain, rain forests, nuclear waste, accidents, VDT radiation, surveillance, genetic patents, genetically modified (GM) crops, human cloning, etc. A grievance can grow in generality; a fade-out might live on in other topics. But Down's main observation remains valid: the cycle is not determined by the severity of the grievance.

A recent model stresses the Eigen-logic of issue cycles (Newig, 2004). A cycle moves from one phase to the other as actors pursue their own interests. Citizens, mass media, politicians and special interest groups all 'satisfice' their bounded rationality of action. Public attention follows levels of aspiration, changing severity of a problem and the visibility of solutions. Public attention increases the stakes for politicians: though risky, there is also something to gain from acting now. Any policy response will reduce attention, but so do competing stories and the declining value of 'an old story'. The issue history leaves a trace. Issues with a history are more easily revived; so are grievances that were only responded to symbolically rather than with substantive remedies.

Issues flow between different media cycles; and phases are defined by the preponderance of certain media and their functions (see Strodthoff, Hawkins and Schoenfeld 1985). Issues are initially framed in specialist media, in niches such as web communities, newsletters or scientific peer communication. The issue is disambiguated within a niche of a like-minded community, not without conflict. In a second stage, gatekeepers and media issue scouts give attention and legitimacy to an issue. Thus the issue moves from the niche community to the general public, the grievance gains clarity, recognition and routine coverage. The issue is routine when a 'correspondent' gets a beat on the environment or on IT. Then the news changes from occasional coverage to full public framing. Wider public attention stimulates renewed specialist attention, a positive feedback that prepares another cycle of public attention.

## Assimilation: Diffusion, Spiral of Silence, and Expert Consensus

Once the community is constituted, it faces the problem of dealing with newcomers, both to expand and to sustain any level of partisanship. Newcomer partisans need to assimilate and be made to fit the existing mould. There are several influence processes by which this assimilation can be achieved. Diffusion assumes a 'natural course of things': New ideas, products and practices spread by imitation in the community, initially slowly then accelerating until approximating the upper diffusion level. A key parameter is the time it takes to reach the turning point of a 50% adoption rate. The adoption of a new idea is thought to depend only on contact with previous adopters. They are imitated because of their prestige of setting a new trend; after the turning point decisions are induced by conformity to a 'critical mass' (see Rogers, 1983).[4] Communication provides awareness and cues to imitate. Diffusion is a very successful idea, but less because of its realism (see Boecker and Gierl, 1988), than through its power to instil confidence in controlling the process. It takes ideas for granted, cannot map changes over time nor deal with refusals, retractions and ambivalence. Diffusion research thrives on a bias for innovation which assumes that the world will be a better place if most people adopt the solution. What is already perfect needs no further attention. Diffusion is part of a technological fix (see Bauer, Harré and Jensen, 2013; Rogers, 1983, 92).

The drama of social conformity is elaborated for mass communication by the idea of a *spiral of silence*. People express opinions, but when asked in public and by a pollster, they fear nothing more than dissenting from the putative majority view. People form views of what the majority position is by taking their clues from newspaper, radio or TV. A position in the mass media thus becomes the benchmark. People with 'deviant' views will become 'silent', continue to hold private views, but publicly go with the *climate of opinion* (see Noelle-Neumann, 1974). Poll data will show that mass media is driving public perceptions, but not by persuasion instead by silencing dissent. Hegemonic media are most likely in crisis moments like war when conformity pressure is not only on individuals but also on the mass media themselves. Spirals of silence have occurred over biotechnology in the US of the 1990s (Priest, 2000; Scheufele, 2007) and on nuclear power in Germany between 1965 and 1985 (Noelle-Neumann, 1991).

It is by no way certain that partisan opinion resonates with public sentiment; partisans and the wider public might be far apart. Here the key variable is *expert consensus*. Partisans need to demonstrate expert consensus. Consensus brings people around because it is convincing. Elite disagreement polarises public opinion along established cleavages. When elites divide, the public tends to split along ideological predisposition and the more politically aware, the more split. This holds for the Vietnam War in the 1960s. The expert consensus effect is strongest among the moderately informed; for the highly educated, consensus might be dubious. The most educated are sensitive to countervailing information and pay attention to even faintly voiced minority views (see Zaller, 1995, 8f and 99ff). Rothman (1990) observed the consensus effect on techno-scientific issues, admonishing that during controversies the parading of expert consensus is key to sway public sentiment. But this tactical concern of the techno-scientific movement clashes with journalistic values of 'objectivity' and 'fairness' which gives equal space to counter-expertise however strong the consensus. This is of course a source of tension between scientists and journalists (Neidhardt, 1993; Peters, 1995) and exploded over the issues such as a climate change when the media are accused of parading discredited opinion rather than verified facts.

## Accommodation: Priming, Cultivation, and Education by Conflict

Assimilation of newcomers is no longer working when the dissenting minority views have gained credibility. The idea of powerful media projecting and legitimating minority views is elaborated in the model of *priming*. Over time, determined activism is able to draw attention to an issue, and in consequence the performance of actors will be assessed on this issue. Once global warming reached the top of the agenda, the performance of governments and industry became accountable on such questions as pollution and global warming. The hierarchy of issues will position public actors either in a positive or negative light whether they like it or not.

*Cultivation* explores the power of mass media, particularly of television, to shape people's worldview. Popular television genres such as news, soap opera and chat shows are a major source of issue framing. Programs are highly competitive and chase the same audiences, which makes them homogenous: many channels pedalling the same stuff. The system produces spurious product differentiation converging on particular worldviews. This model was tested in the US on patterns of violence, gender roles, family life and public affairs. Frequent scenes of violence on TV news and drama fuel belief in a 'generally mean world', the more people watch (Gerbner and Gross, 1976; Morgan and Shanahan, 1997). Cultivation suggests that particular mentalities can be traced to media exposure like feeding on an unhealthy diet. *Mainstreaming* refers to situations when beliefs homogenise with high exposure. High exposure to a 'homogenous diet' of contents generates an unhealthy common sense. The mentality varies only among those with low media exposure; dissenting requires media abstinence. *Resonance* by contrast refers to issues where beliefs diverge among those with high exposure. For example violence resonates among inner city dwellers who are more likely to experience real crime; violence on TV accentuates their fears more than in suburbia. With high media consumption the unified diet resonates or fails to impress depending on real-life experience. Apparently, the German press cultivated a negative image of techno-science between 1965 and 1985. An agitated 1968-generation entered media careers. However, less technical and scientifically educated, they turned techno-science into front-page political news. Science news moved from specialist sections to the political front page where controversies are covered. Short of a 'conspiracy theory', this observation explains the trend in public opinion by partisan cultivation. Kepplinger (1995) argued that the press created an *'artificial horizon'* that lead public perceptions and legitimated anti-nuclear policies.

A cultivation effect of mass media is also in evidence over biotechnology. In Europe of the 1990s, separation of genetic engineering into RED biomedical and GREEN agricultural emerges in the elite. This split also appears in public perceptions. Where the press distinguishes, public attitudes diverge; where the press homogenises, attitudes converge; but this is so among press readers only (Bauer, 2005). This colour coding of genetic engineering created a new strategic situation for the life sciences (see Chapter 8).

With mounting news on a new techno-science, the more educated will be more attentive and a *knowledge gap* along levels of education will appear (Tichenor et. al, 1970 and 1980). Media coverage skews the circulation of information towards the more educated, unless two conditions mitigate this effect. First, public controversy reduces the gap, everything else being equal. Debates create wider relevance, push issues into public outlets and motivate people to seek information; thus controversy educates society. Second, universal and competitive mass media reduce the gaps, ceteris paribus. However in a new media system that is fragmented into ghettos with few links, the controversy will only resonate inside the niche. When global newspapers are replaced by blogs and Twitter for niche audiences with little inter-connections, knowledge gaps are likely to increase. The controversies over GM food and cloning in the wake of Dolly improved the knowledge of genetics across Europe. The more gene news there was before 1996, the more knowledge was skewed on education. However, the stronger a controversy a country saw on these issues in later years, the smaller the gap became; though an entrenched left-right political polarisation increased this gap (see Bauer and Bonfadelli, 2002). Increases in biotech coverage in the late 1990s decreased knowledge gaps within and between countries (see Bonfadelli, 2005).

## CONCLUSION

In this chapter I argued that it might be enlightening to think of techno-scientific developments as *quasi-social movements* that mobilise resources to face up to societal challenges as part of their ecology. Such an approach avoids a number of theoretical dilemmas, and can make use of the conceptual apparatus of social movement and social influence research.[5]

Movement requires a future projection, and a social scandal to address. Public attention is an essential component of any movement that considers and creates a context of opportunities. Social movements have no fixed membership, but a fluid core of activists who deal with large and diverse streams of public opinion in various arenas. Without a sizable membership, movements compensate by focussing and recruiting a much larger number of volunteers and supporters for the core effort, and for this the symbolic work with language, images and value identities is crucial. Movements operate with a small ratio of a few dedicated activists to many more supporters, and while never exhausting the pool of potential support they seek to maximise it with the available means in order to continue.

A quasi-social movement of techno-science needs to organise three basic processes; and SMOs will take the lead. *Normalisation* stands for the process of forming an initial project that serves as the frame of reference, the joint-intentionality for future engagement. Naming the project 'Manhattan', 'War on Cancer', 'the Nuclear Society', 'Life Sciences', 'Nanotechnology', 'Synthetic Biology', 'Human Brain Project' or 'Age of the Robot' and thus projecting images of a better future are key symbolic activities in this.

*Assimilation* refers to dealing with newcomers to the project and dealing with dissent. Assimilation work mobilises conformity and authority. Conformity pressures arise from large masses of public support and the authority of prestigious leadership; it includes mass media work along the lines of diffusion of information, creating a spiral of silence, and parading of expert consensus. The extension of the project into public opinion and politics requires taking into accounts the Eigen-logic of these arenas. Public perceptions, the mass media and the polity can only be engaged on their own operational terms and values. Existing milieus of opinion, arenas of the public sphere and platforms of public engagement demarcate where social movements seek to mobilise resources to sustain their projected future. However, then assimilation reaches its limit and conditions are favourable, a cycle of accommodation will set in. In *accommodation* mode the movement responds to the challenge of resistance by changing its operations, its future vision; the movement is learning. Accommodation involves strategic adaptation and thus a change of project. Accommodation of a techno-scientific project involves responding to neglected issues such as safety and environmental sustainability (see Chapter 4), the cultivation of alternatives such as renewable energy (Chapter 3), and the spread of knowledge in society as a consequence of controversy and debate (Chapter 8). Secrecy might not the modus operandi of the movement, but collective learning can be blocked in manyways (see Chapter 6).

This conceptual apparatus of movement, normalisation, assimilation and accommodation efforts will allow us to analyse and map the effects of resistance on techno-scientific developments. On the historical examples of nuclear power, information technology and genetic engineering, I will explore the impact of resistance on developments and the conditions thereof. It will become evident that public resistance is a necessary condition, but in and of itself not sufficient to change the course of science and technology (see Chapter 3).

## NOTES

1. HBP press release, 28 January 2013. It is no surprise that this news made the European headlines end of January 2013; see 'Die Zeit' (31 January 2013) and NZZ (26 January 2013). Neuroscience has for some time enjoyed growing interest in the Anglo-Saxon media (O'Connor & Joffe, 2013).
2. See also our previous attempt at conceptualising this 'movement' on the occasion of biotechnology in Bauer (2001 and 2002) and Bauer and Gaskell (2002).
3. A classical study of social psychology shows how psychoanalysis found a different reception among urban Liberals, Catholics and Communists in 1950s France. Reception varied along a gradient of resistance; the Liberals were least and the Communists most resistant. In 'social representations' it is important to consider both process (re-presenting) and outcome (representations; see Moscovici, 2008).
4. Note that 'critical mass' is a nuclear power metaphor. It defines the amount of fissile material that is needed to sustain a nuclear fission process.
5. My intellectual home turf is Social Psychology, where these topics have traditionally been elaborated under group dynamics and social influence. But my sense of history increasingly convinces me that the concepts and concerns of this useful canon need extension and to some extent recovering of historically lost ground to accommodate the reality of techno-scientific developments in the 21$^{st}$ century; this book is an attempt to contribute to exactly such a discussion with disciplinary loyalty (see Bauer, 2008 and 2013b).

# 3 The Atom
## Bombs and Power

Atomic bombs and the production of nuclear power are watershed events of modern societies. What ended WWII and was deemed to be the energy source of the 21$^{th}$ century—'too cheap to meter'–came to a grinding halt in the 1980s. In this chapter I will trace this trajectory by addressing the following questions:

- How did nuclear power come about?
- How has the nuclear imaginary changed since the 1950s?
- How and why did public support for nuclear power fade?
- How did the nuclear community respond to public resistance?
- What is the legacy of this resistance?

Figure 3.1 shows the number of atomic warheads and of nuclear power installations since 1945. Neither case fits the famous sigmoid diffusion model of a slow start, turning point of no return, followed by near-complete diffusion. The nuclear arms race started in the 1950s, plateaued in the 1960s, then took off again only to decline after peaking in 1986. Bombs are now being decommissioned without replacement in an exclusive club of nine countries. The civil nuclear project started in the mid 1950s as 'Atoms for Peace', took off in the 1970s and stalled in the 1980s. It currently produces a sixth of the world's electricity in 30 countries; a declining tendency both absolute and relative. The coming and going of both civil and military atomic power and their inter-dependence is the puzzle which we want to explore, and what role did public opinion play in this ragged trajectory?

The world's nuclear debates are richly documented (see Ruedig, 1990). Nuclear power was the first big test of the environmental movement with repercussions into the present. The social sciences are still living off this debate. Many concepts which are currently in use were forged in the nuclear debates. Notions of 'risk perception', 'public understanding' or 'public engagement' originate in responses to the challenges of public resistance to this large-scale techno-scientific projects.

Since 1945, the techno-scientific world blessed humanity with two nuclear achievements: atomic weapons and nuclear energy for base-load

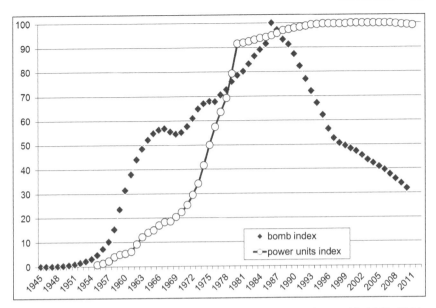

*Figure 3.1* Atomic bombs and nuclear power stations in the world, 1945–2011. Index 100 represents the peak year in each case (1986 = 70,452 warheads; 2002 = 444 nuclear power units). *Source*: Norris and Kristensen, 2010; IAEA statistics various years; Schneider, Frogatt and Thomas, 2011.

electricity and to propel submarines and ships. Medicine uses radioactivity to treat cancer, to sterilise instruments, to trace substances through organisms and to induce variation in genetic experiments. Industry uses uranium by-products to construct hard materials and to test the quality of pipeline work. Nuclear power plants deliver heat to local industry and desalinate seawater. Retailers offered to irradiate food to extend shelf-life (not a consumer preference). Ideas of geo-engineering to reshape the landscape with nuclear explosions also did not carry (Inglis, 1969).

## Physics and the Bomb

Nuclear power was discovered by the new physics of the early 20th century. It revived ancient atomism, discovered radioactivity and Einstein's theory of 1905 predicted the energy of small masses at high speed: $e = mc^2$. The challenge was to control the chain reaction of atomic fission. The breakthrough came with the WWII effort. By late 1942, Fermi's team controlled a uranium chain reaction in their Chicago Pile 1. The Manhattan Project, (partly) hidden at Los Alamos in the New Mexican desert, invented 'big science', i.e. large techno-scientific efforts mobilising science and engineering at the frontier of uncertainty. The result evidenced on 6 and 9 August 1945 over Hiroshima and Nagasaki in Japan:

a blast of unprecedented *yield, heat wave* and firestorm. *Radioactive fall-out* killed 130,000 people instantaneously and thousands more in the months to come. *Radiation poisoning*, inhaled as dust, is an insidious killing machine.[1] The bombs were each delivered by a single airplane. The targets had been reserved by the military so that the impact of the blasts would not be masked by previous destructions (Edgerton, 2008, 16). The Japanese surrendered within days. This impressed the USSR. The US wielded a new weapon. What began as a tactical weapon became the basis of a new strategic game.

That the US would be first to control the atom was no historical destiny. Most of the theoretical work originated in Fascist occupied territory.[2] The US mobilised skills of exiled, often Jewish, scientists who were determined to end the Nazi regime. First building the bomb, then abolishing it, became a way for scientists to resist.

The bomb made physics the lead science of the post-war period, attracting the young science-educated elite. But the A-bomb and later the H-bomb left an uneasy legacy. Subatomic physics faced a lasting demarcation struggle with nuclear engineering for which J. R. Oppenheimer became a symbol.[3] In the Goettinger Manifest (1957), PUGWASH conferences (founded 1955) and the *Bulletin of the Atomic Scientists* (founded 1945) scientists blew the whistle on nuclear arms and its ramifications. At CERN (founded 1954) and ITER (founded 1985) scientists continue to perform subatomic research without making bombs. All in all, the nuclear complex offered a quarter of all physicists' employment (Rothschild, 1980).

## THE MILITARY-INDUSTRIAL COMPLEX AND THE CLUB

The nuclear project ran by many names, each highlighting different aspects of its mission: the *nuclear project*, the *atomic complex* (see Goldschmidt, 1980), the *atomic institution* (see Langer, 1995), the *Club*, the *nuclear industry*, the *nuclear barons*, the *atomic establishment*, the *nuclear community* (IAEA, 1994), the *nuclear priesthood* (Mandelbaum, 1983, 16) and the *nuclear enclave* (in India: Perkovich, 1999).

Nuclear power started as a state-defence project, shrouded in secrecy, jealously guarded, and generously funded (the Manhattan Project cost $2 billion, or $21 billion in 1996 values; see Rhodes, 1987; Schwartz, 1998). Henceforth, the nuclear community faced a dilemma of national security and universal science. The latter implied the open sharing of knowledge, according to Merton's ethos of universality, the former required secrecy and strict access control. This made for difficult relations among war allies, and prime targets of espionage. The nuclear project sealed the bilateral 'special relationship' between the US and UK, which guaranteed some sharing after everyone else was excluded from knowledge and uranium supplies by McMahan's law of 1946 (see Baylis, 1997).

Post-war politics created three international atomic networks. The dominant one was the US, UK and Canada; Belgium and Brazil sourced uranium and other materials. The second network comprised France, who wooed Sweden, Norway and the Dutch. And finally there was the USSR and Eastern Europe (Goldschmidt, 1980).[4] By 1960, 30 countries pursued bomb-making projects, including neutral Switzerland. A combination of parochial defence and universal science unleashed the powers of the atom and formed a *Club (of nuclear powers)* that extended membership not by invitation but by fait-accompli. Hiroshima positioned the US in 1945, Russia joined in 1949, Britain followed in 1953, France in 1964 and China and India in 1974; Pakistan (1998) and North Korea (2006) are informal members. Israel's capability is a 'policy of opacity'. South Africa, Libya and Iraq were discouraged and Iran preoccupies international diplomacy in the 2010s. Many others keep a latent aspiration.

## Nuclear Strategy: MAD or Unilateral Disarmament

Militarily, nuclear bombs were revolutionary, as Napoleon's people's army was in the 19$^{th}$ century and industrial warfare in the first half of the 20$^{th}$ century. They created a new strategic game: Mutually assured destruction (MAD) to dissuade the deployment of 'nukes' was the 'third way' between an infernal Third World War and the utopia of disarmament (Mandelbaum, 1983, 22f). The Cold War (1947–91) became an arms race: developing the nuclear ordnance, missiles for delivery, and orbiting satellites for guidance, detection and surveillance. The space race developed these systems with a civilian face at NASA (founded1958). Upon leaving office in 1961 US President Eisenhower warned of the influence of a *military-industrial complex* seeking ever-larger defence spending for a garrison state, where liberties and democracy are threatened.[5] The nuclear 'priesthood', the techno-scientific system of 100,000 plus military and civil technicians involved in designing, guarding, deploying and thinking about nuclear warfare (Mandelbaum, 1983, 16), formed a 'new man of iron character', a personality type with a heavy toll on individual and family life (Weart, 1988). Secrecy and denial created an elite culture, albeit meritocratic, of paranoia and worst-case scenarios (see Nash, 1980): too worrisome, nuclear logic is best kept out of sight and out of public mind. Policy became a priestly secret (see Williams, 1980).

Nuclear testing was essential R&D. It intensified in waves in the 1950s, in the 1960s, and again in the 1970s to end in the 1990s. Explosions were conducted above ground in Kazsakhstan (USSR) and on Pacific Islands (US, France and Britain). After the Test Ban Treaty (1963), tests moved underground to avoid the global fall-out.[6] The missile agreements, SALT I (1972) and II (1979) paved the way to real disarmament after the stakes had been raised with 'limited nuclear war' and 'Star Wars' scenarios during the 1980s. Strained defence budgets undermined the MAD strategy, and new possibilities alarmed the public: successful interceptions would make a local nuclear war possible again.

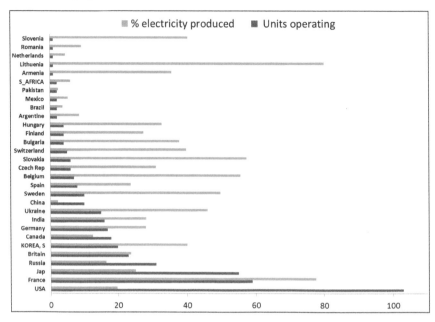

*Figure 3.2* The number of operational nuclear power stations and the electricity generated in percent of the national total in 30 countries; figures are ordered by number of units. *Source*: IAEA Statistics 2006.

By 1986 the US and USSR mustered 70,000 warheads, each hundreds of times as powerful as those of Hiroshima and Nagasaki. This stockpile, reduced to about 22,000 units by 2010 (Norris and Schneider, 2010), is still capable of eradicating the world several times over, and not only MAD, but clearly mad to any common sense.

The move from fission A-bombs to fusion H-bombs alarmed scientists.[7] Fusion unleashed energy beyond predictions. The quest to control fusion is ongoing. Many concerns over nuclear power were first voiced in the late 1940s: radiation fall-out, nuclear winter, waste disposal, economic viability and safety assessment. Some scientists started a counter-movement. Organizations like PUGWASH (founded 1957), The *Bulletin of the Atomic Scientists* (founded 1945) and CND (founded 1958) co-ordinated this dissent.

## 'Atoms for Peace': Creating a Civil Nuclear Spin-off and Trying to Separate Civil and Military Nuclear Power

Ways were sought to separate civilian from military nuclear projects and to convert weapons into ploughshares, to create a civilian spin-off of the bomb. The atom was envisaged as a cheap substitute for fossil fuels of coal, oil and gas. However, the Acheson-Lilienthal report (1946) had concluded that atomic bombs and nuclear energy were linked in the fuel cycle from mining uranium to managing radioactive waste; and in a world of rivalry any separation was bound to fail. Nonetheless, US President Truman signed the Atomic Energy Act (1946), which nominally placed the emergent nuclear power under civilian control but kept it under military secrecy (the Law of McMahon of 1946). In December 1953, Eisenhower started to talk of 'Atoms for Peace'; and the Geneva conference followed in 1955. A proposed idea was that an international bank should give access to nuclear materials to nations who renounced the bomb; this idea was later reinforced in the Non-Proliferation Treaty of 1968. Since 1957, the International Atomic Energy Agency (IAEA), in Vienna, monitors this process and builds technical capacity and public enthusiasm. Even European integration is based on a nuclear vision: CERN (1954) conducts peaceful research, EURATOM is part of the Treaty of Rome (1957), and the Atomium (1958) in Brussels symbolises the unified Europe as elements spinning around a core.

These beginnings were rhapsodic: energy too cheap to meter, every backyard its own atomic boiler, atomic cars and vacuum cleaners, deserts turn into green and fertile lands (Ernst Bloch). This rhetoric appealed in post-war deprivation and to technocratic dreams in a way that is difficult to understand today. In 1956, young Queen Elizabeth inaugurated the world's first civil nuclear energy plant, Calder Hall, in the North of England, and by 2006 442 stations operated in 30 countries, mostly built between 1965 and 1980. The UK and later the US led this roll-out; France, Japan and USSR joined in. In some countries electricity production heavily depends on nuclear power, such as in France, Korea, Sweden, Belgium and former Eastern countries such as Slovenia, Slovakia, Hungary, Lithuania, Armenia and Ukraine (see Figure 3.2).

## The Drivers of the Nuclear Project

What drove this technology? Nuclear energy is concentrated in industrialised countries within the former EAST-WEST constellation. In hindsight, nuclear energy was to close the gap between supply and demand. However, this gap had little salience in the 1950s and 1960s, energy demand is found on the way; it did not stimulate this development (see Radkau, 1983).

*Starting conditions* often determine design paths. British preference for graphite moderated gas rectors, MAGNOX and AGR, arose from a lack of access to natural uranium, graphite was a familiar moderator. Gas cooling was safer and put less constraint on sites. By contrast, the US path, derived from submarine propulsion, used enriched uranium and light water, had to be near water and 50km from conurbations. The commercial designs of General Electric (PWR) and Westinghaus (LWR) followed this path (Williams, 1980, 34ff).

*Competition* among suppliers drives costs down. However, military developments do not make for much competition and military R&D stifled innovations with limited end-user control (Kaldor, 1980). Commercial reactor were spin-offs from technically suboptimal military designs. Not even scientists of the Goettinger Manifest (1957) pushed for better designs that prevent proliferation and minimise waste; thus unsafe light-water reactors with a dual-purpose fuel cycle became the standard consistent with a military agenda (Radkau, 1983). By the 1970s, most countries stopped reactor developments and opted for US commercial designs on hyped economic viability (see Hecht, 1998, for France; Wildi, 2003, for Switzerland). The UK and Canada sustained other designs (heavy water cooling and graphite moderation; see Williams, 1980). Russian technology (graphite moderation) dominated Eastern Europe. Technological *determinism* is the spectre of nuclear power; society needs to adapt to the fait accompli. However, choices were made against experimentation with alternatives. The civil nuclear project was unduly rushed. By 1974, at the peak of the build-up, designs closed on the light-water version with enriched uranium, less efficient and more wasteful than others (SIPRI, 1974).

The most likely driver of the nuclear project was the military-industrial complex, a Public-Private-Partnership (PPP) of *state defence and defence industry*. Nuclear power was a 'quasi-weapon system' (Kaldor, 1980) that innovated like the military do: expert planned and shielded from market signals. The nuclear project became a 'Machiavellian machination' (Latour, 1988) fighting on many fronts, searching for a market rationale amid growing public doubts, and seeking political and industrial allies; not least through offering career moves for Navy and Air Force officers, and revolving doors between industry, government and regulatory bodies.

Efficiency gains in oil, coal and water energy continuously threaten the economics of nuclear power. During privatisation of nuclear energy in the 1960s, the existing energy sector was to take over but warmed to it half-heartedly. The real costs of nuclear power continues to pain the community. Nationalism and independence powerfully justify state subsidies: what business logic discourages, ideology makes imperative. As long as nuclear power is uneconomical, nations will acquire capability for other reasons (Mandelbaum, 1983). The importance of nationalism for nuclear power points to *symbolic factors*, ideas, imaginaries, visions, enthusiasm and doubts that drive the nuclear future. These social representations of the 'atom' and their change over time will concern us now.

## THE NUCLEAR IMAGINATION: TECHNICAL FIX AND CHEAP ENERGY

The atom captured people's imagination. The cycles of public attention and the imagination of the nuclear project are both stimulus and context for development of the technology. The atom entered popular culture as a topic of literature, cinema and drama, of news and reportage in radio, TV and print, and of everyday conversations. Considering the sudden stalling of nuclear power in the 1980s, the key question is, when and why did public imagination shift from a predominantly positive to a negative attitude towards nuclear power?

## Technological Utopia: Reft as Well as Light, Progressive and Reactionary

Strategic technologies engender a genre of sociological writing of axial transitions to a 'new' society, often exaggerated and not always based on the most significant development (Edgerton, 2006). This also brought us the 'atomic society' and the 'nuclear society'. Atomic power is extrapolated into visions of future society. Titles like the 'atomic society', 'nuclear age' or the 'atomic revolution' denote such axial transitions. Nuclear hopes were tied to dreams of national revival and greatness, world revolution and general human progress. For example Darcy Ribeiro (1972), Brazilian anthropologist and philosopher, offered an account of civilization culminating in the 'thermonuclear revolution' that would resolve the energy problem. He was not the first to offer such visions,[8] but his account throws light on their functions. At the time the Brazilian military regime (1964–85) was building nuclear capability with German help, the economic rationality of which was never established. To the contrary, Brazil risked being in possession of defunct technology with no value elsewhere (Galvan, 1983). Ribeiro's rhapsody on the path of civilization buttressed the Brazilian nuclear project with a vision of history taken out of military hands. But Brazilian plans, as earlier Argentinean ones, were vetoed by US proliferation anxieties.

*Nationalism* continues to offer support for a nuclear future. In France the nuclear project compensated for the decline of empire and revived influence in the world. The 'rayonnement de France' doubly signifies the radiance of France's Grande Nation and its capability to mobilise radiating atoms. The nuclear future was a national consensus among the technocratic elites raised in Grandes Ecoles to service a centralised state. This was even supported by Communist trade unions as an act of anti-capitalist defiance counting among their ranks scientific pioneers such as Joliot-Curie. The French consensus thus supported nuclear weapons and power on the right, and nuclear power on the left. The poem *Reft and Light* is one of salvation, redemption and liberation. The atom saved France from economic disaster and redeemed the shame of occupation and collaboration, secured independence on the nation's own resources, guaranteeing liberty in the world.[9] Nuclear sites are the new Cathedrals and Chateaux by which the centre ties the provinces with modernisation, employment and civilization of otherwise 'savage' places. Normandy and Brittany shall rejuvenate in an atomic future. The French CEA (Nuclear Energy Commission) and the EDF (French Electricity) mobilised strategic communication as early as the 1950s, including films and books for children (Hecht, 1998, 201ff). The images become ingrained and France is unable to heed dissenting voices. The few voices of dissent are fractured into Catholic, dissident Communists and Poujadiste voicing anti-materialist, anti-communist or anti-centralist, anti-technocratic sentiments with little resonance in the great consensus (Hecht, 227ff).

Germany performed a similar 'Technikeuphorie'. In 1955 the war occupation ended and restrictions on research lifted. Nuclear energy became the 'hope of the new republic'; it symbolised new beginnings and the economic miracle of the post-war years. Fission came back to its origin and reversed the drain of intellectual and industrial resources. With the slogan 'we will live by the atom', the industry hired nuclear physicists to advice on efficiency. The philosopher Ernst Bloch rhapsodised, albeit in the East, on the *Principle of Hope* (1957) and how small amounts of Uranium or Thorium suffice to turn the desert lands of Sahara, Gobi, Siberia, Greenland and Antarctica into an Italian Riviera.[10] The SPD conferences erupted in 'Kern'ergie, schoener Goetterfunken' (on Beethoven's 9th choral symphony). Atomic power was a Promethean act of liberation. The research centres of Karlsruhe and Juelich embody these hopes (Radkau, 1983, 78ff); this euphoria was as much the effect of propaganda as it was hopeful expression of a people who suffered wartime and post-war hardship. Germany became a leading advocate of EURATOM—the atom would unify Europe as the railway had unified Germany a century ago (Rusinek, 1993). By contrast France saw the EURATOM as a new Germany hegemony (Hecht, 1998, 144).

The 'atomic revolution' and 'atomic society' suggested not only hope for a better future, but a historical necessity, an opportunity to be grasped, a societal imperative, and a risk to be overtaken by history. This is succinctly captured in the 1950s expectation of electricity 'too cheap to meter'.[11] Baby-reactors in back-gardens would produce energy practically free of charge; nuclear cars (never to go beyond the design stage) were in the cards.[12]

The *'nuclear state'* encapsulated the dystopian future. Technologies tend to construct their users by conditioning. The atom and strategic bomber command required a type of personality: the diehard, emotionless, machine-like, error-free operator. The handling of nuclear bombs was no task for emotional characters with social bonds nor was the running of nuclear power stations (Weart, 1988). Rational minds were in demand; what Charles Taylor (2007) later calls the emotionally and socially 'buffered individual' was sine-qua-none. Kubrik's film *Dr. Strangelove* (1964) caricatured their penchant for paranoia and neurotic ticks. The quest for disembodied rationality creates its own irrationalities.[13] Unrealistic social psychology was one element of nuclear governance. The *Nuclear State* (Jungk, 1969) intimated that the security required to run a grid of nuclear power stations would favour totalitarian regimes. The fuel cycle has risky transitions; to pre-empt the deviation of fissile and radioactive materials widespread surveillance of citizens would be necessary. In a culture of fear, the security imperative will trump civil liberties. This popular reading of the 1970s received new relevance after 9–11–2001, when the trading of security against freedom became policy. Radioactive pollution, a dirty bomb and attacks on nuclear installations from the air became real risk scenarios (see Glaser and von Hippel, 2006). Air defence systems were stationed near nuclear installations, and home security and surveillance operate very close to Jungk's premonitions.

## THE ATOMIC CULTURE INDUSTRY

Popular cinema and fiction was the stage on which the nuclear future was performed by 'diegestic prototyping' taking the sting out of the future (Kirby, 2011). In 1955 the launch of *Nautilus*, the first submarine with nuclear propulsion, was a staged media event to demonstrate the powers of the atom other than destruction (Langer, 1995). In the 1950s, governments and the emerging nuclear industry constructed a positive image of the atom in popular culture. *'Atoms for Peace'*, co-ordinated by Abbot Washburn of the US Information Agency, was a Cold War communication effort to win over public opinions. According to secret memos it was ideal psychological warfare:

> Beginning with the positive, substantive proposal on the part of the president [Eisenhower] and continuing to the co-ordinated exploitation of the 'friendly atom' by government and private resources, Atoms for Peace was quite possibly the largest single propaganda campaign ever conducted by the American government probably the largest ever propaganda effort ever in US history (Osgood, 2006, 155f).

The objective was 'nuclear numbing', stimulating a symbolic defence mechanism that prevents people from being overwhelmed by fear. The aim was 'nuclearism' or the acceptance of nuclear weapons as part of everyday life (Osgood 2006, 180); civil nuclear power was instrumental to this strategy. Walt Disney became a key partner in the effort. The animated film *The Atom Our Friend* (book 1956; film 1957) is the fruit of this collaboration. The 'atom' became a key attraction of Disney's 'Tomorrow Land' epitomising human progress and America's leadership in and through science and technology. Theme parks, films, books and TV tied the culture industry to the nuclear complex. Considering that the arms race had three angles— fission and fusion capability, missile delivery, and satellite technology, one might even count NASA and Kennedy's 'man on the moon' as public relations for defence projects that would otherwise run the risk of public scrutiny. Indeed, outside the war effort, NASA finds it difficult to sustain the funding which puts later projects at risk of failure (Vaughan, 1996). Whether 'Atoms for Peace' was a true commitment to a Biblical 'swords to ploughshares' or gigantic total war propaganda continues to preoccupy historians (see Weiss, 2003).

*Hollywood* played its part in 'nuclearism' with dystopian science fiction. All through the 20th century the genre of 'atomic cinema and television' elaborated topics of irradiation, mad scientists, subversion paranoia, secret agents, aliens, mutants and monsters, waiting for war and the aftermath of disaster.[14] Filmmakers seem to have been particularly in touch; their dramas anticipated the changes in public opinion. Monsters like *Godzilla* (1954), *Tarantula* (1955) and *The Fly* (1958) play on fears of genetic mutation from irradiation. *On The Beach* (1959), sequels of *Mad Max* (1979) and *The Day After* (1983) imagine the aftermath of a thermonuclear Armageddon and the totalitarianism arising from the ashes. *Dr. Strangelove* (book 1958, film 1964) ridiculed the paranoia of 'hard men', and James Bond movies (the first, *Dr. No* was released in 1962) offered irony on proliferation, secret agents and special relationships. *China Syndrome* (1979) and *Silkwood* (1984) dramatised nuclear meltdowns arising from imperfect designs, operational incompetence and cynical cover-ups. Reality imitated fiction in an uncanny manner, when *The China Syndrome* was released just days before the TMI incident (see below).

## Framing the Arguments for and Against

Arguments need to resonate to mobilise public opinion. Resonance arises from 'packages of meaning' that fit both public sentiment and media practices. Frames are sponsored by actors who seek advantages and organise alliances by offering striking images and metaphors (see Chapter 7).

One such frame of meaning is the idea of (1) *progress*, which dominated the debates until the 1970s. Nuclear power is frontier science and signifies progress writ-large. A force for good and evil, society has to choose. This view favours a pro-nuclear position. The oil crisis of 1973 raised concerns about energy security and (2) *independence*—an alternative frame. The world realised its dependences. Car-free-Sundays brought bicycles onto motorways to save energy. Nuclear power reduced the risk of being hostage to Arab Sheiks. Consumer organisations and other groups highlighted the lack of (3) *public accountability* and the abuse of market positions. The nuclear industry operated outside public interest, with dishonesty, arrogance and in secrecy. The sector needed to be called to task, scrutinised and monitored closely. When comparing (4) *cost-effectiveness* with other sources of energy, nuclear does not score well. Further investment could mean throwing good money after bad. Scientists and economists tended to think like this. The (5) *soft path* highlighted fundamental choices. Nuclear is a hard path; centralised large-scale technology is blind to ecological consequences. Ecologists show soft alternatives in locally sustainable solutions and 'small is beautiful': sun, wind, waves and water are the future. (6) *Runaway* captures the fatalism that enters a later debate: nuclear energy is a runaway train, a 'time bomb' we must live with, like it or not. Beyond any cost-benefit analysis, a runaway force cannot be contained, and the monster strikes unpredictably. Only humour and sarcasm help, and cartoons on the theme indeed proliferated. The idea of a (7) *Devil's Bargain* emerged after the TMI accident (1979; see below). The benefits of an inexhaustible source came at a terrible price. And the more we use of it, the harder it is to get rid of it (so far see Gamson and Modigliani, 1989).

Since the debates of the 1980s two new frames have appeared. The nuclear industry and its international institutions weathered an adverse climate of opinion with the frame of (8) *keeping the options open* (IAEA, 1994). The idea was to lie low, operate safely, improve efficiency and recruit new blood to ensure that essential know-how did not disappear. This policy keeps the nuclear option until a future renaissance (it would be too costly to start anew). With growing awareness of climate change a new opportunity arose in (9) *clean energy*. With a low carbon footprint, nuclear might be a 'sustainable energy', part of a decarbonised economy. After progress and independence, clean energy is the third positive frame of nuclear power in public conversations.

## Archetypes of Nuclear Fear

The shift in public opinion from positive to negative attitudes was conditioned by resonance with fears and deep-seated images. Weart (1988) identified these as archetypes of transmutation, hard/soft paths, and elitist technocracy.

*Transmutation* through death and rebirth achieves renewal in a novel state of affairs. But this necessitates a last battle of nuclear Armageddon, to prepare the 'second coming' which will liberate the few who toed the line. Clearly this apocalypse is not a theme of sober physics or engineering textbooks, but resonates with a Gnostic eschatology, an old expectation of the end as the new beginning for the selected few. Hence the few might usher in an 'apocalypse now', which is clearly the fear for all those who do not think themselves elected. This provides themes for films and novels: forbidden secrets, punishment with abandonment, destructive rage, victimisation by authority, suicidal urges and related guilt, struggles through chaos, heroic triumph over peril, miraculous regeneration, survival into joyful communion. The mushroom cloud of atomic explosions becomes an icon of the new age. In folklore the mushroom is a symbol of poison with transformative powers: it feeds and kills, but opens minds; it is associated with witchcraft and shamanism, enhanced life, the control over life and death, new life arising from the rot (Weart, 1988, 401ff).

Nuclear power fosters the hard in humans. The men of Strategic Bomber Command must not have emotions that interfere with their task. They are extensions of somebody else's will. This 'hardening' of subjectivity makes for cyborgs, predictable human machines capable of controlling their own destructive potential. This character type became a source of anxiety and revolt, contrasted in the soft approach of 1960s 'flower power': loving, spontaneous and peaceful. *Hard and soft* are collective lifestyle choices.

Hard technology bulldozes the world mercilessly into shape, and humans become machines. The soft path is adaptive, the machine adapts to human measure and desires. Information technology played out this duality in the concepts of hardware and software (see Chapter 6). The nuclear power programme, civil as well as military, was shrouded in elite secrecy. This came up against the idea that techno-science must be a public affair, a common good under public scrutiny. It seems inconsistent to mobilise people into the scientific ethos which is then conducted as an esoteric-gnostic exercise. This contradiction is only absorbed in a vision of technocracy which holds that for good reasons the world shall be ruled by scientifically trained elites. In Veblen's (1964) terms, the 'instinct of workmanship' is not to be contaminated by tribal loyalty (community), seniority principle (respect for elders) or magical and animistic efficacy (i.e. religion). However, in the democratic ethos, scientific competence is just one among several sources of legitimacy. Democrats abhor *technocracy*.

*Nuclear fears* are no constant in public conversations. People cannot live in constant fear, hence try to ignore and repress the uncanny. Thus the fear factor arises at certain times, disappears at others. To explain the cycles of sentiment towards nuclear power, it is therefore important to consider those factors that keep fear in check; anxiety is a latent possibility remembered by myths. Like Sherlock Holmes in Silver Blaze, we have to explain why the dog did not bark, the abnormal silence. A key variable to reduce public fears is the relationship people have to authorities. Can people trust authorities to keep restraint in the handling of nuclear weapons? Can people trust the safety regimes of the nuclear installations, or is safety sacrificed on the altar of costs-saving? Lack of trust in authorities and their systems reveals the unease of the nuclear age (Mandelbaum, 1983).

## News Cycles and Attitude Change

In newsprint we observe the ebbs and flows of *nuclear sentiment*, as shown in Figure 3.3 for UK, which shall be our baseline. Nuclear news peaked in the mid 1950s, stayed high into the 1960s, then declined to rise briefly into the 1980s. Nuclear news never recovered 1950s levels, but increased massively into the 2000s marking the 'nuclear renaissance' with peak in 2004. Similarly, US nuclear news never recovered the levels of the 1950s (see Nealey, Melber and Rankin, 1983).

Each article is rated for news tone. Good news comes in three waves: in the mid 1950s, the mid 1970s and again after Chernobyl 1986 (see below). News was bad immediately after WWII and in the late 1960s and the 1980s. The oil crisis of the 1970s and the later 1980s saw more positive news. Before 1975, intensity and positivity are correlated ($r = 0.38$). After 1975, more news meant more bad news ($r = -0.27$).[15] The ambivalence of public discourse is found in the variability of tone, which is large in the mid 1970s, after Chernobyl 1986, and in the 1990s with the budding renaissance.

The nuclear news from 1953 to 1964 captures the euphoria of 'Atoms for Peace', 'energy too cheap to meter' and modernisation for a better future. The atomic age is a success of British engineering. Into the 1960s, Britain led the world with the largest civil nuclear power programme. The loss of this leadership became an index of British 'decline' (Williams, 1980). The counter current is marked by news of nuclear dust from afar.[16] It highlighted the fourth parameter of nukes after yield, heat and radioactivity: fall-out. Nuclear winters with crop failures, food shortage and anthropogenic climate change would follow a nuclear explosion anywhere. Globalisation entered as a shock idea. Nuclear testing worried the public and scientists alike. It revealed the split reality of the nuclear revolution: 'Atoms for Peace' here, thermonuclear fall-out there. When testing went underground after 1963, it had alerted people to the globalisation of pollution (Edwards, 2012; also Chapter 4).

The *change* from more positive to more negative nuclear news had a different timing in different places. Britain performed its nuclear hype in Harold Wilson's 'white heat of technology' speech at the party conference of 1963. Shortly after, mainly behind closed doors, started the second nuclear power programme, scaling up reactors and making crucial design decisions.[17]

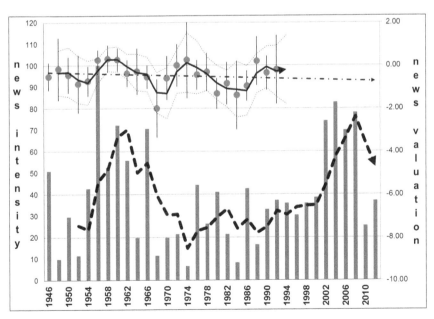

*Figure 3.3* The index of nuclear news in the UK Press 1946–2012: numbers of articles, bi-annually, and eight-year moving average (1955/56 = 100; left scale). News slant with upper and lower bands; rating of discourse of 'concern' (-4) and 'promise' (+4); valuation data is available until 1992 (n = 502), news intensity until 2012.
*Source*: Bauer et al. (1995) and Lexis and Google Trends (see Appendix 2).

British nuclear power was scrutinised by the coal industry: energy pricing excluded the costs of safety and of the dismantling of coal culture. Investigations tried to clarify the situation, and the press reflected this scrutiny. The oil crisis of 1973 put nuclear energy temporarily in a positive light. High oil prices improved the position of nuclear energy relative to fossil fuels. Britain had its nuclear power debate between 1975 and 1978 over the Windscale Fast-Breeder Project, a plutonium recycling machine which the *Daily Mirror* anticipated would make 'Britain the World's Nuclear Dust Bin' (headline 21 October 1975).[18] The British Nuclear Industry countered with the 'Power for Good' exhibition and many others since, stressing to be an energy solution. At the time anti-nuclear attitudes were explained by immature public understanding, biased media, and misused engineering jargon (Williams, 1980, 261ff). In the US the crossover from good to bad press had occurred already in 1967 (Weart, 1988, 387). Germany saw a rapid change after 1973: less than 1% of news had previously printed critical assessments of nuclear technology (Kepplinger, 1995, 374; Radkau, 1995, 339). Thus the age of nuclear hope ended when nuclear fears returned and got the upper hand in the 1970s.

## The Impacts of Nuclear Accidents on Public Opinion

Absolute safety was essential for nuclear power, but accidents are a normality of complex systems (see Perrow, 1984), and as nuclear power is a complex system, safety posed a paradox: an impossible necessity. There had been serious incidents at nuclear units (see Appendix 2), often understated or covered-up like the South Ural event of 1957. However, the events at Three Mile Island (1979), Chernobyl (1986) and Fukushima (2011) were historical. Let us explore their cumulative impacts.

*Three Mile Island* (TMI) near Harrisburg (Pennsylvania, US) operated a pressurised water reactor (PWR). On 28 March 1979, around 4 am, malfunctions and operating errors nearly led to a disaster. A leak in the primary cooling system went unnoticed. At 9 am an explosion occurred within the containment, and it took six hours to regain control. The result was a partial meltdown and leakage of radioactivity into the environment. On the incident scale TMI rated 5 (max 7). The cause of the valve failure was never determined, much analysis focussed on the handling of events. The effect of the leaked fall-out remained inconclusive (see Mangano, 2004). Subsequent studies claimed that the main 'fall-out' of TMI was the stress caused by evacuation of 50,000 households (144,000 people) within a 15-mile radius. People with higher education and households with small children were more likely to move out (Houts, Cleary and Hu, 1988, 18ff). The short-term costs to individuals and businesses were $90 million and shareholder losses were $800 million by 1987; the utility company was saved only by state subsidy. The clean-up of the site lasted 14 years and cost another $1 billion (Walker, 2004).

TMI made media coverage and opinion polls. Figure 3.3 above shows how UK news picked up TMI, but the cycle hit bottom very soon. The arrival of Margaret Thatcher in the UK made different news. At the time, three questions were asked to gauge public sentiment, the first two asking about general support for nuclear energy and support for nuclear units in the local area. Locality only opposition or NIMBY (not in my backyard) tended to be larger than general opposition or NIABY (not in anybody's backyard). NIMBY opposition was thought to be amenable to compensation for taking risks (see Wolsink, 1994).[19] A third question gauged those who were happy to live with existing installations, but would not support any extension of the 'devil's bargain'.

TMI brought home an event of 'very low probability'. The safety regime of nuclear energy fell into disrepute as it emerged that warnings on TMI's operations had been dismissed. A 'conspiracy of silence' resonated in popular culture.[20] Science writers recall TMI as watershed event which mutated their work ethos from idolatry to iconoclasm (Franklin, 2007). However, public perceptions had started to shift already in the 1970s. US support for nuclear power was 60% in the early 1970s. After about 1975 support declined. All through the 1970s about 20% remained 'unsure' on the issue, by 1978 this was at 40%, and opposition increased to about the same figure. TMI boosted opposition. By 1979 polls recorded 60% opponents, 35% supporters, and 5% 'unsure' (score = 64). With TMI the previously unsure became opponents (Nealey, Melber and Rankin, 1983; Rosa and Dunlop, 1994).

TMI had an effect across Europe and the US with support dropping 10% as shown in Figure 3.4. US polls observed a 'rebound effect', support for nuclear power recovered but only to the long-term downward trend. Europe varied, and some countries went into opposite directions. In 1978 public opinion in the UK, Italy, Denmark, Ireland, and Luxembourg supported nuclear power, while France, Germany, Belgium and in particular the Netherlands did not. In Germany, public support had eroded after 1975 (Kepplinger, 1995). By 1982, the French, Germans and Dutch showed more positive views; the French and Germans now even supported nuclear power, while the Danes, British and Irish had become markedly more negative. In Germany and in the UK, much of this change comes from high ambivalence (i.e. neither supporting nor opposing) where opposition had reduced in Germany and support in UK. France and Germany, both with large nuclear projects, succeeded in outsourcing the blame: calamitous devices are elsewhere, home-grown technology is safe.

*Chernobyl* in Ukraine was an altogether different story. The accident was a 'worst case scenario', point 7 on the incident scale. During a safety training routine in the early morning of 26 April 1986, reactor unit 4, a Russian RBMK light-water graphite moderated design, got out of control, became critical and went into meltdown. A massive explosion ignited a fire that threatened other units. Large radioactive fall-out spread with wind and weather. Because of

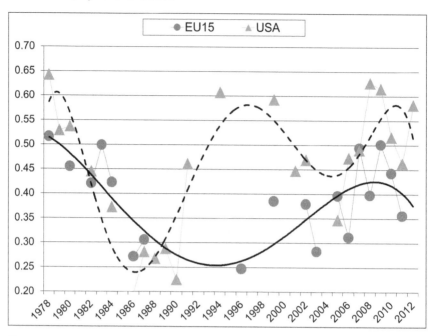

*Figure 3.4* European and US sentiment on nuclear power with fitting trends. The score is calculated as acceptance = $\%_{supporters}$ / $[100 - \%_{Unsure/dk/na}]$. A score of 0.50 means a 50:50 balance of support and opposition. Support means nuclear power is 'worthwhile', 'acceptable risk', or 'generally favored', depending on the poll. *Sources*: Nealey, Melber, and Rankin, 1983; Rosa and Dunlop, 1994; Eurobarometer surveys; PollingReport Energy; and recent polls from NOP, Gallup, YouGov, etc.

official denials, the evacuation of the population was delayed. Some 40,000 moved on rumours, before finally 130,000 people were evacuated. Fall-out was reported in Finland on 27 April, in Sweden on 28 April; they raised international alert to an unknown source of radioactivity. The USSR government despite 'Glasnost' (transparency policy) struggled to inform. Radioactive dust dispersed into Poland, East Germany, Czechoslovakia, Hungary, Austria, Southern Germany, Switzerland, Northern Italy, and to Cumbria in Britain. Iodine-131 and Caesium-137 were of concern for public health. The radioactive fall-out exceeded that of nuclear tests in the 1950s; 100 times the radiation of Hiroshima and Nagasaki was released. This brought a global affair home again; radioactivity ignored national borders, even heavily militarised ones, as it crossed Europe. Locally, 28 heroic firemen and plant operators died from radiation illness within months; 134 suffered acute radiation sickness. 109,000 people were permanently resettled outside an exclusion zone, which remains 20 years later. Seventy per cent of the fall-out descended on Belarus. In order to prevent fall-out from reaching Moscow, the Russian Air Force drained clouds over the area of Gomel, 150km north of Chernobyl, which has seen a ten-fold increase in thyroid cancer among children (OCHA, 2000).[21]

The world continues to count the dead, diagnose the sick, and assess designs after Chernobyl (Chernousenko, 1991; Ramberg, 1996; *Nature*, 2006).

By 1984, overall public opinion had shifted less in Europe than in the US; attitudes were entrenched by strong local controversies; however the 'Chernobyl effect' was stronger across Europe. Finland, Italy, Austria, Luxembourg, Germany, and Denmark reacted with negative shifts of 20% and more. The UK, Greece, France, Netherlands, Sweden and Spain reacted less strongly. In West Germany, France and the UK a favourable majority evaporated after Chernobyl; only in Sweden opponents remained a minority (see Renn, 1990). Shifts in public opinion were mostly unrelated to local radioactive fall-out ($r = 0.05$; $n = 14$; Rüdig, 1990). US opinions, with no real fall-out, remained unaffected; opinions had deteriorated through the 1980s and this trend continued. In Japan, likewise without fall-out, opinions shifted but remained favourable. Greece, with large fall-out, showed little shift, as they had always opposed nuclear power. According to Kepplinger (1992, 1995) this disproportion of fall-out and public opinion was consistent with a theory of 'artificial horizons', i.e. a conspiracy of hijacked science news by the 1968 generation. Humans, without any sense organ for radioactivity, must rely on official reports, rumour and mass information, and the mass media had reported the situation in a 'responsible, accurate and fair manner' (Renn, 1990, 161). Again after Fukushima 2011 (see excursion 3.1 below), public opinion withdrew support from nuclear power, but did not change the structure of opinion formation (see Visschers and Siegrist, 2012).

Figure 3.4 above shows the long-term 'rebound effect' after Chernobyl, fast in the US, and slower across Europe. The rebound was stronger in countries with low fall-out, where people had less obvious reason to stick to their entrenched positions on nuclear power (Renn, 1990, 158). Little is known on opinions in former Eastern Europe. Before 1989 there were few if any opinion polls. However, data from former Yugoslavia showed that in the wake of Chernobyl opposition had doubled to 80% and rebound to 60% later (Van der Pligt, 1992). Eurobarometer reported in 2006 and 2008 that Croatia and Slovenia opposed nuclear power, while Bulgaria, Czech Republic and Slovakia, Hungary and Lithuania strongly favoured it. Here, Chernobyl faded away with the totalitarian past, while Estonia, Latvia, Poland and Romania retained sceptical publics.

TMI and Chernobyl led people to reassess the past. Earlier incidents came to new light: Kyshtym in Southern Ural (1957), Lucens in Switzerland (1969) and Fermi-1 near Detroit (1966). In Cumbria UK, *Windscale* had provided plutonium for the British H-bomb project 'Hurricane'. From 6 to 12 October 1957, a heat-release was delayed and overheating lit a fire that released radioactive materials into the air, level 5 on the incident scale. The authorities admitted local fall-out, but it was the radioactivity recorded in Norway and the Netherlands which created a diplomatic incident. The consumption of local milk was banned; farmers were compensated. The official inquiry blamed the operators when the causes of the accident remained inconclusive; operating staff continue to seek vindication. Five studies assessed the fall-out cancer burden. An initial null-increase became one of 90 to 248 additional cases, of which 10 to 100 were to be fatal (Arnold, 1992). The site changed name: 'Windscale' became 'Sellafield', but mistrust of official authorities persisted, as documented in a case study of how local knowledge challenges scientific expertise (Wynne, 1992).

Eurobarometer and other data showed that after 2000 support for nuclear power continued to be lukewarm at 30%, but opposition declined, and ambivalence increased. The nuclear 'renaissance' had been taking hold in public opinion, in the US as in Europe, by making opponents more ambivalent (see Figure 3.4 above). This realignment was particularly clear in Germany, Italy, UK, Sweden, Switzerland and Finland, where nuclear debates had reopened. Finland had restarted a nuclear programme with public consent demonstrating that societal risk assessment was temporary (Litmanen, 2004). Many European governments were anxious to follow Finland's lead. Then, on 11 March 2011, a massive earthquake in the Pacific was followed by a gigantic tsunami and the meltdown at Fukushima power plant in Japan.

Figure 3.5 shows the differential effect of Fukushima on world opinion. Opinions in Lithuania swung the most, from widely positive to totally negative. Strong shifts also occurred in Switzerland, Italy, Sweden, Germany, India, US and Canada. Unsurprisingly, Japan experienced a major

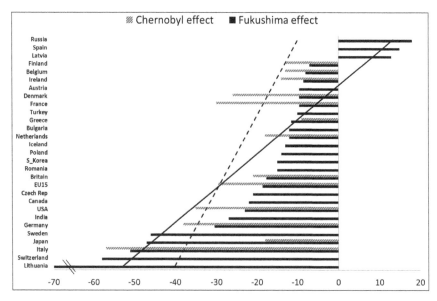

*Figure 3.5* The Chernobyl and Fukushima effect on public opinion in comparison. The index shows the differences in the balance of opinion (favour against disfavour) two years before and two years after the events of 23 April 1996 and 11 March 2011. The two effects are correlated (r = .65; n = 13). *Source*: own collation of various opinion polls, e.g. Gallup International; Ipsos Global @divisor, June 2011.

turnaround in public sentiment. A country that hitherto supported nuclear power, turned against it. The data available shows three countries where public opinion improved after Fukushima: Russia, Spain and Latvia. While there is some correlation between public reactions to Chernobyl and Fukushima, the great puzzle remains the diversity of responses: a single trigger event, and a wide spectrum of reactions. The pragmatic meaning of Fukushima is clearly constructed in the local circumstances including the legacy of resistance to nuclear power.

*Excursion 3.1* Once Bad luck, Twice Suspicious, Three Times a System Failure

*On Friday 11 March 2011, mid afternoon, a very large earthquake of extraordinary magnitude 9.0 off the eastern coast of Japan caused three nuclear installations at Fukushima Daiichi to shut down automatically; three others were already shut for maintenance. One hour later a devastating 14m tsunami of the earthquake made landfall and hit the installation. Constructed to withstand 5–6m waves, the flood destroyed the emergency diesel electricity generators of the plant, and a catastrophe unfolded. Without pumps and the cooling system no longer operative, reactors overheated, went critical and into meltdown, and caused explosions and emissions of radioactive materials in days to come. At 7 pm that day, the Japanese government declared a nuclear emergency. In the days after, the population within 20km of the power station was evacuated; people further away were advised to take precautions. The government considered evacuating Tokyo at 250km from the nuclear installation, but did not have to go as far as luckily, the winds blew radioactive materials the opposite direction, into the sea.*

*The massive tsunami itself killed 15,000 people and left 3,000 missing. In Fukushima, two nuclear workers died in the emergency, and 100 to 1000 people are expected to die of cancer as a long-term effect of exposure to Gamma radiation, isotopes of Plutonium and Strontium, and Caesium-137 and Idine-131 in the environment. 100,000 were screened but nobody was found acutely affected. The decontamination of the power plant area is estimated to last until 2052. 210,000 people took flight or were evacuated from the area.*

*On 7 June 2011 Japanese investigation concluded: that a) TEPCO, the private electricity operator, was insufficiently prepared to cope with a critical nuclear accidents, b) after the batteries and power supply boards were inundated, almost all electricity was lost, c) the operations did not consider such a power failure or any kind of prolonged power loss and d) the company thought that in a serious incident, controlling the reactor would be possible, but made the error of assuming that traditional power sources were reliable. A normal system failure as it appears. A week before the disaster, TEPCO had considered a report of the uncontrollable consequences of a large earthquake and tsunami on the plant. Though a known scenario since 2008, no action was taken, as it was a very low probability risk. Regulatory capture is a major condition for this disaster: the relations between the government, TEPCO and the regulatory authority (NISA) were too close to really work; the industry was a lucrative career which no regulatory employee wanted to jeopardise. After the disaster the Japanese government nationalised the Fukushima nuclear plant as no private company can survive with this legacy (Sugiman, 2014).*

*Initially the Fukushima events were recorded level 3, 'nuclear incident', which on 11 April 2011 was corrected to level 7, 'major accident', the same level as Chernobyl 1986. This uncertainty in the assessment is indicative of problems between TEPCO and the regulatory authority, and the Japanese and world public, which only later came to the fore. Subsequent investigations revealed a series of underestimations of the radiation emissions into the sea (the largest in history) and into the atmosphere (estimates are at 40% of Chernobyl). Two years after the disaster, it is unclear whether the reactor is stable. Five thousand people are working on reducing the damage of Fukushima, 500m³ of ground water mixes daily with radioactive material and needs to be contained, but partially leaks into the sea where fish continue to be contaminated. Two million people are being monitored for health effects of small dosage radiation exposure (source: NZZ, 6 March, 2013).*

*The radioactive fall-out, serious as it was, might in the end be contained, but the political and cultural fall-out of this third serious nuclear accident occurring in a highly technological nation is globally consequential. The so-called 'nuclear renaissance' is seriously in doubt, with many nations reversing 'reversed' nuclear policy within weeks of the events in Japan. Germany decided the exit nuclear energy, and so did Bulgaria, Switzerland and Italy. Even the US, Japan and France are for the first time calculating the real costs and a future without nuclear energy (Lovins, 2013). The Japanese have lost their traditional trust in technology and the system. While, Korea, China and India stick to their nuclear plans after in-depth and costly reviews of operational procedures (Ramana, 2013).*

---

## THE MOBILISATION OF RESISTANCE

The demonstrated long-term shifts in public sentiment over the last 60 years provided an opportunity to mobilise public resistance to nuclear power as much as it is the outcome of such mobilisation. What started as local discontent over armament and local power stations in the 1960s became a global movement by the 1980s. Anti-nuclear protest marched separately and at times in concert.

## LEVELS OF RESISTANCE

Resistance is mentality, attitude and action. Resistance is political participation. This might include, with increasing disruptiveness, the signing of petitions, street demonstrations and sit-ins, interrupting public events, litigations against corporations and governments, civil disobedience, rioting and violence against installations and persons. Such actions are likely to be reported in the mass media, and thus provide the data to assess the intensity and 'styles' of resistance. Germany, Switzerland, Denmark, the US and Austria see the largest numbers of people in protest, more than in Scandinavian or Mediterranean countries. Japan and France fall off, and

Belgium and Canada report least public resistance (Rucht, 1995). Most Western European countries see some anti-nuclear mobilisation after 1975, but with different timing. Sweden and France peak in 1976 and 1980; Germany mobilises mainly in the mid 1980s. Switzerland, the US, and the Netherlands had their peaks of protest in 1978, 1979 and 1981 respectively (Van der Heijden, Koopmans and Giugni, 1992; Jasper, 1988).[22] The protest cycle defines the *politicisation process* (Jasper, 1988). Before the large protests, the political arena was very limited public debate. Protest becomes a measure of debate intensity and culminates in a decisive event. Oregon (US), Sweden, Austria and Switzerland voted in referenda; Netherlands and Taiwan (1988/89) staged national debates. The issue is temporarily closed by a democratic decision.

French and German resistance was different in style: France preferred petitions but saw more violent protest, while Germany preferred street demonstrations, civil disobedience and interrupted hearings. France, Germany and Japan scored highest on disruptive protests (Rucht, 1995; Ruedig, 1990). These differences express political cultures, and the level of violence could be a measure of the frustration over the non-response of the political class as in the case of France and Japan.

Resistance was not only a matter of private persons demonstrating publicly; but states resisting diplomatically. So, for example, Austria and the Czech Republic exchange regular notes of protest regarding a Czech power plant at the border over safety concerns. Similarly the EU made the closure of Lithuania's Chernobyl type power plant a condition of EU membership in 2004.

## The Motives that Mobilised People

What made people join the streets in protest against nuclear power? The initial concern was *radioactivity*. Radiation is not detected by human senses and hence uncanny, but its power to induce genetic mutation and cancerous growth of biomass is well known. While most fission products decay quickly some, like Strontium-90 and Caesium-137, stay around 25 years and longer, and accumulate in the food chain. The dose-effect remains controversial. Is there a *threshold*, below which radiation is irrelevant or even beneficial—the hormetic J-curve (like drinking wine)—or is there a *direct dose-effect* (like smoking)? Is the dose-effect linear or non-linear, and of which the slope is a parameter (see Wigg, 2003)?

Specification of this detail is the key to model the disease burden arising from exposure to the nuclear fall-out. People worry about uncertainties in the models which underpin official reassurances. The radiation concern gained public profile with the *nuclear fall-out* debate that followed the nuclear tests of the 1950s. The thermonuclear H-bombs dispersed radioactive dust over long distances. The late 1950s cycle of anti-nuclear protest cantered on fall-out and ended with the Test Ban Treaty of 1963 (see chronology in Appendix 2). Protest was carried by scientists who had pioneered the nuclear effort but now put their minds on a different future; Britain had its Committee for Nuclear Disarmament (CND, founded1957). The Easter marches counted on writer J.P. Priestley, philosopher Lord Russell and historian E.P. Thompson. Fall-out raised awareness of globalisation: the 'nuclear winter' defined global pollution in a shocking scenario (IPPNW, 1991; Edwards, 2012).

Governments tried to absorb the nuclear threat in *civil defence measures*, built concrete shelters and trained a resilient population to 'survive' nuclear attacks; risk communication was born. School children were instructed to 'duck and cover' under the desk. What preoccupied observers were unrealistic assumptions about nuclear war: a population capable of absorbing retaliation would embolden the military on the first-strike option (Churcher and Lieven, 1983). Civil defence perversely increased the likelihood of a nuclear war. Slips in nuclear constraints were also more likely because of progress in missile technology. Missile defence systems in big cities made the first-strike option feasible, as the counter-attack would be absorbed by local defences. Steps that undermined deterrence worried common people but also defence experts, who stimulated protests against missile defence systems in the US in the late 1960s. Citizens of Utah and Nevada mobilised against inter-continental missile silos in their vicinity. Such installations had detrimental effects on the water supply and the local community would become a strategic target. Various anti-ballistic missile (ABM) treaties banned missile defence (Mandelbaum, 1983). Similar concerns mobilised globally against the 'Star Wars' idea in the 1980s.

In Europe, NATO's decision (1979) to station missiles with 'neutron bombs' that killed people but not equipment revived the Peace Movement. The fear of a tactical nuclear war in Europe, and doubts about the restraint of the US government—in 1981 Ronald Reagan was elected on a war mongering ticket—fuelled protests in 1982. Expectations of a nuclear attack within a lifetime rose from 10% to 33% of the population (Thompson, 1985, 11). To press for *disarmament* was the purpose of the Peace Movement in the UK and elsewhere. MAD was considered literally mad, and so was the idea of tactical nuclear warfare; pressure from the street could encourage governments to seek ways out of the nuclear pile-up. It was unclear was whether this should be a unilateral move or a mutual commitment. At the same time, atomic energy had become the battleground for the green movement. The idea of building nuclear power stations close to large cities was perceived to be similar to stationing nuclear missile systems, be that in Long Island (US), Britain's Greenham Common or German cities. Attempts to prove the safety of nuclear installations brought to light the uncertainties and galvanised unprecedented anti-nuclear protests, military and civil (Radkau, 1983).

Nuclear energy installations sourced military projects from the beginning. The mechanisms of 'Atoms for Peace', the IAEA and the treaty of 1969 have not entirely failed to contain *proliferation*. The Bomb Club has grown, and the countries with declared projects declined from 23 in 1960 to 10 in 2006. Proliferation remains a major issue since Acheson and Lilienthal in 1946 declared it in the nature of things. Nuclear capability is a symbol of power, signifies achievement, membership of a prestigious Club of the few. As long as the economic rationale of atomic energy remains doubtful, its military and symbolic value will have to justify the effort (Cirincione, 2007). The nuclear fuel cycle remains risky, and proliferation creates its own 'chain reaction'; countries are anxious about their neighbours (Mandelbaum, 1983, 81ff).[23] The credibility of civil nuclear power hinges on the proliferation issue.

Nuclear energy has to be *safe and to be seen to be safe*. Safety has two aspects. The reactor is safe when redundancies are built in to contain radioactivity that might accidentally escape. The station must also have a safety culture in daily operations and in dealing with radioactive waste. Probabilistic risk assessment, scenarios of the maximum credible accident (MCA), the incident scale (INES 1–7), regular emergency drills and international co-ordination of staff training to report and monitor incidents contribute to a safety culture. However, past and present nuclear incidents cast a shadow of doubt despite the rational reassurances of comparative risk assessments.

Reactors produce *radioactive waste* of various degrees of danger, which waste management stores, reprocesses and recycles. Materials often are transported to the disposal sites from afar. The depositories might be scientifically determined, but the local inhabitants are often unhappy to be the damp of high-tech civilisation. Storage and transfers are risky: material might end up in the wrong hands, pollute the environment and endanger the population. At the end of its lifetime of 25–30 years a reactor site will be contaminated for years to come. The safety of operations and waste disposal made the *siting* of nuclear power controversial. Initially a de-central solution to energy supply that would foster local autonomy, size and strategic value, atomic energy became the dominion of central powers. Local communities saw themselves disenfranchised, and concerns over safety and waste galvanised NIMBY and locally unwanted land-use (LULU) positions (Wolsink, 1994). What seemed a golden apple became a hot potato.

The *real costs* of nuclear power are a perennially controversy. As long as the projects were military, the real costs were hidden in defence expenditures. But privatisation of the sector brought full-cost accounting, including R&D, construction, operation, insurance, waste management and decommissioning. Comparisons of unit costs for fossil fuels, nuclear power and renewable sources are the battleground. Full economic costing is the precondition for the renaissance of nuclear power in Britain as elsewhere, but it remains to be seen how this 'condition' will be implemented. Capital costs depend on the requirements for licensing, safety standards, liabilities and consultation exercises. It might have taken 3–4 years to build a nuclear power station in the 1960s; it takes 12–20 years to build one in the 21st century in Europe, though much quicker in China.

Investment uncertainties are high. The competitiveness of nuclear energy hinges on high oil prices; and any new commitment to nuclear power distracts from fully developing renewable sources. A factor of full economic costs is the *burden of liability*. How much of the risk is taken by the operators and how much is the taxpayer's through direct and indirect subsidies. Citizens end up subsidising nuclear power and free-marketers recognise distorting state interventions.

Not everybody is equally attentive to all of these issues. However, motives give these issues personal significance and they resonate in different constituencies. Different motives inspire acts of resistance (Van der Heijden, Koopmans and Giugni, 1992). The *localist* sees in nuclear power a threat to the community, its traditions and autonomy. The technocratic agenda infringes the proud local autonomy. The *anti-technology* mind reckons that the technology is, despite reassurances, unsafe. Nuclear power is the very problem to which it offers itself as the solution. Social choices need to consider the 'soft path'. The *anti-industrial* sees the environment, health and civil rights sacrificed on the altar of a private industry and profit; leading to private greed and public squalor. The *anti-authoritarian* worries about the military's control of atomic energy. The 'nuclear state' envisages total government surveillance and control where security is traded off against civil liberties. The *anti-capitalist* combines several of these motives with a (neo-)Marxist view of history. The nuclear power industry is the attempt of monopoly capital to stem the terminal crisis of capitalism. The *pacifist* harks back to disarmament and opposes atomic power because of the link to nuclear weapons. Civil or military nuclear power is no choice. The joint fuel cycle proliferates unavoidably; nuclear energy sustains the nuclear capability; being against war means to oppose nuclear energy. Finally, the *rationalist* soberly considers the costs-benefit question and shuns all ideology other than calculus. They want to compare energy options on full information, and bemoan accounts that externalise costs. In the early 1970s, local and anti-technological motives dominated; later, anti-capitalist, anti-authoritarian and pacifist motives gained more salience. After TMI and Chernobyl the economic rationale gained its full weight: nuclear power was neither safe nor economically viable.

## The Social Basis of Anti-nuclear Protest

Who were the people who resonated with these images, expressed oppositional attitudes and participated in anti-nuclear protests? Women were more likely concerned than men and reacted more strongly in the negative to TMI and Chernobyl. Education and family income influenced the likelihood of changing attitude but not the level of support for local power stations. Support and opposition was higher among the more educated. Supporters and opponents of nuclear power were equally knowledgeable of the issues. The less educated were more ambivalent and uncertain, and more likely to say 'I do not know'. TMI had a greater effect on the less educated; educated minds were 'inoculated' with knowledge that made attitudes resistant to change (see Eagly and Chaiken, 1993, 305ff). Education had no consistent effect on NIMBY attitudes. However, people with higher incomes were convinced: they were more supportive, though some also opposed, like people with lower incomes. Less income meant more uncertainty. No relationship existed between income and NIMBY attitudes; this surprised observers because nuclear sites brought employment opportunities. Age is complex: the young and the older were more opposed than the middle aged. The younger and the older also responded to TMI (Nealey, Melber and Rankin, 1983).

Geography mattered. Across the US, attitudes varied between states. For example, Missouri remained more favourable than Oregon. In 1980, Oregon accepted a moratorium on nuclear power with 53% yes-votes. The response to TMI was stronger and more lasting in the Northeast, where it happened. The South generally supported nuclear power. Similar geographical disparities persist in Europe. France always had regional differences: Normandy and Brittany favoured nuclear power, while other regions opposed it in the 1980s (Hecht, 1998, 248); clearly there was large disparity across Europe, as discussed above. By the late 1980s, a good predictor of anti-nuclear attitudes in Europe was the left-right self-positioning. The more left-leaning, the more likely respondents opposed nuclear power; the right tended to supported nuclear power. The falling out of left-leaning voters with nuclear power is surprising if one considers the revolutionary representations it had enjoyed back in the 1950s East and West; clearly something had happened in the meantime (Van der Pligt, 1992, 7; also see Chapter 4).

In summary, at the height of anti-nuclear mobilisation of the 1970s and 1980s, the recruiting ground of anti-nuclear opposition in the US and elsewhere were women, often with young children and both younger and older citizens; the more educated are on either side. Those with lower incomes and the less educated often kept an open mind longer and thus responded more strongly to TMI and Chernobyl. Long-term changes in attitudes and the realignment of the atom on the left-right spectrum followed secular changes in values and the emerging environmental sensibility.

## EFFECTS OF RESISTANCE ON THE NUCLEAR PROJECT

The key question about the nuclear power debate is: what is its effect? How does the nuclear project of 2012 compare to that of 1970 or that of 1955? Did the mobilisation of resistance make any difference for the *nuclear project*, locally and globally? We will explore some impacts attributable to the anti-nuclear mobilisation and end with a cautionary comment on a possible cause-effect relationship between resistance and new technology.

## Delayed Diffusion and Differential Uptake

Construction of nuclear power units peaked in the mid 1970s. In the US all construction stopped after TMI (1979), and mounting anti-nuclear activism across the world brought some nuclear projects to a halt; others rescaled, some remained unaffected. Public opinion resisting nuclear energy is one of the factors which explain the development of a global affair, others include: a stagnating stock since the mid 1980s at about 450 units, renewals not compensating for decommissioning, a declining share of electricity production, and the concentration of piles among an existing group of countries. Figure 3.2 shows the situation by 2006. The US operates the most units (104) that produce 20% of the country's electricity. France (59), Japan (55) and Russia (31) follow. After 1989, the dissolution of the USSR created several nations who rely heavily on nuclear electricity: Lithuania (1 unit; 80% of electricity supply), Ukraine (15; 46%), Slovakia (5; 57%) and Armenia (1; 35%). The world's most nuclear countries are France (59; 78%), Sweden (10; 50%), Belgium (7; 57%) and South Korea (20; 40%). Sweden was bound by popular referendum (1980) to phase out by 2010, but retracted this in 1996. The former nuclear leader Britain operates at lower levels (19; 20%), and struggles with decisions on what to do next. Easy access to energy from fossil fuels and water explains an energy mix, in part. But do anti-nuclear sentiments have anything to do with it?

## Encouraging Disarmament

Through three cycles of the late 1950s, late 1960s and early 1980s, public opinion on nuclear weapons accentuated and promoted arms controls, opposed the stationing of tactical neutron bombs, delayed the arrival of Pershing missiles in the 1980 and kept the US and Europe from drifting apart on strategic matters. Public opinion resonated with arms issues whenever people lost trust in governments to show restraint in using nuclear weapons, thus reinforced that very restraint and avoided nuclear disaster. To that effect, public opinion mattered (Mandelbaum, 1983). Also, the anti-war movement of scientist intellectuals kept open non-governmental communication across the Iron Curtain. These contacts eased tensions by staging highly informed opinions and encouraged 'dissidence' on either side. An anti-nuclear movement inspired by survivors of Hiroshima-Nagasaki sustained Japanese pacifism and kept the country from nuclear proliferation. After Fukushima this position seems strengthened in Japan, not least as the previously envisaged massive expansion of nuclear power is cancelled and cannot serve as a cover to develop weapons capability.

## Modulating Expectation of Future Capacity

An effect of resistance is the moderation of expectations. The promotion of new technology involves the projection of future opportunities. The 'past futures' of nuclear power can be mapped by comparing the projections of key actors at different times for a fixed point in the future. Figure 3.6 plots past nuclear futures. It captures the rising expectations in the US in the late 1960s from 900 to 1,500 gigawatt electricity or GW(e). SIPRI's 1974 projections for 2020 were for 15,000 GW(e) global capacity. In the same year, the OECD expected 4,500 GW(e) for 2000. In 1977 this figure had shrunk to 1,200 GW(e); a few years earlier the Atomic Energy Commission expected 1,500 GWs for the US alone. In the late 1970s projections rapidly adjusted to about half of the OECD's, to a seventh of the US' initial projections. By 1991 we reached the expectation of 275 GW(e) for 2000 across the OECD, very close to reality.[24] In 1974 IAEA expected that by 2000 50% of global electricity and 80% of all added capacity would be nuclear (SIPRI, 1974). Currently nuclear electricity is declining at about 15% of global production or a third of the high-time expectations. Before Fukushima expectations had risen again: MIT (2003) predicted 20–35% nuclear electricity by 2050; but this again was corrected downward quickly.

It was projected that nuclear would be an energy source for the developing world. By 2000 more new capacity would be installed in the developing world than in the developed world. This prediction came true, as the developed world saw a moratorium on nuclear build since 1980. However, few Latin America and Asian countries who envisaged in the 1970s

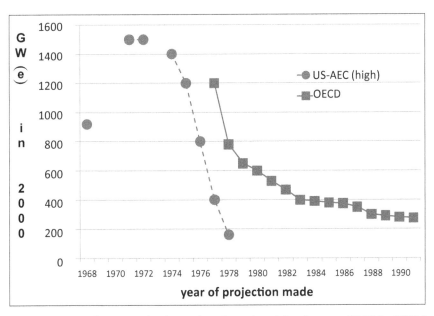

*Figure 3.6* Declining projections of nuclear electricity for year 2000 in GW(e) produced made in different years by AEC (for US) and for OECD countries. *Source*: Lovins, 1979; various IAEA reports.

that they would take up nuclear power did effectively do so; new building of nuclear power is led by China, India and both Koreas. On the whole, the global nuclear project did not live up to its projections of the 1970s. Accommodations of reality set in before TMI and Chernobyl, and followed the widespread anti-nuclear mobilisation in the late 1970s and early 1980s. This would indicate that other factors than public opinion kick-started the adjustment of future expectations of industry and agencies, while clearly public opinion reinforced the recurrent adjustments of 'past futures'.

## Project Alterations

Public resistance creates attention and evaluates nuclear power, and is able to induce alterations to the nuclear project. An index that measures the alteration of the nuclear project helps to assess this impact.[25] Figure 3.7 shows the realisation index for 28 countries. Belgium, Israel, Japan, France, Russia, Czechoslovakia,[26] India and Hungary have by 2006 exceeded their plans of the mid 1970s. South Korea and China (not in the graphic) exceeded more than 10 times (Liu and Smith, 1990). Finland, Canada, Taiwan, Bulgaria and Romania went more than 50% beyond plans. These countries' nuclear projects seem little affected by anti-nuclear sentiment. Brazil, Sweden, Mexico, Slovenia, Argentina and the UK stayed well on target. Spain, Germany, South Africa, Switzerland, the US, Pakistan and the Netherlands materialised their plans only in part. Twenty-seven other countries (not in the graphic: Austria, Denmark, Ireland, Greece, Italy, Luxembourg, Norway, Portugal, Poland, Venezuela, Jamaica, Chile, Honduras, Turkey, Egypt, Libya, Algeria, Morocco, Bangladesh, Philippines, Singapore, Indonesia, Malaysia, Vietnam, Thailand, New Zealand and Australia) had nuclear ambitions, but none materialised by 2006, though due to different circumstances. Some had completed construction, but never started up (Austria and the Philippines). In these contexts public sentiment must have played its part. The Zwentendorf power plant near Vienna is now a museum: the boiler was never turned on after a popular referendum. Many countries retracted plans before build even started: Denmark, Greece, Ireland, Italy, Luxembourg, Norway, Portugal, New Zealand and Australia (on Australia, see Macleod, 1995). Other ambitions were cancelled at an even earlier stage (Turkey, Bangladesh, the Philippines, Vietnam, Singapore, Indonesia, Malaysia and Thailand). Egypt, Algeria and Morocco were discouraged under proliferation anxieties; Libya and Iraq were stopped by force in 2000 and 2001, with similar pressures applied to Iran and North Korea.[27] Some countries stalled nuclear projects under diplomatic pressure rather than by responding to public opinion.

Public resistance influences technological futures in some contexts, but globally it is neither a necessary nor a sufficient factor to alter any project. To assess the impact of public opinion on this variation, one can directly relate

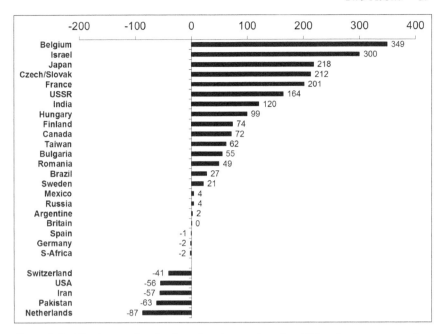

*Figure 3.7* The degree of realisation of nuclear projects in additional percentage to plan. The index measures 2006 realities against 1976 projections: 0 means on target; +100 means doubling capacity, -100 means null capacity realized. China and South Korea are excluded; their earlier plans were exceeded more than 10 x (>1000); 27 other countries which are not in this graphic had plans for a nuclear build up but never even started, hence their degree of realisation would be -100. *Source*: Rüdig, 1990 and IAEA, 2006.

the intensity of resistance to the degree of realisation of a nuclear power programme in that context (see Ruedig, 1990; Rucht, 1995). The correlation is small, because there were many countries where despite public resistance nuclear build-ups pressed on as if nothing happened. Countries with strong resistance and over-realisation of ambitions are Belgium, France, Japan, and India; these countries often have a strong elite consensus. On the other hand, resistance is likely to have had a direct influence in the US (-56%), in Switzerland (-41%), in the Netherlands (-87%) and in countries which abandoned the idea altogether like Italy, Denmark, Greece, Ireland, Portugal, New Zealand and Australia. In these countries there was considerable anti-nuclear mobilisation. In Switzerland several referenda stopped projects at Graben (1978) and Kaiseraugst (1988). Sweden voted (March 1980) to phase out nuclear power by 2010.

Other factors than public resistance played a role. External pressures over proliferation contained nuclear ambitions and are functionally equivalent to internal public protests. There are indeed also many contexts with no or little resistance and rescaled programmes due to diplomatic pressures over proliferation anxieties, mainly in the developing world. There are also countries where ambitions are fully realised, because there is limited political culture to conduct a public conversation over nuclear energy, look at China or Russia.

Context plays a double role: as mediator to allow resistance to manifest itself, and as moderator for resistance to have an impact. A functioning public sphere is a condition for resistance to manifest itself *and* to have an impact. Public resistance is not sufficient to stall nuclear ambitions; an elite consensus on technological ambitions is able to ignore public sentiment.

### Clustering of Installations in 'Nuclear Parks'

According to spatial models, resistance leads to the geographical clustering of innovation (Hagerstrand, 1967). Nuclear power was envisaged as a flexible energy source to the point of being mobile. Compared to water, coal, gas and oil, the supply of uranium was small in weight and volume. The efficiency per fuel unit did away with large-scale logistics, allowing the production of energy anywhere. Nuclear enthusiasm predicted an 'atomic cooker' in every backyard, but also in remote polar region and deserts. Little of this became reality (Goldschmidt, 1980, 252).

Economy of scale, complex public consultations and tightening safety requirements pushed nuclear technology to large units, initially below 500 megawatt electricity, or MW(e), now above 1 GW(e) capacity on average. Local protests forced reactor blocks into fewer locations where resistance could be placated. Once a deal with locals was reached, the industry expanded these locations, rather than taking the battle to new shores. The waste problem was also contained at geologically feasible locations. The new Finnish programme is based on this idea of 'nuclear parks' (Litmanen, 2004). Public concerns over safety and radioactive waste led to a virtual geography of nuclear futures. Public opinion contributed contours of improbability by raising the hurdles on safety and waste disposal. Rather than small, flexible and ubiquitous, the nuclear project became centralised parks, where combined units base-load the electricity grid remote from conurbations, and where waste is deposited locally. Nuclear power is by far less flexible than initially envisioned, and public opinion brought this reality about. There are new attempts to make nuclear electricity flexible again, on floating ships (Russian ideas) and in modular underground piles (US ideas) of small capacity below 500 MW(e). But public opinion is recognised as the difficulty (source: *Financial Times*, 15 February 2013 'flexible fission').

### Raising Safety Standards and Construction Delays

Much mobilisation against nuclear power was motivated by doubts over the safety of nuclear installations. TMI, Chernobyl and Fukushima brought home the safety issue even to the uninformed and least interested. One accident might be bad luck, two makes for suspicion but three looks more like a system problem. Formal risk assessment lost public credibility in the face of this reality (see Chapter 7). The safety reputation of the industry is based on performance and its public perception. The technology appears obsolete, badly maintained and subject to regulatory capture even in high-tech contexts. The safety of nuclear power installations is dubious, but is a necessary condition of any renaissance. The nuclear safety regime comprises the entire fuel cycle, from procurement of fuel to enrichment, operations, waste disposal, reprocessing and decommissioning of defunct sites. But safety is costly and makes nuclear power ever more expensive to build, operate and decommission (IAEA, 1994). The longer the construction time, the more costly electricity will be. The nuclear controversies have brought about licensing regimes involving extended public consultations. Thus, the planning horizon for a plant has risen from initially 3–4 years in the 1960s to 11–18 years in the new millennium (IAEA, 1994; MIT, 2003). The future of nuclear power, particularly in places with a legacy of confrontation, hinges on public opinion. It will define 'how safe is safe enough'. Public opinion and the costs of nuclear power are directly related. Promoters of nuclear power will want to constrain the statutory influence of public opinion on licensing and safety standards, or work on public opinions in the field of communication. Public opinion clearly matters.

## The Limits of Liability and the 'Risk Society'

After precursors in the early fall-out debate of the 1950s, Chernobyl brought home what 'globalisation' means: global consequences of local actions. The idea of a 'risk society' brings this to the point of liability: the success of modern technology undermines its own foundations as its consequences are no longer insurable (Beck, 1992). Since the 18$^{th}$ century, human ventures are protected from unintended consequences by insurance schemes; individual risks are covered collectively at a smaller cost to individuals as in health, employment and life insurances. However, nuclear power risks of meltdown and fall-out are not insurable; no market consortium will guarantee the sums involved, estimated at 1,000$ billion for a Chernobyl-like scenario (in Switzerland). Who can absorb such risks? A state guarantee becomes another cost factor hinging on public consent. How much is the taxpayer prepared to burden themselves and future generations? Taking away the liability from the nuclear industry is a subsidy that distorts the competition in the energy market. Liabilities leave the nuclear future in a dilemma: only a subsidised industry is viable, but subsidies distort competition. The nuclear controversies made visible hidden costs, secret deals and liabilities which makes the nuclear future a difficult one.

## Raising Attention and Unfreezing Decision Procedures

By the late 1970, observers noted that public protest had limited impact on the nuclear power projects. Nuclear commissions that planned constructions seemed immunised from public influence by secrecy and the defined expert base bound to consider only bureaucratic and expert information. However, politicians with election cycles started to respond to the concerns of the streets. Resistance to nuclear power raised awareness in the context of wider environmental issues, and this steadily translated into the polity (see Kueppers and Nowotny, 1979; see also Chapter 4). The old nuclear regime, based on secrecy and expertise only, was increasingly challenged by what came to be known as 'extra-parliamentary opposition'. Many countries revised how their nuclear affairs were run along lines of transparency, accountability and expertise. This legacy of public involvement in science and technology policy was later to be taken for granted in debates over information technology (1980s) and genetic engineering (1990s).

## Discovering the Rationality of an 'Irrational Public'

Public resistance challenged the 'nuclear priesthood' of the technocratic elite. An elite subscribing to technocracy sees itself in a Platonic light: philosophers shall govern affairs of state as only a scientific and technological elite is capable (Bucchi, 2010; Luebbe, 1978; Veblen, 1964). Public opinion seemed an irrelevant nuisance.

In the course of dealing with public resistance since the 1960s, it dawned on many that public opinion is capable of judging complex technical matters. Critical opinion was in most cases no expression of 'nuclear phobia'.[28] Many of these decisions are not technical anyway, but on matters of real uncertainty, values, priorities and interests. In deconstructing the logic of risk assessment, the social sciences rediscovered 'common sense' and its wisdom on a wide range of issues, i.e. the cultures of risk. The discussion over risk perception is predicated on an 'unscientific' distinction between expert and lay judgment which is dubious in the first place (Van der Pligt, 1992, 12f; see also Chapter 7). But, public opinion is not only a matter of competence, but also of due procedure. Deliberation is necessary to form opinions. This gave rise to what became aptly known as 'technologies of humility'. Agencies are now regularly and routinely organising public deliberation in many formats. The fight over nuclear power opened new ways of governing, so that later developments in information technology and biotechnology occurred in a more open context. Anti-nuclear resistance opened techno-scientific agencies to institutional learning as they started to experiment with participatory democracy (see Jasanoff, 2003 and 2007; Mejlgaard et al., 2012, and also Chapter 5).

## Changing the Ethos of Science Journalism

In the post-war years science journalism had professionalised as a genre of reportage on science and technology to the wider public. The dominant paradigm was one of service, extension and diffusion. The machinations over nuclear power and the TMI incident (1979) marked a turning point. The experience led many writers to convert from an ethos of triumphalism to critical writing on science and technology (Franklin, 2007). The 1980s brought critical science writing; in 2010 we enter another phase as the new professionalism of PR for science takes hold and science journalism faces a crisis at least in some parts of the world (Bauer and Gregory, 2007; Bauer, 2013).

## Setting the Conditions for Future Resistance

An effect of past resistance is the establishment of conditions for future resistance. Public protest is itself a learning process for individuals and for groups. Skills, competences and structures to support are built and persist. Groups like Greenpeace, Friends of the Earth, WWF or Sierra Club, although founded before the nuclear controversy, found in these debates a 'baptism of battle' as they internationalise local issues. Nuclear protest was the forming experience of many activists and sympathisers who radicalised during the events; nuclear protest was recruitment ground. The environmental movement has in the 'anti-nuclear' its foundation myth which provides a narrative of heroic actions that will be difficult to replace. These organisations grew in the conflict and continue to exert vigilance. The period created a field of skilful moral entrepreneurs who translate the nuclear issues of safety, choice, democracy, environment and equity to new issues such as biotechnology and information and nanotechnology. Resistance thus created entrepreneurial networks and competence. Some of these resources disappeared; others diversified and expanded a growing sector of activism that is capable of mobilising public opinion over techno-scientific developments. By 1994, the nuclear community formally concluded: any renaissance must carry public opinion, the key hurdle of a nuclear future (IAEA, 1994). Such a statement was not even thinkable before the debates of the 1970s and 1980s, evidencing collective learning.

## Mediating Factors

The comparative evidence on the challenges posed by resistance and the responses of the nuclear community suggest a contingent relationship. If we consider resistance as the independent variable and responses as the dependent variable, we must conclude that the linkage is not a direct one. There are techno-scientific changes without resistance, and there is strong resistance without any techno-scientific response. The responses are generally not proportional to the intensity of resistance. In other words, the techno-scientific response is contingent on other factors, which however does not make resistance irrelevant.

One of the main contexts that appear to mediate the impact of resistance is elite consensus. Consensus in the polity across parties and ideological boundaries is able to 'filter' resistance; it allows the ignoring of resistance, or even dismissing it as irrelevant: no further action required. Where the political elites are in agreement, public resistance from the streets is likely to be frustrated. France is the classical case over nuclear power: the French polity on the left and right backed nuclear power as a tool of French national independence vis-à-vis the US (see Hecht, 1998; Rucht, 1995; Flam, 1994).

64  *Atoms, Bytes and Genes*

We might conclude a rule: without a split in elite opinion, public opinion has difficulties in being heard. When political elites split, one part will seek advantage in aligning itself with the resistance and thus translate anti-protest into the policy arena. For example, in the UK the nuclear establishment was deeply split over design options—to pursue British designs or to buy US technology. Anti-nuclear feelings thus found the entry point and established a moratorium, though did not achieve a reduction of the nuclear project. Similar situations arose in other countries in the 1980s. Elite conflict is a context that renders resistance effective.

Single mindedness and national consent is not only in techno-scientific policy a buffer against doubts and alternative actions, though it carries risks of being proven wrong in the long run. Ignoring resistance carries opportunity costs. Being successful but wrong is the dilemma of new technology, as it mobilises overpowering hyperbole for a 'better future'. Here lies the functional significance of resistance: to alert to and overcome this dilemma by showing the path of a sustainable future.

**THE LEGACY OF PAST CONFLICT**

In this chapter I traced the path of military and civil nuclear power since WWII through public opinion and beyond. The nuclear Club had two sections: one for bombs and energy, another for energy only. The passage between the rooms, though policed, remains desirable. The civil and military build-up came to a sudden halt in the 1980s. Mass mobilisation and two serious accidents brought to light that neither the safety nor the economics of nuclear power were as clear-cut as claimed.

Attitudes and opinions on nuclear power reflect the ambivalent imagery of this technology: global destruction and abundance of cheap energy. Nuclear news, opinions, attitudes, frames and archetypes layer the public imagination in several cycles of change. In this symbolic context fear and anxiety are a constant possibility. Their containment needs to be explained. TMI, Chernobyl and Fukushima were significant events where this containment broke down. One accident is bad luck, two draw attention, three mark a system failure. Nuclear power is 'stigmatised' by catastrophe, as accidents reappraised earlier incidents, like Windscale in the UK and Fermi-1 in the US: nuclear safety was never controlled as claimed. Since the 1990s, overall public opinion had rebounded. Opposition declined and ambivalence increased. For the knights of the nuclear future, global climate change puts nuclear power into a new light of clean energy.

Public resistance impacted nuclear projects globally. However, the techno-scientific responses depended on context. Elite conflict and competing designs were entry points for public resistance to be effective. It delayed diffusion of nuclear plants, concentrated piles in nuclear parks and lowered the future expectations. Projects stalled or were dropped entirely; safety threshold were raised which at present delay construction and massively increase costs; the limits of private liability became public knowledge as 'risk society'; the ways of techno-scientific decisions were challenged; the rationality of an 'irrational' common sense was rediscovered; science writers changed their ethos; and resources were built up to translate resistance and controversies to other developments. All this left an important legacy: *public opinion is a hurdle to be reckoned with.* Public opinion will have to be persuaded that nuclear power is safer than at TMI, Chernobyl and Fukushima; that nuclear waste can be disposed of safely; that nuclear electricity is economically viable; and that regulations over licensing, liabilities, safety regimes and investments are not captured by private interests. All hinges on public opinion granting legitimacy for future decisions.

The issue of proliferation remains equally pressing. Can military and civil nuclear power really be separated, or does it remain an illusion? Is it just coincidence that Britain renews its nuclear defence, the Trident submarine, at the same time when nuclear electricity is on the agenda? Defence observers are wary of ruminations that Japan is considering dual capacity and former Cold War strategists see proliferation getting out of control (see Weinberg, 2004; Cirincione, 2007). The revival of nuclear energy stirred nuclear ambitions in China, India, Russia, Eastern Europe, the UK, Finland, Italy, Germany, Switzerland and the Netherlands. Many countries that had abandoned such ideas under pressure of proliferation diplomacy are reconsidering: Brazil, Argentina, South Africa, Kenya, Indonesia, Malaysia, Vietnam, Turkey, Syria, Iran and elsewhere. However, what the nuclear community and the IAEA see as a new dawn, defence observers warn of *new risks of proliferation.*

The third legacy of resistance to nuclear power is the opening of collective decision making to institutional learning. The *experiments with public deliberation* that came to characterise the governance of biotechnology in many places took inspiration from nuclear to overcome the secrecy and technocracy of the military-industrial complex.

Will a nuclear renaissance face up to the new legacy of public consultation or will it restrict the scope of such procedural obligations? The machinations over the nuclear dawn, for example in Britain, are not entirely hope inspiring.[29] What looked like a learning outcome with lasting effect, might relapse into the old learning difficulties of technological hyperbole and single-mindedness in Britain and elsewhere.

## NOTES

1. In November 2006, Alexander Litvinienko, a former KGB agent, was poisoned by polonium -210 applied in minute dose to his tea in a London Hotel. He died three weeks later of radiation poisoning, the first victim of 'nuclear terrorism'. The case remains unresolved, but soured relations between the UK and Russia.
2. The German nuclear project is the stuff of drama. The play *Copenhagen* (Frayn, 1998) dramatises a meeting of Niels Bohr and Werner Heisenberg

during the WWII. Confronted with several versions of this encounter the audience is left with the question: was Heisenberg signing up Bohr for the Nazis, or was Heisenberg delaying the Nazi effort and signaling as much to the Allies via Bohr? This encounter remains an ambiguous episode in the history of physics.
3. J. Robert Oppenheimer was a theoretical physicist and leading figure in the Manhattan Project, which produced the A-bombs for Hiroshima and Nagasaki. When he developed doubts about the wisdom of developing the H-Bomb after 1945 his rival Edward Teller took over and Oppenheimer was investigated for 'anti-American' activities (see Rhodes, 1987). His case became the stuff of drama as in Kipphardt's (1964) 'In the case of J Robert Oppenheimer', or Durrenmatt's (1962) 'the physicists'.
4. Argentina wanted to join the act early on, but was discouraged from doing so with an open US threat in 1947 (see Cabral, 1990, 12). Brazil started organising its science base to muster the atom to secure national independence. Attempts to defend its relevant material deposits ended in murder in 1949 (Cabral, 8f).
5. Eisenhower addressed the nation before leaving office after having resisted constant pressures to increase US defence spending in response to an apparent missile gap between US and USSR. Secret information showed a different reality of which public opinion was not aware. There was a clear danger that military threats were unduly amplified in public perceptions by interested parties: 'this conjunction of an immense military establishment and a large arms industry is new in the American experience ... the total influence—economic, political, even spiritual—is felt in every city, every statehouse, every office of the federal government ... in the council of government, we must guard against the acquisition of unwarranted influence, whether sought or unsought, by the military-industrial complex. The potential for disastrous rise of misplaced power exists and will persist ... to endanger out liberties or democratic processes. We should take nothing for granted' (Ambrose, 1990, 536f).
6. Statistics on nuclear testing are available at: http://www.johnstonsarchive.net/nuclear/nuctestsum.html (accessed 4 December 2006).
7. See note iii.
8. Similar atomic visions can be found earlier for example in Rubinstein (1954) writing in Stalin's USSR: nuclear power could fulfil its potential only under communist rational planning; capitalism would stifle its potential.
9. This vision of France is expressed in the inscription for Charles DeGaulle on Champs-d'Elysee in Paris: 'il y a un relation secular entre la grandeur de La France et la Liberte du Monde'.
10. Nuclear euphoria was shared evenly across the Iron Curtain. This grip of technology on post-war minds urged Gunter Anders in 1958 to call for a 'social-psychology of things' (see Anders, 2002, 58ff), which would elucidate the blockages of thinking about technology across economic systems.
11. The origin of this statement '*It is not too much to expect that our children will enjoy electric energy in their homes too cheap to meter*' is controversial. One attribution is to Lewis Strauss, then chair of the US Atomic Energy Commission. He seems to have dropped the line at the National Association of Science Writers' Founders Day (16 September 1954; see Cohn, 1997.; or Wikipedia 'too cheap to meter'). The phrase was rehearsed by nuclear advocates like Alvin Weinberg (US) and Walter Marshal (UK) to highlight the technology's promises; later it became the key example of exaggerated expectations of the period. That is how the phrase mainly survives today.

12. For US car manufacturers of the 1950s—for example Ford's 'Nucleon' of 1958 would run more than 5,000km on a single charge—atomic car projects were a way of demonstrating work at the frontiers of science.
13. The 'Die Hard' (since 1988) action movie series, stars Bruce Willis, who is stoically in pain, but indestructible on his mission to save his family and thus the wider world from evil terrorists, seems like a post-Cold War re-visioning of the 'iron man' of Strategic Bomber Command, but this time round 'incarnated' in pain and with family obligations.
14. Broderick (1988) lists over 500 movies, of which two-thirds are US productions, which project the ebb and flow of nuclear imagination onto the screens. The genre comes in two waves, peaking in the US in the late 1950s and again in the early 1980s. Europe's high is the late 1960s; the second coincides with the US.
15. This correlation of news intensity and bad news was observed and generalised into the 'quantity of coverage' hypothesis by Mazur (1975, 1984 and 1990).
16. The tested H-bomb yielded five times the total Allied bombing during WWII, blasted Bikini Island in March 1954 and threw up a cloud of radioactive dust that rained down on the inhabitants of Rongelap Atoll and on the unhappy Japanese the fish trawler *Happy Dragon* 150km away.
17. The UK nuclear establishment agonised over gas, light water, heavy water and fast breeder designs, and their economic viability (Williams, 1980). However, the take-off of commercial nuclear in the 1960s was with US technology, light water, Westinghouse) and pressurised water reactors (PWR, General Electric). Most countries in the West abandoned ambitions and imported US technology, even the French; Canada stayed with heavy water and the UK with gas-cooled designs. With the quasi-US monopoly on nuclear technology, its value as symbol of national achievement and scientific leadership faded.
18. A prediction that sounded still true in early 2013, when Cumbria's local democracy finally refused to be that 'dust bin' at Sellafield, where a consortium of British army, French AREVA and US URS maintain a storage site, the decommissioning of which is estimated at $100 billion (source: NZZ, 6 February 2013).
19. The emergence of the NIMBY opinions contrasts the 1950s, when regions were competing for nuclear power stations in search for modernization. In the UK Labour MPs were demanding nuclear stations for their constituency. For example in 1956 HR Mr Gower, Member for Glamorgan, urged the government not to leave out Wales as potential nuclear site, as Northern Ireland, Scotland and England had already been considered (source: Hansard, 555, 4 July, 1329).
20. In an incident of reality imitating fiction, the accident occurred just two weeks after the release of the thriller movie *The China Syndrome*, which fictionalised the meltdown of a nuclear power station (starring Jack Lemmon and Jane Fonda; released 16 March 1979).
21. A BBC4 production of April 2007 broadcasted an interview to that effect with a Russian military pilot who had received a military medal for his sorties over Belarus at the time.
22. In Germany nuclear protest became an annual ritual of collective memory: once every year protesters disrupt this nuclear transport by train. To transfer the waste to the Gorleben depository, thousands of policemen line the railway tracks to fend off disruptions.
23. Currently, 49 countries have the know-how though not necessarily a project to make nuclear bombs; these defines the circle within which a new arms race

could develop (see Isaacs and Downing, 2008, 288). In an irony of history, former exponents of MAD strategy are now calling for total nuclear disarmament. George Shultz, William Perry, Henry Kissinger and Sam Nunn are called the 'four horsemen of the apocalypse' (see *Economist*, 15 November 2008, 74f).
24. Recent projections are optimistic again: MIT (2003) anticipates 3,500 GW(e) by 2100, or 10 times the current 380 GW(e). But Fukushima 2011 has launched a new round of downward correcting future expectations.
25. An index of alteration is the degree of project realization, i.e. the match between ambitions at time $t_0$ and the state of affairs at the time $t_{0+k}$ (modeled on Rucht, 1995). Nuclear ambitions fail in various ways: reduction of capacity, retraction of plans, construction but no start-up, and phase out without replacement. These all register as capacity at time $t_{0+k}$ being equal, larger or smaller than projected at time $t_0$. The index is defined: $100 + 100*[GWe_{2006} - GWe_{projected\ in\ 1974}) / GWe_{projected\ 1974}]$.
26. Note that Czechoslovakia split in 1993 into two states, the Czech Republic and Slovakia.
27. There are a number of countries who maintained and might still maintain reactors for research purposes without producing energy: Australia, Austria, Chile, Columbia, Denmark, Egypt, Greece, Indonesia, Iran, Iraq, Norway, Philippines, Portugal, Turkey, Uruguay, Venezuela, Vietnam and Zaire (SIPRI, 1974). Israel sustains a nuclear reactor to supply its nuclear weapons capability, which is not officially acknowledged and kept a state secret; blowing the wistle in 1986 to the British press brought Mordechei Vanuatu 18 years of imprisonment for 'spying'; he was released in 2004.
28. Psychologising opponents with pathological categories such as 'phobia' is an eristic-rhetorical trick to disqualify and undermine opponents in public debate. Notions like 'nuclear phobia', 'radiation phobia' or Jungian psychoanalysis of dissenting scientists apparently suffering from disturbed mother-child bonding, suggesting unconscious pathological motives of anti-nuclear positions, had widespread currency in nuclear debates (see Weart, 1988; on notion of 'technophobia' see Bauer, 1995 and chapter 5).
29. A British court rejected the public consultation for the new nuclear programme 'as flawed and unfair' in light of the legal obligations (source: *Daily Mail*, 10 February 2007), and the consultancy firm responsible for the exercise was found falling short of the professional standards of even the British Marketing Society. But changes in UKplanning law are introduced with the Planning Act 2008 and subsequent legislation which restricts the rights of local communities, and courts might no longer able to test cases of technocratic decision making.

# 4  Environment, Safety and Sustainability

Once the equivalence *technology* <=> *Progress* has become dubious, independent criteria are needed to evaluate the contribution of techno-science to social progress. I will argue that the past 50 years have brought the rise of two benchmarks against which new technologies are evaluated: safety and sustainability. These benchmarks were carved into 'common sense' by two social movements, the ecology movement urging sustainable lifestyles and the consumer movement raising concern over product safety and market transparency. I will chart the new common sense with focus on sustainability by showing how representations of 'nature' have changed and how activism that started locally and became global created the present 'green' mentalities.

The way most people now think about 'nature' differs from the ways of the 1950s, or back in the 19$^{th}$ century. Changes in attitudes to the environment are grounded in long-term changes of representation of nature. The shift from a mechanistic to an ecological worldview underpins even my own 'pain model' of resistance to new technology (see Chapter 5). Morris Zapp, in Lodge's comic novel *Changing Places* (1975), lectures on the 'influence of the 19$^{th}$ century on Shakespeare'. Nonsense, one might say: Shakespeare lived in the late 16$^{th}$ century, later events cannot influence earlier ones; post-hoc cannot become propter hoc. Can we not trace the influence of environmentalism on climate change? Nonsense, climate has no politics! However, how a strict separation between 'nature' and 'culture' has lost some of its credence will become clearer in this chapter.

## REPRESENTATIONS: FROM NATURE TO ENVIRONMENT

The way we see, feel and think the world is no historical constant. The historian Keith Thomas (1983) chronicles the changing sensibilities towards the 'natural' world between 1500 and 1800. From the Judeo-Christian Bible, the Western tradition inherits a dilemma between action and admiration. The natural world is on the one hand man's dominion, created for man's sake and sole benefit. Human authority over animals and plants is absolute and the cultivation of useful animals and plants is an action imperative, imitating God's own act of creation. On the other hand, God's creation mirrors the 'perfect beauty of God'. Beholding this beauty requires a sensibility which conflicts with the imperative of cultivation and exploitation. This dilemma between design and observation, vita activa and vita contemplativa, has a single source. From shooting and looting we moved to observing and admiring nature. Nature used to terrorise its inhabitants; the vagaries of climate brought devastation, hunger and pestilence to the land. Forces erupted in storms, landslides or earthquakes, but cultivating and 'civilising' the land increased the security of life and livelihood and fostered human resilience. As droughts lost their terror with ample stocks of water, so did stormy rain mitigated by river corrections and flood plainst. But, the cruelty of nature needed no admiration, nor does every useless plant need eradication. Besides classifying animals and plants according to human utility, the new sensibilities encouraged a view of nature in-and-for-itself, and beyond human purposes. By the 18th-century, these new sensibilities were cultivated by a travelling leisure class with a romantic gaze, and by a distinctly Anglican clergy, Deist in theology, who through 'botany, books and bottles' is preoccupied with Natural History.

These attitudes to 'nature' form over centuries and persist in present dilemmas between town and country views: the preservation of wilderness versus its cultivation and exploitation; urban sensitivities and sentimentality against industrial farming, mass killing in meat production and unapologetic going for the hunt. This dilemma is typified on three levels: in historical changes of attitudes, of concepts and of metaphysics. What appears a historical sequence is also a diversification of simultaneous possibilities and a struggle over relative preponderance.

## Changing Attitudes: From Fear and Trembling to Modern Stewardship

The original attitude might be characterised as 'being in awe of Nature'. The capital letter for nature is apt, as the forces are often personified. Humans confront an enchanted and overpowering nature with fear and trembling, and with a stubborn determination to stand the ground. Thus, sea, desert, jungle and mountain are awesome adversaries, but open to negotiations for particular expeditions. Mythologies of the sea, desert, jungle or mountains express this attitude of 'being-in-awe' and exhort advice of how to conduct one's business in the face of overpowering forces. Myths and magic promise some control of the situation. Before one goes to sea, Poseidon, god of the sea, must be placated by a sacrifice or the good word of an ally on Mount Olympus.

This attitude of hopeful respect is later superseded by an attitude of confident *domination and exploitation*. The world is a resource to exploit, a free good to consume, a human dominion. Indeed, the Biblical passage of Genesis 1:28 'multiply and dominate' grants authority over the biosphere. Natural resources distinguish rich from poor nations, and the battle over resources such as coal, iron ore and oil becomes a major driver of history.

Later appears the attitude of stewardship, of *sustainable* management of nature. Human actions are within the ecology where success can undermine its very basis by over-using the resources. If fossil fuel reserves are depleting and their emissions warm the atmosphere, basing human prosperity on fossil fuel is self-defeating. Our activities are not sustainable because they do not renew the basis on which their continuation rests. The stewardship attitude highlights the consequences and turns 'dangerous nature' into a risky environment: interventions can go wrong. What hitherto were natural dangers are henceforth risks of human actions. The former terrors of nature are now man-made: deforestation brings soil erosion and flooding; air pollution damages the trees that protect against avalanches; $CO_2$ emissions lead to global warming and climate change which put into jeopardy the safety of the world.

Thus, our attitudes to nature developed from being in-awe-of, via domination, back to the stewardship of sustainable management. This goes hand in hand with a semantic shift from 'nature' to 'environment' in common parlance.

## Changing Concepts: Organic—Mechanistic—Cybernetic

Moscovici (1977) offers a schematic history of this changing mentality in terms of changing root metaphors. Organic, mechanistic and cybernetic concepts of nature frame very differently the potentials of 'nature'. Not only is culture a part of nature, but the 'culture of nature' is historical.

*Organic nature* was the location of activity of craftsmen and artists, who shape materials with their hands. The craftsman is methodical; he diligently follows a plan that gives form to material, for example by rotating clay into a pot. The clay does not add value; the 'essence' is the plan which remains fixed as it materialises. The workshop was the place where skills and knowledge were handed down from generation to generation. Man is the measure of all things, at the centre of this universe, doing the work; without intelligent design chaos reigns. In this ancient world, mind informs substance. Form is eternal, substance is transient. Change is teleological, unfolding a 'telos' through the cycle of birth, flourishing and decay. There is an analogy between the microcosm of human purposes and the macrocosm of the God's universe (see Collingwood, 1945, 8). Politically, this view inspires the 18th-century quest for 'good government'.

The Renaissance, so goes the story, broke with this ancient model. Inspirational for the *mechanistic worldview* was a growing stock of instruments to make things, and a focus on the anonymous forces of causality. The instrument refocussed attention from designed things to the process of making them. As products became secondary, the process took priority. Mechanics and later engineers became the social elite who defined this new take on things. Invention changes its meaning from that of a penal offence, to a heroic act that substitutes and enhances nature (Godin, 2010). Extending the human body with mechanical devices no longer imitates nature (mimesis), but exceeds it. Thus clocks, mills, pumps, lenses and telescopes are the new achievements. The stellar universe was an automaton that ran eternally and with predictable regularity like clockwork. How this mechanism arose was a secondary question, to grasp its causal logic sufficed. Mind and matter were separated by strict dualism. Deism was the theological version of this view: God created the world and then withdrew to observe it. Work, previously a human capacity, is henceforth the moving mass over a distance. Machines, horses or humans work equivalently. Scientists observed the laws of nature and avoided the 'prejudice' of purpose; and engineers harnessed these laws to utility. This view extends to living things. Robert Bakewell (1725-95) is an English agricultural pioneer whose effigy in the London Portrait Gallery carries a famous quote: 'a sheep is a machine to turn grass into mutton'. Harnessing mechanisms allowed for the design of animals and plants to maximum profit. The mechanistic outlook made available natural causes to human purposes of cultivation, while purifying nature from all purposes. This seems very much in line with Judeo-Christian views of the bounty of nature laid out before Adam and Eve for their use and manipulation, politically only tamed by the 19th-century fight for equality and the redistribution of income.

With the *cybernetic worldview* we reached the 20th century, whose embodiment is the modern scientist. Innovations now are focussed on synthetics, the construction of new materials in chemistry; atomic fusion and fission mutate one material into another; genetic engineering mixes and constructs new life forms. Substances not found in nature proliferate. Matter itself is designed. The key is structural development. Matter moves by information and feedback. It is directed, regulated and evolves. The process of nature has become multi-layered and multi-causal, and the science of cybernetics elaborates this shift in emphasis. Time and space are relative; everything has rhythm and is set in a field of forces. Invention and creativity are general human potentials and no longer the privilege of genius. Human work becomes process control, the monitoring of signals with a view to keeping production going. Autonomy and sustainability is the new orientation: a machine is monitored to avoid breakdown; human interaction avoids certain topics to continue; business accommodates the market in order to continue profiting; fishermen catch fish only in numbers which reproduce year on year. A sustainable nature follows the motto 'the show must go on'. Thus the 'environmental question' entered the polity in the late 20th century with a kind of 'hold your breath' trepidation as limits were reached and balancing the books becames more complicated.

Humans participated in an enchanted universe of myth and magic, then, were strictly separated from nature's causality to maximise its use, and finally, the environment is a function of human culture and co-evolution (see Collingwood, 1945; Cranach, 1987; Luhmann, 1990a). Thus, human views of nature are effective and not epiphenomenal. Guided by our views we prosper or walk into disaster; thus we are part of nature's development. Anthropogenic climate change brings home that very idea: take care of what is not a cold automaton but a warm lifeline.

## Shifting Metaphysics: Closed, Open and Self-active Systems

As *system thinking* takes hold, our metaphysics change, and 'nature' becomes 'environment'. Environmentalism invites us to think differently. Ecological thinking presumes processes in a system-environment relationship. Luhmann (1984) builds on this basis a theory of three paradigms worth rehearsing.

Closed systems operate as wholes-with-parts. The picture is more than the sum of its pixels. However, a single pixel can contain the entire picture as in portraits that are made up of mini-portraits of the person in the portrait. Units can represent the whole in a desire for unity. The relations are the key, as the parts are interchangeable. A melody is identical whether played in G or in D# even when most notes differ. Equally, a bureaucracy combines a set of roles and rules, while real persons are replaceable. Bureaucracies, melodies and pictures are closed systems. What counts is their structure, ideally in the perfect order of a 'one best way'. Closed systems succumb to entropy; they fade like a candle that runs out of wax. Similarly, a bureaucracy burns out its staff; an endlessly repeated melody bores the listener. Closed systems consume themselves without renewal.

Open systems by contrast develop in an environment. The flow of inputs of energy, matter and/or information is processed into outputs. Open systems stabilise an interior milieu by selective perception and adapting movement. An observer can note that different internal structures perform equally well (i.e. equifinality or functional equivalence), and that environments favour certain structures (i.e. contingency). Open systems are cruising projectiles, set on a target, navigating circumstances with sensors and course corrections. The overall direction, however, is set by design, command and control. Open systems consume and grow by prospecting ever new environments to expand their efficacy.

The third system paradigm refers to *self-activity* (also autopoiesis) with a focus on self-organisation and self-reference, thus radicalising the thinking one more time. The 'image' is the basis of all operations, and this includes a or self-other difference. Self-active systems are autistic and closed to information—relevance and meaning is produced only in self-reference—but in coupling with other systems through exchange of matter and energy. Living is an instrument flight; it can only react to irritations in terms already given. The issue is 'how to stay connected'. This paradigm refocusses operations from command and control to sensitivity and autonomy. 'Nature' is not represented once and for all, but a relevant environment continuously constructed on resistance and irritations. The environment is inside as well as outside; and it irritates. The system suffers lack of sensitivity and learning difficulties in the effort to self-perpetuate. 'The show must go on' brings autopoiesis to the point. Self-active systems consume by renewing their environment in response to signals of unsustainability.

This history reveals the diversity of attitudes, concepts and metaphysical assumptions. Concrete events show how this diversity is not superseded but engenders current environmental conflicts. Out of history comes the current confusion.

Traditional mentalities left little room for thinking in alternatives. The 'heretic' was not just disagreeing with the 'orthodoxy', but calling upon all the wrath of Poseidon: 'Mother Nature' was a matter of life and death. Everybody participated in this universe without an opt-out clause. Deviating from orthodoxy was not a private matter but endangered the entire tribe. This gave formidable reasons for conforming to collective pressures (see Taylor, 2007). Modern mentalities tolerate contingency within the public and the private: things can be different, and indeed they are, and are known to be, lived and experienced differently. The common sense of 'nature' references not only an objective world, but also expresses a community and appeals to a certain future. For social psychology, knowledge of the world is thus a function of the triplet of object-it, subject-self and project-other (see Jovchelovitch, 2007; Bauer and Gaskell, 1999 and 2008). The modern world no longer deals with heresy and excommunication, but absorbs simultaneously viable views. Among religions this parallelism of 'truths' is envisaged in peaceful dialogue. This tolerance might be a model for the scientific worldview, when the historical diversity of actual nature comes to the fore in unusual cases. This is illustrated by the 'Waldsterben' and the Shetland oil spill.

*Excursion 4.1* The 'Waldsterben' and the Shetland Spill

*I remember well the commotion during my university years known as 'Waldsterben'. In the early 1980s German foresters observed worrisome tree tops: unusual colouring and widespread loss of leaves and needles in the crown areas. Such damage was concentrated near industrial sites which pointed to air pollution as a potential cause. The alarm resonated with public opinion: 'der Wald stirbt'* (*'the forest is dying'* cried the German magazine Der Spiegel *in autumn 1981*). *Apocalyptic scenarios painted a future where trees failed to secure the snow and avalanches made the Alps uninhabitable. European history would rewind back to the 11$^{th}$ century. Forestry research obtained unprecedented funding to assess the damage and its causes, while freedom-loving automobile lobbies fended off attempts to police cars to reduce air pollution. By the end of the 1980s observations and experiments had shown that tree tops were not a reliable indicator of damage nor caused directly by air pollution; trees and public opinion were otherwise under stress, not from air pollution (see Newig, 2004).*

*The Alps span France, Italy, Austria, Germany and Switzerland, but ironically the tree issue concerned only the German speaking world; the Swiss followed the Germans one year later. The French and Italians used the German term 'Waldsterben' to mark what they viewed as a curiosity. In German romanticism the 'Wald' is the locus of spiritual renewal, and most forests are publicly accessible via footpaths. This resonated culturally: the 'Waldsterben' is not only a matter of Alpine safety, but threatens to exhaust the spiritual fountain of the nation. What resonates for Germans does not necessarily resonate with the French or Italians, thus providing evidence of different representations of nature (see Graumann and Kruse, 1990).*

*Another illustrative episode is the BRAER oil spill in the Shetlands Islands, 200km north of Scotland, in January 1993. Gervais (1997) observed the aftermath, the invasion of some 1,400 journalists and the performance of different representations of 'nature'. Among Shetlanders the spill brought to the surface an organic, a mechanistic and a cybernetic view of sea and coastline. The event and the intense media attention challenged the local identity, and brought fault lines to the open. Gervais' study showed how an ancient organic-participatory and a modern systemic view fought off a mechanistic and exploitative view that was epitomized by the leaking oil tanker and its abstract business logic. The distress of the 24,000 islanders expressed itself in talk of 'invasion', 'murder' and 'rape' of their nature. 'Invasion' was already a trope in the local history of this marginal place in the North Sea.*

*'Waldsterben' and Braer demonstrate how environmental sensitivity is part of local culture. Different concepts of nature are anchored in society, and this diversity comes to the fore when unusual things happen. The idea that cultural values structure both our sensitivity and what we consider risky has spawned an entire research field. A typology of cultures—hierarchical, egalitarian, individualist and communitarian—explains the differences in what people care about, worry about and in how they view their relation with nature (see Douglas and Wildavsky, 1982). In the US, for example, egalitarians and communitarians worry about global warming, pollution and nuclear power, favour gun control, and expect anti-terror war to fuel terrorism; hierarchists and individualists do not, or much less so (see Kahan et al. 2007). However, some of these concerns also rest on achieved affluence: Indeed, opinion polls in the 1980s showed that concerns over air and river pollution, industrial and nuclear waste, deforestation and threats to sea life were correlated with higher incomes. The better off individually and as a nation, the more people are aware and ready to do something for the environment (European Commission, 1982, 28; and European Commission 1986, 64).*

## THE RISE OF ENVIRONMENTALISM

Semantic changes like that from nature to environment, or changes from a mechanistic to a cybernetic view of nature, from exploiting to preserving an environment, do not come from outer space. These secular changes are sponsored by protagonists with a repertoire of public mobilisation. During the latter part of the 20$^{th}$ century, an industry of social movements emerged which brought ecological awareness centre stage in a polity that hitherto had put equity and the distribution of resources central.

### The Institutionalisation of Green Issues

The story of the green movement is often told as a series of phases in which ecological awareness moved from the fringes to mainstream politics. Jamison's (1996) and Radkau's (2011) account of these phases is useful. Pre-1968 is the *awakening phase*. Conservationist concerns reach back into the 19$^{th}$ century when groups first formed to struggle for the preservation of natural environments. These organisations endorsed a romantic admiration of 'natural beauty' and cultivated a critical view of destructive industry. Organisations like the Sierra Club (founded 1892 in the US) set up nature reserves to preserve landscape and biosphere in a 'state of nature' shielded from destructive cultivation. In Britain the Royal Society for the Protection of Birds (RSPB, founded 1889) formed in protest against the impending extermination of the Great Crested Grebe by fashion and fur trade. Similar movements appeared across Europe: the 'Bund fuer Vogelschutz' (Germany, 1899), the 'Natuurmonumenten' (Netherlands, 1905), and the Campaign to Protect Rural England (Britain, 1926).

These movements mark conservationism's long past, while the short history of environmentalism is said to begin with a bombshell book, Rachel Carson's *Silent Spring* (1962). This book found resonance in a public which was already prepared by the nuclear fall-out debate of the 1950s. Many people recognised the similarity between showers of Strontium-90 and chemicals sprayed on crop fields (see Watkins, 2001). *Silent Spring* was serialised by the middle class *New Yorker* in June 1962, published in book format in September, and by April 1963 it became a CBS television documentary. It charts the devastating consequences of excessive use of pesticides like DDT in industrialised agriculture. The evidence of collateral destruction of wildlife and of health hazards arising from residuals in the human food chain made scandalous news that stirred a propaganda war over regulations of pesticide use (see Kroll, 2001; Wang, 1997; Lear, 1992). In conjunction with the Thalidomide—the drug against morning sickness that caused deformed limbs in babies—and the nuclear fall-out debate, *Silent Spring* alerted the American and international public to sinister aspects of technological progress and to the hazards emanating from scientific laboratories. Pollution had been a standing concern of conservationists, but only the public health alarm reached a larger audience and accomplished public awakening. It pitted a soft, warm and female ecology-biology against a hard, cold and patriarchal chemistry-industry. Organisations like Nature Conservancy, World Wildlife Fund (WWF, founded in 1961), a revitalised Sierra Club, and the International Union of the Conservation of Nature (founded in 1956) henceforth used the opportunity to bring ecological issues to the attention of the financial and business establishment.

By 1968 Ehrlich's book, *The Population Bomb*, also reached massive public resonance with a Neo-Malthusian doomsday scenario of overpopulation and hunger crisis. The thesis was taken up by the Club of Rome (founded 1968) whose report on *The Limits of Growth* (Meadows et al., 1972) broadened the message of unsustainable exploitation beyond food to all natural resources, and in particular the non-renewable energy sources.

This defined, between 1968 and 1974, environmentalism's revolutionary *organisation* phase. Friends of the Earth (FoE, founded in 1969) and Greenpeace, founded in 1971, today stand for global environmentalist activism. These organisations with an activist core set themselves apart from the traditional conservation movements with whom they clashed over nuclear power. FoE split the Sierra Club, and gathered support in the anti-capitalist and non-conformist student movement of the late 1960s. Whatever the merits of '1968' at universities in the industrial world, a lasting influence of this revolt is the new social movement that formed around ecological issues. Its more sinister legacy is urban terrorism as in the German *Rote Armee Fraktion* or the Italian *Brigate Rosse*, which erupted in the 1970s and 1980s.

The years 1975 to 1980 see the *formation of a social movement* through political action and widespread mobilisation on the anti-nuclear issue. On nuclear energy, a matter of peace, safety and waste disposal, the green movement gains both public profile and initiation of battle in mass protests with various degrees of violence. In these years most Western industrial countries experience large-scale demonstrations and public happenings that challenge nuclear power, spearheaded by growing green activism.

This leads in the 1980s to the *professionalisation of activism*. This means on the one hand the professional spearheading of protest. The effectiveness of groups like Greenpeace to take up and create a public issue became their trademark. Greenpeace International operates globally on deforestation, whaling, industrial pollution and global warming, Antarctica, nuclear power and biotechnology. They fund research, stage media stunts and raise funds through donations that pay for global campaign logistics of ships, helicopters and troupes of highly trained stunt men and women who dive, climb or jump to create television coverage. With a professional approach Greenpeace and similar organisations walk a tightrope between effectiveness on public issues and alienating their basis of support.

## 78  Atoms, Bytes and Genes

*Excursion 4.2*  The Risks of Professional Activism

---

*Occasionally a campaign backfires like the one on BRENT SPAR oil platform of 1995. The British government had licensed Shell to dump a defunct platform into the Irish Sea. After international protest that included a boycott of Shell petrol stations, the platform was finally dismantled at shore in Norwegian Stavanger. It later became evident that this operation at shore posed higher risks to the environment than deep-sea dumping would have posed. Greenpeace won the battle for the hearts and minds of the public, but they had got the facts wrong. This damaged their reputation of being both scientifically correct and politically right. Professionalisation can create reputational risks for activism.*

*The other side of this professionalism is the development of a new service sector that offers environmental assessment for public and private projects. Environmentalism achieved changes to planning regulations which obliged planners to anticipate and evaluate environmental impacts. Potential negative impacts turned into calculated risks as the assessment moved upfront in the planning cycle. Power stations, road building, tourist developments, water barriers and industrial operations now require environmental risk assessment before works can start. This created a market for advice provided by statisticians, geographers and civil engineers who developed methodologies for conducting environmental risk assessments.*

*Much expertise accumulated and translated 'sustainability' into steps to avoid future damage by realistic cost accounting. The oil spills of Exxon Valdez (Alaska, 1989) and Braer (Shetland, 1993), the BP deep-sea oil drilling disaster in the Gulf of Mexico (2010), the chemical pollutions at Seveso (Italy, 1976), Bophal (India, 1984) and Basel (Switzerland, 1986), the Space Shuttle Challenger disaster (1986) and the nuclear accidents of TMI (1979), Chernobyl (1986) and Fukushima (2011), highlighted the environmental burdens, the public health hazards and the limits of safety regimes of large technological projects. The clean-up and the loss of biosphere from an incident must now be part of corporate liability and of social responsibility.*

---

The late 1980s saw the *internationalisation* of the issues on a large scale. The anti-nuclear and environmental protests of the 1970s moved from local protests to trans-national co-ordinations and global campaigns spearheaded by Greenpeace and FoE. The world's insurance markets increasingly recognised their limitations to cover large events like nuclear accidents and more frequent natural disasters. International climate watchers under the patronage of the UN, which was already engaged in population control, stepped in with the formation of the Intergovernmental Panel on Climate Change (IPCC, founded in 1988). Further milestones of internationalisation were the Rio summits of 1992 and 1997 on biodiversity, and the Kyoto Convention of 1997 and the Bali follow-up of 2007. The IPCC establishes the scientific facts of climate change and maps the consequences in a sequence of high-profile reports. Kyoto established a market for carbon trading on agreed emission targets. Those who emit $CO_2$ beyond their target can trade their surplus with those who emit below their targets. This creates a system of tracing and of trading emissions against capital which sets financial incentives to avoid pollution on a global scale. Save for opt-outs, global carbon trading operates in a similar manner to late mediaeval indulgences: accounting and paying for your sins (i.e. trace and trade your carbon emissions) encourages moral behaviour (i.e. less polluting) and raises funds for a good cause (i.e. development for the less well off; see MacCulloch, 2003, 120ff).

## The Social Basis of Environmentalism

Social movements reach beyond a core group of activists. They thrive in a social reservoir from which they draw financial support, recruit members and mobilise sympathisers to mass demonstrations, the public opinion that levers in policy changes. Not everybody is equally attentive to ecology. So, we have to ask: where is the social location of an ecological consciousness and the recruiting ground for activism, and has this changed over time?

In the industrial world environmentalism is situated in the sector of new movements including peace, civil rights, feminism, gay liberation, alternative lifestyle and other issues. New movements are characterised by identity and culture politics that focus on lifestyle and aspirations in contrast to the traditional movements whose focus is class, occupation, grievance and hate. In the repertoire of activity, new protest superseded industrial strikes (Hechter, 2004; Touraine, 1985).

The social basis of new movements is heterogeneous, but in contrast to traditional worker and peasant movements mainly middle class, both in activists and sympathisers. Environmentalists are initially recruited at the intersections of the intelligentsia, the service sector, and the socially marginal. But there is a cleavage within the new middle classes between educated professionals loyal to an 'ideal community' and managers loyal to 'business and markets' (see Kriesi, 1998). Activists and sympathisers are disproportionately university educated, often in the social and historical sciences. In particular, green activists and their sympathisers are disproportionately employed in education, social, health and mass media services. This *human service sector* was about 10% of the labour market in Germany by 1980 and grew quickly. Activism extends to the socially marginal, such as unemployed teachers and social workers, part-timers, people on social assistance and among all, particularly the young. The sympathisers include *locally affected populations*, the NIMBYs who fight against particular airports and nuclear power stations in their neighbourhood (Raschke, 1988, 414ff). This selective recruitment base has given rise to doubts about the degree to which the environmental movement represents the common interest, and how under a green agenda, in particular in developing countries, different interests might be able to co-exist (see Forsyth, 2007).

Large-scale surveys in the 1970s and 1980s demonstrated that the rise of environmentalism is best explained by changes in social values (see Inglehart,

1990; Pakulski, 1993) and less by economic interests. While the materialists are concerned with law and order, inflation, economic growth, and national security, the post-materialists value democratic involvement, freedom of speech, ideals, beauty and friendship. This shift from predominant 'materialism' or 'post-materialism' in one generation occurred during a period of prolonged prosperity, rising levels of education and peaceful life for an entire generation post-WWII. Research in Europe shows that post-material values are neither an achievement of the wisdom of age, nor explained by economic cycles, but an effect of the post-war cohort of baby-boomers born before 1963. This marked the shift from the politics of distribution to a politics of life quality.

Membership of the ecology, anti-nuclear or peace movement is best explained by post-material values, better than by the left-right orientation, age, family income, religiosity or party affiliation (Inglehardt, 1990, 385ff). Post-material values fostered membership of environmental organisations; UK studies showed that green commitments were spreading beyond the original 'middle class radicalism': to women more than men, to rural more than urban dwellers (Norris, 1997; Witherspoon, 1994). Similar observations apply in Australia and the US (see Skrentny, 1993).

In the UK, membership in environmental organisations increased from below one million in 1971 to over five million by 1997 or about 15% of the population (Cowe and Williams, 2000, 21). While this increase seems universal, the level of memberships varies widely across the globe (see Dalton, 2005), concentrating in Western industrialised countries. The late 1980s, following nuclear accidents and global campaigns, saw a swell of people signing up to the green cause.

It is notable that the German Green Party became a major political force with government responsibilities in the 1990s. The collective action repertoire is conventional including signing petitions and street demonstrations and creative including stunts and theatre. The new ways of gaining public visibility are expected to be mainstreamed by the market system as 'guerrilla marketing' (see Amann, 2005).

In Germany between 1974 and 1989, 24% of public protest was ecological, while this figure is between 13% and 18% for the Netherlands, France and Switzerland. France persisted with traditional labour protest. Ecological protest was more conventional in Switzerland, using signatures and petitions, while more confrontational in Holland, Germany and France. France saw the highest numbers of violent anti-nuclear incidents. All four countries experienced a peak in protest in 1980 and again in the mid 1980s. In France, protest faded with the presidential victory of the socialist Francois Mitterrand in 1981 (see Van der Heijden et al, 1982).

## Ecology Moves from Right to Left: Modernised Conservatism

A feature of environmentalism is the mix of tradition and radicalism, the pursuit of conservative aims with progressive methods. Rooted in romanticism and localism, environmentalism emerges from a desire to protect the simple life from the grip of a noisy industrial society and its bleak chimneys. This conserving impetus made the movement initially lean to the political right. *Rural and local conservatism* defending a traditional way of life is a more direct ancestor of environmentalism than the urban and internationalist radicalism of Marxist-Leninist or Liberal provenience, which embody the rationalist worldview (see Mannheim, 1986; Devall, 1985; Sieferle, 1984). The prominence of European aristocracy, Dutch and British princes, as ecological figureheads testifies to the former.

However, things got confused in the 1970s, when ecological awareness mixed with new radicalism, which shifted the action to the young political left. The old left of the labour movement and of Marxist-Leninism were still tied to the scientific-technological revolution and dismissed environmental awareness as a petit-bourgeois or reactionary distraction. Free-market liberals equally showed learning difficulties, dismissing it as an interventionist, even totalitarian temptation. The trope of an eco-dictatorship, to sacrifice democracy in order to save the planet, continues to enter political debates. Indeed, the paradoxical heritage meant that the green movement has to fend off 'entrism' of neo-fascists and xenophobic nationalists, 'Lebensraum' ecologists before their time. The over-population frame resonates across the political spectrum. These and other contradictions characterised newly forming green parties, where 'deep ecologists' or 'Fundis' sought system change and clashed with reformers or 'Realos' who sought change within the market system to save the planet (see Naess, 1973).

Not only did the ecological awareness and its agenda move from right to left, but also up from local to global politics, and thus beyond 'reft and light'. Into the new millennium, the environmental agenda had become mainstream politics and ecology the *ideological master frame* with wide public resonance. It solved the paradox of being both modern and conservative by defending the life-world against bureaucracy and technocratic rationality (see Eder, 1996).

## RESPONSES TO THE CHALLENGE

The achievements of this broad mobilisation since the 1960s are contained in three outcomes: a series of cycles increasing issue salience and rising public awareness; changes in public attitudes and commitments; and the introduction of international regulations of corporate conduct.

### Issue Cycles: Salience and Awareness

Considering environmental news since the 1940s, the British mass media show four cycles of ups and downs of environmental news, each highlighting a different mix of issues (see Figure 4.1). The first cycle, post-war into the early 1960s, focussed on air pollution, landscape and road planning (see Brookes et al., 1976). In 1956, Britain passed the Clean Air Act in response to the great smog of London of 1952 which caused several thousand deaths mainly among the very young and the elderly. The act created smoke-free zones and encouraged alternatives to coal fire heating of households. In the same period, the nuclear fall-out debate with scenarios of nuclear winter, widespread crop failure and hunger crisis caused by the dust cloud of an atomic explosion provided an early motive to think globally. But the actual radioactive fall-out of the Windscale (now Sellafield) fire in Britain of 1957 was treated as a local issue, despite the global news event and diplomatic protests from Norway, where the emitted radioactivity was registered.

Pollution from agro-chemicals, over-population and the limits of energy resources dominated the second cycle between 1962 and 1980 (Carson, 1962; Ehrlich, 1968; Sandbach, 1978). However, in contrast to population and resources, pollution remained a local issue. The 'Waldsterben' and acid rain, despite in evidence across national borders, remained a local issue. While the Norwegians complained about the British acid rain coming over with the prevailing winds, where the German and Swiss saw a 'catastrophe in the making', the British, French and Italians wondered what the hysteria was all about.

The ozone hole, deforestation in Amazonia and the 'greenhouse effect' of $CO_2$ in the atmosphere finally created a global drama, and a third cycle of attention between the foundation of the IPCC in 1988 and the Rio Summit of 1992. Here global synchronisation of opinions is well documented (see Mazur and Lee, 1993; Einsiedel and Cochlan, 1993). In this cycle environmental news became a routine beat of specialist reporters (Warren, 1995, 52): exaggerated and scientifically unwarranted claims were made (Bell, 1994); scientists opened the debate but then relinquished it to political actors (Trumbo, 1996); local dramas such as 'poisoned seals' in the North Sea, Prince Charles speaking out on aerosols (Warren, 1995), Chernobyl 1986 and the 1988 assassination of activist Chico Mendes in Brazil pressed the green issue into public attention (Mazur and Lee, 1993); Greenpeace successfully focussed public attention with a series of high profile campaigns (Hansen, 1993).

Whatever the quality of coverage, the global alarm of the late 1980s created a political agenda and legitimised the formation of the IPCC in 1988 (Mazur and Lee, 1993). In the US awareness of a 'greenhouse effect' increased from 39% in 1986 to over 80% by 1992; it stayed on a high level into the new millennium, around 90% by 2008 (see Figure 4.2). As environmental concerns lost public priority in the 1990s, the IPCC secured the attention. The IPCC put up a strategy of piecemeal media events, reporting in 1990, 1995, 2001, 2007 and again 2013 on the science base for global warming and anthropogenic climate change. This activity ushered in the fourth news cycle after Kyoto 1997. The evidence of anthropogenic influence on the

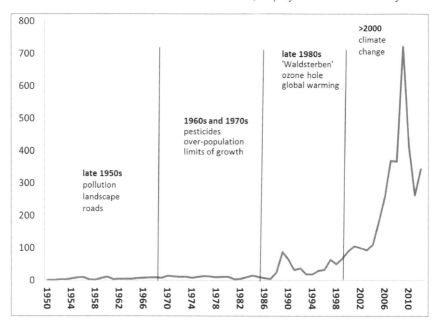

*Figure 4.1* Environmental news references in the UK press, 1946–2010. Index combines overlapping data series [2002 = 100 or 458 articles in the British *Times*; peak 2009 4,217 items; compare 'credit crunch 2008' = 5,320 items; 'Gulf war 1990' = 4,185 news items]. *Source*: Brookes et al., 1976 (1950–1970s); Bauer et al., 1994: 1946–1992; Warren, 1995: late 1980s; Tennant, 2012; and Guardian News Archive: 1999–2006.

climate emerged from long-term observations that compared climate models with and without human influence and projected future scenarios. Graphics showing the marked rise of global temperatures in recent years (the 'hockey stick' graphic) visualised the problem with iconic status (see Maslin, 2004, 47). Other visuals depict polar bears threatened by melting ice, and gigantic icebergs breaking off the Greenland and the Antarctic shelves. Despite global information, the awareness of climate change varies across the globe. Gallup polls in 2007 and 2008, before the Copenhagen Summit of 2009, showed awareness above 90% in Europe, Australia and North America (as in Figure 4.2). In the BRIC economies awareness was lower between 50–70%, while across Africa and the Arab World awareness remained below 50% of the population, and as low as 25% in Benin, Niger, Liberia and Burundi.

Global warming became a meta-frame that merges scientific, political and cultural drama, and puts nuclear power into a different light. The Kyoto Protocol 1997 and the controversy over the US and Australian opt-out until and beyond Bali 2007, created a huge attention cycle on 'global warming' and the 'carbon economy', the international system of tracing and trading

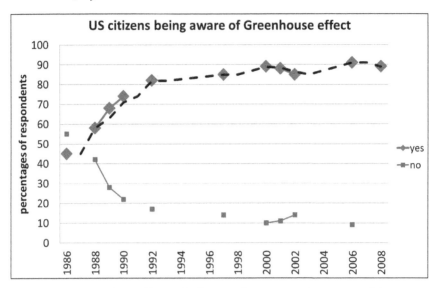

*Figure 4.2* US opinion polls on whether people are aware (yes or no) of the 'greenhouse effect' and 'global warming' between 1986 and 2008. *Source*: Nisbet and Myers (2007) and BBC poll (16 July 2008).

$CO_2$ emissions. But an apocalyptic drama is not sustainable for long. A question of 'what happens when prophecy fails' is lurking. The media logic of heightened drama runs the risk of a sudden reversal: when disaster is no longer news, the absence of disaster is the news, global warming might be good news after all (see Weingart, Engels and Pansegrau, 2007). Preaching apocalypse is a risky business. It might create a global community of concern, but if the doomsday scenario does not materialise, science will be seen to have acted in its own interest and credibility will be lost.

Here I return to the paradox of how the greens might have influenced climate change. Since the 1960s changing public attitudes have focussed attention and mobilised resources to ascertain the facts of global warming and the deterioration of 'nature', and to disentangle endogenous cycles from man-made contributions. Changing attitudes raise public attention and push resources to issues previously thought to be unimportant, such as climate change. Nowadays we live with icons that visualise global warming and motivate actions to mitigate it. These images were not available before the 1980s because of insufficient scientific attention. Green mobilisation legitimated this research. Thus environmentalism indeed influenced how we know about climate change.

## Attention, Attitude Change and Behavioural Commitments

Public attention is often short lived. Figure 4.3 shows how during the third attention cycle public salience changed the public's commitments to protecting the environment. The percentage of the population who regard the environment as a serious issue peaked at 30% in June 1989 and 1990, then declined again. Public salience follows the press coverage. High and low press coverage is reflected in high and low public salience of this concern; it appears that the audience loses interest whenever the press does. The one follows the other, though it is not clear who leads whom.

Surveys of public commitment show a different trend. People watch environmental programmes, give money to activism or participate in public demonstrations. The trend shows rising behavioural commitment into the 1990s. Respondents with five or more environmentally friendly behaviours increased from 15% in 1988 to 30% by July 1991, only to decline to 23% a year later. At the same time the consumption of lead-free petrol increased from 12% to 37%, today it must be close to 100% and the rate of 'green consumers' in the UK increased from 20% to 40% (Worchester, 1993). The use of energy saving light bulbs increased in the UK from 16% in 1993 to 72% in 2007 (Defra, 2007). While news attention and public salience comes and goes, some new habits seem to stick.

Sorting household waste, worrying about wildlife and thus taking part in public protest would have been ridiculed in the 1950s—into the new

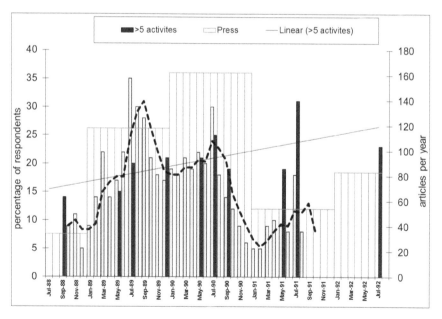

*Figure 4.3* Peak of the third cycles of environmental news between 1988 and 1992 (annual average, right scale); the percentage of respondents for whom the issue is important (monthly light bars; left scale) and percentage of respondents reporting five or more behaviours supporting sustainability in Britain (dark bars and linear trend, left scale). *Source*: IPSOS Mori polls & EC, 1992.

millennium it is good citizenship. The burden of proof has shifted: behaviour that earlier needed justification, is now taken-for-granted. However, the taken-for-granted needs no attention; awareness might fade. Saving the planet and avoiding apocalypse has become a way of life. Some will sceptically argue it has become a quasi-religious commitment: carbon trace is a sin and carbon off-setting is the monetary indulgence by which we pacify our modern anxiety about our salvation.

## International Policy and Corporate Conduct

Several cycles of public attention brought alarm and changes to national and international policy. The greening of politics occurred in the 1990s, when most political parties adopted an environmental rhetoric and at best a green platform. Green parties, with few exceptions, remained marginal despite their success in setting the agenda. In Germany the Green Alliance, founded in 1980, achieved a significant role in coalition government between 1998 and 2005 when Joshka Fischer, a former radical, became foreign minister.

With the IPCC since 1988, the Rio summits 1992, the Kyoto Protocol 1997 and the follow-ups in Bali 2007 and Copenhagen 2009, the 'environment' reached the global agenda. How to save the planet is no longer the fringe concern of nerdy conservationists, but central to politics and economic planning. The debate is no longer on whether deterioration is happening, but on options and commitments for remediation and self-constraint. The Stern Review (2006) diagnosed a gigantic market failure to provide remediation, and proposed a 1% of GDP investment to mitigate climate change—a figure that Stern considered too low six years later. Clearly this is not a conflict-free zone as it includes the allocation of resources and costly regulations. Economic interests motivate climate change scepticism (see Monbiot, 2006). These machinations exploded in 'Climategate', when on 17 November 2009 Russian hackers accessed an e-mail exchange among climate scientists from East Anglia University that apparently showed a collusion among scientists to suppress data that was inconsistent with global climate change (see House of Commons, 2010). Climategate significantly influenced an already polarised US public opinion over anthropogenic climate change (Leiserowitz et al., 2010).

What dominates the post-2000 cycle of news is the mobilisation of the corporate world into a green agenda within the market logic. Events like the World Economic Forum in Davos, occurring annually in Switzerland, mobilise global business commitment. Sustainability has entered the language of corporate social responsibility (CSR). Green investors demand audits of the corporate 'carbon footprint'. The industry responds to green regulations, green investors and green consumers. Conflict over the merits of regulation, deregulation and self-regulation to achieve a sustainable economy absorbs much energy and is the focus of much academic research.

However, climate change denial and disillusionment about the persistent gap between word and deeds, about sustaining the unsustainable modes of production and the simulation of changed behaviour (performative simulation), give rise to 'post-ecological discourse' (Zeyer and Roth, 2013).

## Toward a New Common Sense: Sustainable Lifestyles

Acting sustainably so that the consequences of action do not undermine the conditions for future actions has become the new horizon against which, and within which, affluent industrial countries evaluate new techno-scientific developments (Jackson, 2003). It brings together various old and new strands of thinking. Private and public watchdogs monitor the application of the new benchmark. Many people and corporations have already changed their habits. Some countries are more determined, for example the Dutch, concerned with predicted rapidly rising sea levels. Other countries are more reluctant, avoiding the costs of remedy, or they genuinely lack public awareness. But be that as it may, it clearly has become difficult to promote new technology without considering its environmental impact. Nuclear power and biotechnology are cases in point. The nuclear power movement had to face up to this question in controversy and conflict. Equally, the environmental impact disrupted the trajectory of biotechnology in the 1990s and looking forward throws a dubious light on nanotechnology. What makes the criterion pertinent is the emergence of new configuration of human being: the green consumer-investor.

The green movement has been successful in raising awareness, in pressing the political agenda, first nationally then internationally, and in changing policy, to set incentives which encourage sustainable economic production. After the fall of the Berlin Wall in 1989, an invigorated capitalism was averse to statutory regulations. Consumers and investors became the lever of change by demanding sustainable business models as well as sustainable goods. The fourth cycle of environmentalism sees the marriage of environmental and consumer concerns in the 'green consumer-investor'.

*Excursion 4.3*   Consumerism and Product Safety

---

*The consumer movement has its own protracted history of conflict. It started with fights against price-fixing monopolies and food and drug adulterations in the early 1900s, followed by a struggle for objective information and product labelling against the rising tide of advertising and marketing claims-making that has exploded since the 1930s. These issues of price extortion and health risks continue to loom large in many countries—note the BSE meat crisis in the UK in the 1990s, and the milk adulteration scandals in China of the 2000s or product adulteration in the 2013 scandal of 'enrichment' of beef products with cheaper horse meat across Europe. Where individuals cannot check the quality of products at the points of sale, independent consumer organisations, if they exist, seek to make this information available through testing of products and the publication of results. This logic made consumer organisations wary of GM food when they appeared on the market in the early 1990s (see Chapter 8).*

*The push for product safety and corporate liability is a second phase of this consumer movement that started in the 1960s. Many of these fights were spearheaded by local personalities like Ralph Nader in the US (see Nader, 1965). The fight for product safety spilled over into concerns over safety engineering of technological installations as in nuclear power plants, or dams that hold back water to produce electricity.*

*Trade unions and the international labour movement had achieved earlier successes in implementing measures to guarantee safety at work. For example, in the UK, the 1974 Health & Safety Act was landmark legislation on occupational health, drawing conclusions from epidemiological research on the handling of toxic substances. It defines procedures and responsibilities, and made possible redress and compensation claims in cases of accidents and work related illnesses. In this context 'asbestos' was recognised as a dangerous substance that, despite useful insulation and fire-proofing characteristics, was no longer viable as building material.*

*The consumer movement has seen its own internal debates between radicals seeking system change to achieve equity and for whom big business is the culprit, and the reformers who fight for effective regulations to deliver quality goods that are efficiently priced. The fight is against rogue actors and monopolists within the market system. This dilemma between system change for equity and reform for efficient markets continues to split the consumer movement over the choice of tactics, which include changes to the law, appeals to self-regulation, forming cooperatives, calling and organising consumer boycotts and campaigns, and filing litigations and court actions (Mayer, 1989).*

---

In the 'green' investor-consumer-citizen these two movements, environmental and consumer concerns converge: an issue coalition formed over agricultural biotechnology and GM food in the 1990s. Some speak of the 'ethical consumer' and mean 'green consumer'. The affluent and green investor assesses the carbon footprint of a company before investing his or her money in stocks. Banks now offer green investment funds and monitor indices with companies that are vetted on their green credentials. Green consumers are wetting the reputation and avoid brands and products without green credentials. Green investment is now a sizable market segment and green products cover the entire livelihood from energy to clothing, travel, leisure, holidays, food and appliances. However, the consumer is confused by a plethora of labels and certification which signal the credentials of products, services and businesses; and certification like money is prone to devaluation due to proliferation of false pretence.

The British pollster, Sir Robert Worcester (1993), had claimed that by 1992 around 40% of British were 'green consumers', probably an overestimation. By 2006, this new lifestyle was embraced by about half of British adults, who recycle household waste, consider ethical products, buy organic, think twice to travel far and are prepared to pay a premium to indulge their desires in a sustainable manner (ICM research, October 2006). By 2008, 76% of Europeans recycle household waste, 64% save energy and 55% save water. The older and more educated more so than the younger and the less educated. Environmental awareness makes a difference: 44% of Europeans are prepared to pay a 10% premium for green energy (EC, 2008, 62ff). The British Cooperative Bank has published several 'ethical consumerism' reports since 2000. By 2012, the British *Guardian* offered a regular supplement of 30–40 pages advocating a 'green & ethical lifestyle' with advice on financial services, design and healthy living. For the green consumer-investor-citizen the world needs to be *safe and sustainable*. And henceforth any emergent technology must pass this benchmark or face public resistance.

# 5 Ten Propositions on Learning from Resistance

In this chapter I will elaborate in more abstract terms the idea that pervades every line of this book. Techno-science is a quasi-social movement in search for material and symbolic support. In this context the problem of resistance arises. The social movement literature focusses on the mobilisation of resistance against techno-scientific developments, for example against nuclear power and genetic engineering. This is not the main concern here, but is implied. I will turn the issue on its head and ask: how does techno-scientific mobilisation deal with the resistance that is encountered? This raises the following questions:

- What is 'resistance'?
- How is resistance registered or ignored during mobilisation efforts?
- What is the effect of resistance being registered?
- What is the relation between 'resistance registered' and movement learning?

In a classical note on 'resistance and social movements' Vander Zanden (1958/59) captures the perennial problem. Social mobilisation faces two challenges: the inertia of all movement (movement creates friction and faces doubts: 'don't fix what ain't broken') and counter-mobilisation arising from vested interests and from good reasons and genuine concerns about the future. Movement and counter resistance build on each other, leading to outcomes quite different from the ones initially projected by movement (Zanden, ibidem, 313). It would be a whiggish error of hindsight to attribute the outcome solely to the projected movement, and to resistance only 'disturbance'. Such interaction is modelled in 'war games', which involve guessing opponent's intentions, undermining their resources, deploying propaganda and a common set of rules of engagement (see Zald and Useem, 1987). However, this shall not be the focus of my present analysis. My focus is movement change, or learning in response to resistance—something the literature pays little attention to. In the context of techno-scientific projects, what concerns us here is the *encounter of movement and resistance*. Resistance disrupts the established course of action, is registered and elicits sense making and learning, but within constraints. This basic insight will be elaborated in 10 propositions about the effects of resistance on social mobilisation.

## The Main Thesis Is: Mobilisation Efforts Have Inherent Learning Difficulties

In the extreme, mobilisation is a hot movement, and it is nigh impossible to derail its project and progress. For example, chiliastic religious movements are 'hot' mobilisations that are immunised against contradiction through strong belief in a mission on earth. However, we ask, what role is there for challenge in this effort? Resistance makes a difference under certain circumstances, which we need to understand. The present orthodoxy is the past heterodoxy; the course of history is no straight line. The literature on strategy uses many terms for problems in movement: obstacle, hurdle, friction, irritation, stress, strain, pain, provocation, disruption, recalcitrance, unknowns, uncertainty, reaction and resistance (see Akerman, 1993; Bauer, Harré and Jensen, 2013).[1] Most of these terms have a metaphorical core and conceptual entailment. For the present purpose, I elaborate the entailment that arises from a metaphorical transfer between pain and resistance in movement.

Resistance triggers additional activity in movement, namely sense making and mobilising emotional energy. Reflective consciousness, cognition and motivation determine sustainable activity.

My model is based on a process analogy between pain and resistance. This suggests what pain does in everyday activity, resistance does to social mobilisation: it *monitors ongoing activity and signals that something is going wrong*. This signal has learning potential. My model urges a shift from a 'deficit' to a 'resource perspective' of resistance, fostered by insight into how pain works.[2] This resource perspective has several implications:

(a) We consider primarily the functionality of resistance=pain; dis-functionality arises as a secondary process.
(b) Because resistance=pain threatens the sustainability of a project, it requires an immediate response.
(c) Resistance=pain triggers this response, but does not determine it.
(d) The response determines whether it develops into chronic resistance=pain or not. Likewise, the dis-functionality of resistance=pain is secondary, hinging on the immediate response.
(e) Everyday responses to pain form a cultural set—with threshold and preferred courses of action. Likewise, responses to resistance are symbolically mediated. The theories, concepts and stories of resistance condition the responses to the challenge and are therefore part of a mentality which we need to characterise.

(f) Resistance=pain is more motivator than sensation (see Wall, 1999). The correlation between causes and pain are tenuous; there is pain without injury (phantom), and injury without pain (endogenous modulation); there is registered pain that does not matter (asymbolia, see Grafee, 2001). Pain is modulated by disposition and endogenous control (Melzack and Wall, 1988). Constant are the functional responses and their learning potential. By analogy, resistance=pain is modulated by public discourse which frames responses.

The elaboration of a resource model of resistance=pain probes this analogy by unfolding a number of empirically testable hypotheses. Thus, the analogy is creative. I unfold this framework in the spirit of a 'General Resistology' for the social sciences (see chapter 9).

## WHAT IS RESISTANCE?

Basic features of resistance are surprise and the clash of dynamics, forces, intentions and will powers (see Akerman, 1993)—as such, it creates uncertainty for strategic action. Resistance is the flip side of power. Power implies resistance: Napoleon is credited to have observed 'one can only rely on what resists'.[3] Social mobilisation pursues a particular course of action, for weak or strong reasons, and faces resistance on the way; everything else derives from this basic encounter. Any social movement consists in holding to a particular project invested with the will to change reality. Who resists does so also with reference to a projected future; it affirms the possibility of an alternative, often not yet defined.

### A Formal Definition of Resistance

Formally resistance may be seen in the following logic. The verb 'to resist' is a predicate of conflict ($C_r$) with at least three arguments and one meta-argument: project (P), actors (A), repertoire (Rep) and mentality (M) as the meta-argument, such that: $C_r$ {P, A, Rep, M} and M {$C_r$, P, A, Rep} contains recursively all the elements. This points us to the logical paradox of resistance: it is both element and the set that contains all the elements.

$$C_r \{P_{ts} [ A_{spo}; Rep]; R [ A_{res}; Rep] \} \quad \rightarrow \quad C_r' \{P_{ts}' ; R'\}$$

$$M_{bda} [A_{obs}, A_{spo}, A_{res}]$$

In order to analyse resistance, several elements of this formalism need to be considered. Project ($P_{ts}$) and resistance (R) are in a *temporary* and for ($P_{ts}$) *surprising conflict* ($C_r$). We consider a techno-scientific project ($P_{ts}$) sponsored by actors ($A_{spo}$) who see themselves as 'innovators' and command a repertoire of framing actions (Rep). Others ($A_{res}$) will resist this 'provocation' drawing from a similar repertoire (Rep) of actions, reasons and motives. $A_{spo}$ and $A_{res}$ draw on a common cultural stock (Rep), making use of an overlapping subset of actions, which characterises their particular repertoire. In time, this leads to unexpected transformations: project ($P_{ts}$), resistance (R) and conflict ($C_r$) move on to ($P'_{ts}$), (R') and ($C_r'$). These transitions become part of *discursive accounts which make up a mind set* ($M_t$) that register events from the perspective of observer ($A_{obs}$), sponsor ($A_{spo}$) or resister ($A_{res}$). The mind set ($M_t$) is the meta-argument; it is both an event and presents all the events recursively as discourse. The mind set ($M_{bda}$) operates in anticipation of problems (before, $M_b$), through analysis and comment (during, $M_d$) and in commemoration (after, $M_a$) of 'resistance' ex-post-factum. The commemoration of 'resistance' ($M_a$) has the features of hindsight (see Geyer, 1992) such as fostering identity and capacity building. Resistance does not 'exist' as such, but 'insists' through a mind set before, during and after events; social representation matters to make sense of what otherwise remains opaque.

What for the sponsor ($A_{spo}$) is an *unexpected surprise* is for the resistor ($A_{res}$) a reaction to *provocation*. $A_{spo}$ encounters another 'will-power' of $A_{res}$ and the techno-scientific project ($P_{ts}$) finds itself *resisted*. Who is the proactive and who is the reactive part can only be determined arbitrarily by punctuating turn-taking over time. Techno-science is disruptive and thus a provocation; resistance responds to this provocation. The conflict (C) is a 'conversation' of challenges-and-responses. The analysis can focus on the project ($P_{ts}$), on the resistance (R), or on the conflict (C). The distinction between the project ($P_{ts}$) and the sponsor ($A_{spo}$) allows us to distinguish resisting a project from resisting particular actors. The polemic can be ad-rem or ad-personam, e.g. hackers do not resist IT, but its use for state surveillance. (Rep) refers to the repertoire of activities, reasons and motives. The effects of resistance are manifest in the transformation ($C_r$) → ($C_r'$), e.g. from an acute to an ingrained conflict, in altering the project (→P'); when some features are dropped, and in recursive effects on resistance (→R'), when past experience conditions future action and capacity is built, encouraged or demoralised.

The public discourse of resistance includes episodic accounts and theorising. Conceptual analysis and storytelling are modalities of social representation. Events are narrated and can be studied as discursive mind sets ($M_t$). A narrative has perspective. Stories are told from the perspective of the sponsor, the resister or an outside observer. The same is true for conceptual analysis. We recognise that a 'deficit concept of resistance' is a particular mind set ($M_t$), whose significance lies in constraining the responses to the challenge: causal attribution of an external nuisance. In contrast, the 'pain model of resistance' favours internal attributions: the problem is the project itself and its mind sets. This formalism specifies the basic logic of resistance as a predicate with four arguments and a meta-argument.

We cannot take this for granted. The problem of 'resistance' has been given many farewells—useless for analysis and without prospect of success ('resistance is pointless'), only to return under a different guise. We can identify many disqualifications of resistance. The *innovation mind set* sees only an unproductive delay; as most new ideas fail, it is more important to understand success than failure (Rogers, 1983). To declaring 'X is resistant to change' to explain failure is attributing blame externally, and *rhetorically* a distractive argument ad hominem. The attribution slanders serious actors and their genuine concerns (Hirschman, 1991). Resistance analysis is thus often the 'white-wash' of managerial risk-taking, a form of blame shifting in cases of failure. For the *revolutionary mind sets*, resistance is 'reactionary'—an action bereft of 'true' historical consciousness and thus doomed. Its motives are 'pre-revolutionary' and immature 'Kinderkrankheiten' (Lenin) and must be eradicated. For *post-modernists* resistance belongs to the discourse of modernisation, and since grand narratives of history are obsolete, the concept of 'resistance' is an epistemic error of a transcendental perspective (Bateson, 1972). Despite all these good or bad farewells, I keep the term 'resistance' for what it does: *marking a space where there seem to be no alternative.*

## Taxonomy and Repertoires of Resistance

Resistance has *intensity and scope*. One might ask, how many people resist and how often? This can be observed for example around a nuclear power site that becomes a focus of public gatherings and protest. Researchers and actors use mass media reportage to estimate these events, the scope of actions and the number of participants (see Franzosi, 2004). Resistance might be short-lived, flare up and disappear, or persist for a time and become a social ritual, like commemorative protests against nuclear power sites have become in places.

Resistance is social conduct comprising both *attitude and action*. A resistant attitude expresses critical opinion, views, perceptions, beliefs and arguments. Some argue that attitude is not sufficient. Taking risks in action defines resistance: actions with costly consequences such as confronting security forces, being brought before a court and potentially imprisoned.

The action repertoire varies on the dimension *active-passive*. In passive resistance all violent action is withheld, but this holding-back is highly disruptive and keeps illegitimate powers in check (see Sharp, 1973). Active resistance designates a set of actions that involves escalating violence, which in the extreme does not refrain from killing or maiming other persons. The military strategists explore resistance as 'guerrilla tactics', which, after defeat by an overpowering force, keep open the remaining possibilities (see Stahel, 2006). Resistance to techno-scientific mobilisation mainly takes the form of passive-symbolic action with the occasional escalation into violence, as in the anti-nuclear protests of the 1970s and 1980s, anti-abortion campaigns in the US or in the case of the 'Unabomber' who, in the 1980s and 1990s, sent letter bombs in protest against a 'technocratic conspiracy' (see Chapter 6). Techno-scientific resistance mostly takes the form of attention seeking stunts and camera-suitable event making for which issue entrepreneurs such as Greenpeace have become the innovators.

Finally, resistance sees *degrees of organisation*. At one end of the continuum are spontaneous acts of non-conformity or foot dragging that derail mobilisation. On the other end is the formal organisation, openly or conspiratorially, that co-ordinates acts of resistance involving competent and trained professionals. *Weapons of the weak* allow people to resist powers by foot dragging, co-ordinated only by tradition, storytelling and routines without any formal organisation, and this to cumulative effect (Scott, 1987). Such resistance without leadership is difficult to control; there is no way to disrupt leadership where there is none. Historians distinguish *resistance* from *dissent* (German: 'Widerstand' versus 'Resistenz' or 'Eigensinn' or stubborness). Key is the conspiratorial intent of regime change that is risky because of consequences of life and death. Dissent or 'Eigensinn' are everyday acts of foot dragging, soldiering and non-conformity (Broszat, 1981). Some researchers see risks as the defining criterion, and exclude non-conformity, dissidence, attitudes and beliefs from the strict definition of resistance (e.g. Rucht, 1995).

## Resistance and Opposition

It is useful to demarcate resistance from opposition. Both challenge the dominant mobilisation, but opposition does so through an established role. Opposition is institutionalised; it is expected. Opposition can fail its role, as governments do. By contrast, resistance is unexpected, *comes as a surprise*—its name is uncertainty, unknowns, stuff that happens. It works unexpectedly in time, location and dramatis personae. Resistance is the unexpected friction of strategy that demands 'all change'.[4]

## Causes and Reasons

Resistance might be explained by causes or by reasons to act. Causal attribution includes the search for interventions which the actor who attributes causes can control. The social engineer looks for actionable causes and for policy levers. Attributions of *cause-effect* identify the age, gender, or socio-economic status, mind sets and the incentive structures of those resistant. We learn that resistance is often inversely related to age, and women with children might be more resistant to nuclear power or GM food than men, ceteris paribus. Or the population can be profiled into 'innovators', 'adopters' or 'laggards'. Thus socio-economic positions are 'risk factors' that make resistance more or less likely. Attributions of *why-causes* identify the lack of information, attitudes and false beliefs, erroneous cognition or lack of skills. Here the *notion of deficit* falls easily: information deficits, cognitive deficit, skill and moral deficiencies. Internal and external attributions work for both self as well as others. As the self is situated or disposed, so are others. By contrast the hermeneutic exploration of reasons seeks to understand, justify and legitimise resistance by appeal to a sensus communis.

*In-order-to reasons* are the justifications given to resist, drawing on a discourse of 'natural rights' of moral reasoning and appeal to rights in the rhetoric of resistance. The *right to resist* commits us to oppose the abuse of power, any authority out of bounds and unjust. Age-old myths and folktales justify such acts. Antigone insisted on the law of kinship over the laws of the father to bury her outlawed brother. The moral of the story: common-natural law supersedes statutory-positive law. However, a rule 'not to follow the rule' is a paradox: no code of law can legalise breaking the law without undermining itself. Constitutional thinking seeks for ways out of this paradox. The law needs to tolerate some stringently defined deviance to compensate for imperfections in the system. Dissent within the moral community of the constitution is justified; dissent from outside that community is not (Rawls, 1972). Democratic societies need acts of resistance to test their system of justice, to check the misconduct of their institutions, and to signpost the lacunae in the operations of the law (Rhinow, 1985).

The *rhetoric of resistance* (Hirschman, 1989) points to three tropes often used to derail mobilisation for change efforts. All three arguments challenge an exaggerated sense of command and control. 'Perversity' points out that actions will achieve the opposite of what is intended. 'Futility' claims that the intended results will not be forthcoming because the circumstances are stronger: the more you change the more things stay the same. 'Jeopardy' admits that actions might be laudable but previous achievements are put at risk. These tropes contain a kernel of empirical truths towards which the reforming enthusiast is often blind (Hirschman, 1995, 45ff).

## Consequences

Like all activities, acts of resistance have a dual outcome. Recursive effects comprise further entanglement, solidarity among activists, identity construction and the building of capacity for future action. Stories of 'heroic deeds' become part of the identity kit of a counter-movement. Historians talk of *totemic resistance*: memory widens the scope for retrospective resistance in the re-interpretation of the past. Past resistance becomes a symbol for a new order, present obligations, and new struggles in present society. Scattered acts of defiance develop coherence only in hindsight (Geyer, 1992; Rabinovici, 2008). This totemic resistance is brought to the formula: '*Je me revolte, donc nous sommes*' (Camus, 1951, 432). An act of resistance entails a paradox: *by resisting you get entangled.* There is always the risk of being co-opted and caught in a predefined role. Thus resistance is a balancing act between entanglement and autonomy.

Then there are the immediate effects of resistance in delaying or stopping the project, forcing tactical retreats and instigating adaptive learning to make the future sustainable and to pre-empt troubles in the future. As such, resistance has intended and unintended consequences, whether its reasons are good or bad. The effects of resistance pose an attribution problem as recognised in the 'sleeper effect' or the tragedy of minority influence: the mobilisation agency will be happy to bath in the glory of final success, while the contributing effects of counter-mobilisation are easily forgotten.

An analysis of resistance by consequences (rather than causes and reasons) privileges an ecological analysis of the phenomenon resistance=pain. By thinking of resistance=pain as a resource and learning opportunity, our attention will focus on the *constraints of learning*. The learning difficulties then become a matter of opportunity costs, the costs of not being able to learn from resistance, that need to be analysed.

## THE DISCURSIVE REGISTERS OF RESISTANCE

In conducting a functional analysis we ask: what effect does resistance have on the techno-scientific mobilisation? In principle we might distinguish two kinds of effects. One outcome is *regulating* the mobilisation effort with reference to the established intentionality of the project: tactical adjustments will be needed. Another outcome is *redirecting, and thus redefining* the project in the light of resistance. How will that be achieved? Let us conceptualise a bit more.

I have argued the basic warning function of resistance. Like pain, resistance raises attention, enhances awareness of what is going on; it evaluates, delays and potentially alters ongoing activities, and thus offers an opportunity to learn. These established functions of pain become empirically testable hypotheses of resistance by analogy. However these functions are not hard-wired into the social fabric of society, they are mediated by public discourse and mind sets. The way we give meaning, talk about and thus deal with resistance is a matter of discourse, tradition, and mentality.

This implies that resistance per se does not 'exist'. Its ontology is discursive and its contours are fuzzy. But ignoring, denying or dismissing resistance does not make it go away. Moving actors register 'resistance' in a way that has consequences for what to do next. Discourse for action represents 'resistance' schematically. There will be competing ways of doing so, and all discourse and mind sets are prone to blind spots, error and misrepresentations.

### Framing Attention

We communicate primarily, though not exclusively, to solve problems of social co-ordination: when routines break down we seek orientation and direction by sharing knowledge and by resolving conflicts of interests and values. Resistance signals a breakdown of normality and a conflict of directions. Resistance that goes unnoticed has thus limited effect, but attention is no guarantee for an adequate response. We need to distinguish two empirical problems: the *salience* and the *framing* of resistance in communication. Resistance is a more or less frequent reference. In the continuous stream of project mobilisation, referring to 'resistance' is only one among many topics that are registered. Resistance therefore has an irregular salience, and this can be observed by research.

Resistance is registered and 'appears' already *framed* in a particular way. There is little resistance per-se, except maybe undefined inertia of the project. Everyday thinking and project work elaborates that inertia as 'resistance', who it is, where and when it appears, what causes it and what might be done about it. In the mobilisation effort, formally or informally, a representation will appear: $A_{res}$ does $R$, *because of* $Y$. R stands for the mind set that defines 'events' as-R, e.g. blocking a road, a reactionary challenge, putting up a hurdle, a nuisance to be overcome become instances of 'resistance'. $A_{res}$ identifies the actors, and Y suggests causes/reasons and frame a course of counter-measures to control the 'causes'. By knowing the causes, one generally gains confidence to be able to intervene.[5]

## Narrative and Conceptual Representation

We need to ask, where and how does resistance appear in techno-science mobilisations? The analysis needs to consider different modalities of communication which operate in parallel, differences between channels of circulation and between genres of representation. Resistance appears in more private and *informal* conversations, gossip and rumours about what is happening, but also in *formal* news media. The circulation media offer a multitude of elite, specialist and general media, through which resistance resonates with public attention. And resistance being itself a counter-mobilisation effort, this is also a strategic issue of maximising public attention for an actor ($R_{res}$) and its causes.

Another distinction arises from the difference between *narrative and conceptual* accounts. Clearly protest and dissent appear first and foremost as events narrated as stories. People will tell concrete stories of who did what to whom with what effect. Stories of resistance form a vivid memory of heroism and pride, standing up to authority. Beyond stories, academics develop concepts and models of resistance, which bundle these stories into theories of resistance, its motives, reasons and possible consequences. These modes correspond to episodic and semantic memory; we remember the past either as a series of actions or as a categorical order of actions (Bruner, 1991). Thus narratives and concepts are often a *taken for granted* part of the mobilisation practice and mind set. For example, an orthodox believer knows that dissent is 'heretic' and therefore wicked. Similarly, for the convinced change-manager, resistance is a sign of impending failure; they will do anything to squash it. For the revolutionary vanguard, resistance is 'reactionary' and therefore a legitimate target of violence. Social research might use *abstract notions*, even mathematical models, such as 'logistic models', 'rational choice' or 'false consciousness' to deal with inertia in social mobilisation. All these representations, narrative and conceptual, common sense or scientific, circulate in communication, and constitute an empirical mix, the mentality of resistance before, during and after the events ($M_{bda}$).[6]

## Misrepresentation and Error:
## Recipation, Phantoms, and Promoter Signs

Resistance appears in the mind set of actors in the form of framed 'schematic representation'. As such they can fail us, with error and misrepresentation (see Norman, 1984). The mirror of representation is neither plane nor plain.[7] The basic problem will be the following: how can any confident techno-science mobilisation effort recognise an error? All that is at hand is a risky mind set but no sense of being wrong. Dealing with errors involves learning difficulties, and these manifest themselves as the risks of mobilisation (see Chapter 2).

The first error arises from lack of or misplaced vigilance. Resistance might be registered when there is really nothing to worry about (commission or false positive), or resistance might be ignored when there is (omission or false negative). Errors of *commission* arise from a state of alert with strong expectations that something might happen. In a state of high expectation, fiction can overpower reality. We might call this *'Resipation'* (= resistance + anticipation; an analogy of 'nocipation' as in Iggo, Iversen and Cervero, 1985). The anxious anticipation of resistance lowers the threshold of attention and makes actors prone to false alarm. . In a kind of panic, events hastily become 'resistance', either by category extension or over-generalisation of markers. A faint rumour of resistance might lead to an over-anxious mobilisation effort. False alarms can be used strategically, if the consequences of registering are in themselves desirable. If there is advantage in confusion, one might cry 'fire' independently of whether there is a fire or not; the subsequent panic will create the desired confusion. If there is a perceived slack in techno-scientific mobilisation, identifying 'resistance' as a false positive might just unleash that desired additional mobilisation effort.

*Omissions* are false negatives: relevant events go unnoticed. Omissions occur when distracted by boredom or by the hyper-stimulation of dominant concerns. Boredom lowers the threshold of registration, and over-stimulation masks the relevant markers. The fervent mobilisation effort can lead to over-confidence and complacency. Techno-scientific projects usually make great strides to mobilise strong *'promoter signs'* (Valsiner, 2007) such as 'progress', 'revolution', 'economic imperative' and 'investment opportunity'. These signs leave little space for alternative considerations. It is generally known that pattern recognition operates most reliability at medium levels of activation. False negatives thus occur in conditions of high and low levels of mobilisation with talk of revolution and economic imperative.

Registration errors are not the only forms of misrepresenting events. More subtle forms arise from elaboration and modulation in communication. Resistance is *amplified or censured* in communication. In the case of *'phantom resistance'*, resistance is registered with all consequences, but its base in reality remains opaque. Its opposite in *understatement* is equally possible: communicative references to resistance are suppressed to reduce their importance. The public discussion of resistance legitimates it; many acts of resistance seek 'the propaganda of the act' where public attention in mass media and everyday conversation is both means and objective. Hence the counter-strategy of the mobilisation includes censuring this public attention through news management in order to keep coverage low-key. Finally, symbolic registration of resistance involves *displacement*. Resistance that originated in one location is erroneously attributed to a different place involving the disjunction of action and signal. The discourse of resistance is not bound up with the moment and location of occurring events. Displacement occurs when real 'causal' conditions are replaced by some 'spurious' causes, either by pure association or for rhetorical expedience. For example, resistance in one place can serve as a pretext to intervene at another place. External *attribution* of resistance during mobilisation efforts often constitutes such a displacement of conflict. So for example, local resistance to GM crops in Brazil was attributed to 'foreign influences' (see Chapter 8). Displacement by externalisation is risky, because it becomes part of the problem. For example, deficit notions of resistance make 'them' different from 'us', externalise others as 'ignorant' or 'irrational' people and insult people and reinforce the problem. *Prejudices* are self-serving mobilisation symbols; they are a dysfunctional substitute for a realistic analysis of the situation.

In considering the possibility of error and misrepresentation of resistance we recognise the 'blind spots' of any mentality of resistance ($M_{bda}$) which constitute a *learning difficulty* in the very act of mobilising the techno-scientific future. The basic problem is how to recognise an 'error' when there is little room for doubt, and no sense of being mistaken. How can an alcoholic recognise his or her problem when intoxicated (Bateson, 1972)? Similarly, the thermonuclear, the genetic or Internet revolutions left and leave little space for doubt against the logic of 'revolution''. Only *disturbing a euphoric self-confidence* with resistance will bring the potentially flawed mobilisation effort into a sustainable future. How this might work, we will explore in the following.

## HYPOTHETICAL FUNCTIONS OF RESISTANCE

Resistance=pain primarily alarms and irritates and stands for discomfort and trouble ahead. Resistance is noticed with *a sense of emergency* and diffuse anxieties that can escalate to a specific fear of failure. Experienced theoretical (know-that) and practical (know-how) knowledge will be required to contain and save the situation, hence it is likely to enhance mobilisation efforts.

## Proposition 1: Resistance Focusses Attention Where It Is Needed

Resistance redirects attention during the mobilisation effort to where it is needed, and thus creates problem awareness. The focus of attention is narrow, so paying attention to resistance will be at the expense of other topics and issues. Attending one thing risks missing out on something else that might be equally or even more important. Paying attention to an 'event' makes registration as 'resistance' more likely and this creates an impetus to take action; resistance stands for 'the situation is tense and intolerable'.

Attention that is buttressed by a mind set operates better (Waldenfels, 2005). Attention may freely float but be primed to pick up events as 'resistance'. The issue is to pick up early warning signs, to anticipate and spot signs and signals (monitoring function). Attention notes that *'thing are going wrong'*. This invites diagnosis and interventions on pre-fixed signal schemata (alarm function). Mind sets to the title of 'how to deal with resistance' or 'why do people resist' (e.g. Lawrence, 1954) suggest such fixed patterns of registration and interventions. Diagnosis codes events into defined problem shells (coding function). Markers are recognised and turned into thematic communication. Finally, attention is focussed on certain locations at certain times. In these moments, everything else is irrelevant noise (focus function); focus sorts out a figure-ground or signal-noise ratio of perception. Thus attention triggers alarm, focus on and code particular events as relevant according to primed schemata. These schema are used and rehearsed , thus memory of past events influences present sensitivities. Regular occurrence reduces sensitivity, while surprise captures attention but can also traumatise. That is why surprise is a key feature of resistance. Attention enhances the elaboration of the situation; it interrupts quasi-automatic routines by breaking up the established links between diagnosis and intervention. Rather than rushing into action—like a bullet out of the barrel—attention might open time and space for further thought.

Because these elaborations during a mobilisation effort are unlikely to be unified and consonant, the project will confront conflicting action impulses, and has to sort this out. Public support and expert advice are well suited to help to settle such internal conflicts of interpretation and remedial action, opening a lucrative field of consultancy. Empirically, the attention function of resistance is studied in the co-occurrence of references to 'resistance' with other techno-scientific topics in public discourse. We must ask, how often and in what manner is anti-nuclear or anti-biotechnology resistance a topic of public conversations? Thus, the 'resistance' theme has an empirical salience during the mobilisation effort in relation to other themes and topics. The salience and the framing of resistance is an empirical problem of analysing the mentality of the mobilisation effort ($M_{bda}$).

## Proposition 2: Resistance Shifts Project Self-awareness

With a toothache, my body image is dominated by that hurting tooth. Likewise, we hypothesise that resistance highlights the locations and stakeholders of the techno-scientific project who hitherto went unnoticed or had been considered irrelevant. Resistance marks a newly relevant location for the mobilisation effort.

Mass media reportage of events gives attention to some actors and their themes. Themes and actors often move from specialist or social media forums into the general mass media, and from there onto the political agenda (see Strodthoff, Hawkins and Schoenfeld, 1985). Resistance actions have news value when controversy makes for good news. Protagonists and issues can be dramatised. Resistance is bad news for the mobilisation effort, creating costly delays and opening the spectre of failure. A mobilisation effort confident of success might suddenly switch to one dominated by fear of failure. Success orientation or fear of failure creates very different mind sets of mobilisation. With a focus on success, the actor seeks new opportunities; focussed on avoiding failure, actors seek to avoid responsibility and to shift the blame for the upcoming failure. The focus of attention moves from the project to reputation management.

Through attention of others, actors seek to position themselves as a relevant for a topic. Empirically, the mobilisation effort can be assessed through listings of relevant actors and their discursive position in conversations, memos, documents and public media coverage. We ask: who are the actors that are given a position on the 'techno-scientific project'? How does this list of names and their network of topics change over time? We expect that changes in linking topics and actors are related to the elaboration of resistance, and this enhances the self-observation of the techno-scientific project.

## Proposition 3: Resistance Evaluates the Mobilisation Effort

Resistance=pain puts the current course of action in doubt; resistance evaluates the ongoing activity in the light of process and outcome criteria.

Resistant people reject the provocation of a particular techno-scientific project, e.g. the 'thermonuclear civilisation' or the 'bio-society'. Various elements of this project can be evaluated: science and technology per se can be regarded as good or bad, acceptable or not; the people leading the project have a reputation to be mistrusted; unintended consequences of the innovation may be registered. Resistance evaluates the mobilisation effort, initially diffusely (for example Hamlet's observation that 'something is rotten in the state of Denmark' is only preliminary), and with time this evaluation becomes more focussed, considered and able to name the point of contention. This evaluation becomes an input for subsequent action. It is part of a learning process: how to do things differently in the future. The mere fact of public attention implies alarm, i.e. the anticipation of potential hazard. Intensified attention turns public opinion negative if the techno-scientific project is risky (see Mazur, 1975).

Empirically, the evaluation function of resistance is visible in the mapping of public discourse. The quantity and scope of critical references registers the public resistance. Changes in the content of critical arguments over time will show the move from diffuse to more specific resistance. Resistance itself will differentiate in line with different mind sets and conflict mentalities ($M_{bda}$). For example, the nuclear discourse might be framed as a 'war and peace' issue, 'economic viability' or a 'threat to civil liberties'.

## Proposition 4: Resistance Disrupts the Mobilisation Effort

In pain, we start limping and stiffen up. What might have been a smooth and elegant movement becomes stop and go, erratic behaviour. Likewise, resistance disrupts the mobilisation effort in full flight. Attention is paid to the resistance, which brings to light conflicting tendencies where previously there might have been a 'unite de doctrine'. Consensus breaks up and latent controversies in the social movement come to light; what lied dormant becomes visible in public discourse. Some techno-scientific activists and followers become doubtful and consider alternative futures. The movement might show signs of breaking into factions. The leadership of the mobilisation effort is put to test over how they manage the resistance. The quest for leadership of the new techno-science is reopened. The mobilisation effort is disrupted as limited resources of attention are allocated to evaluate what is going on, making sense of and dealing with resistance.

## Proposition 5: Resistance Alters the Mobilisation Effort

Resistance potentially redirects *and* motivates modifications to the techno-scientific project. Resistance creates a sense of urgency where things cannot continue as before, and the evaluation of things point to alternative ways forward. Changes may range from abandoning the project altogether, to altering the strategy, tactics or operations of bringing about the techno-scientific future. These alterations are conditioned by memories of past events.

### Stop and Bring It to a Halt

The most radical project alteration is abandonment; the mobilisation comes thus to a grinding halt. The nuclear project saw such effects of resistance in the 1980s, when in many countries planned nuclear power units were not built, and some that had been completed did not go into operation (see Chapter 3).

### Moratorium: Time to Think Before Mobilising

A source of anxiety in techno-scientific mobilisation is delay. Much seems to hinge on leading the pack. Resistance can physically obstruct efforts, or force delays in activities of research and developments in courts and through legislative process. Politically successful action can change the licensing regulation, putting considerable burdens of time and costs on the project. Consultation processes and safety measures delay the construction and strain the returns on investment, as happened with nuclear power. A regulatory framework affects the planning horizon of new projects. More regulations generally mean longer delays and greater costs, but once regulations are in place this might also enhance mobilisation efforts by providing a predictable context and closing the issue cycle of 'resistance'. The moratorium can thus prompt a new, faster beginning.

A moratorium on a particular mobilisation effort is either *de facto* or *de iure*. GM foods were not common in Europe even in 2010, despite being perfectly legal; in many countries the build-up of nuclear capacity was stalled in the late 1970s despite being perfectly in line with law. A de iure moratorium disallows the pursuit of a particular project, as in the case of embryonic stem cell research in the US (2002–2009) or the cultivation of GM crops in Brazil (1998–2005). A de iure moratorium often means that resistance has successfully achieved political influence by being able to call up existing laws or change the law. The start or end of moratoria are a key tactical objective of public mobilisation efforts.

### Sense Making and Learning

Psychology considers different types of learning and learning constraints (Bandura, 1969). Learning is cumulative change in habits of mind, emotion and behaviour.. Through operational learning, actors enhance their ability to do things better, to co-ordinate moves with less effort and to greater efficacy, to build a repertoire of skills.[8] The attribution of efficacy is 'internal' to the actor. Through symbolic learning, by observing other actors, the actor enhances the capacity to categorise and make sense of the world by attributing causes also to the situation (external), to pass the buck when things go wrong. Note that both foci of attribution—dispositional and situational 'causes'—are learnt operations of the mobilising effort.

Learning theories further distinguish *modalities*. Primary learning (schema training) means doing things better within a given mind set, within a given project orientation that is not at disposition. Learning here means to make tactical adjustments within a fixed strategy. In the context of mobilisation, primary learning means professionalising the effort in order to mobilise more efficiently. We know what we need to achieve and we seek to do it better. In contrast, secondary learning (reflective thinking) means putting the strategy itself at disposition. With secondary learning the mind set is up for grabs. Action always involves cognition, but reflective thinking opens habits to awareness and reflection, and ways of discourse to meta-communication. Other terminology for these modalities includes regulating versus directing, learning curve versus paradigm change, association versus Gestalt switch and mode I versus mode II learning (e.g. Argyris and Schön, 1978). Learning is cumulative but reversible, and over-learning is possible; beyond an optimal point training is detrimental to the efficiency of movement, as any athlete or piano player will know, and too much thinking can distract from action. Two big dilemmas of a techno-scientific mobilisation effort are: how much to train for efficiency, and when to start and or stop the self-reflective process. Resistance might provide answers.

With *learning constraints*, theory refers to inherent conditions that inhibit symbolic and motor learning (Hinde and Stevenson-Hinde, 1973). Learning depends on flexibility and dispositions to make the most of opportunities. Learning difficulties arise when the actor is 'closed minded' and 'operationally rigid' through overly strong convictions and euphoria (see Kruglanski, 2004). In the context of individuals we talk of being brainwashed or under the influence of drugs. In the context of techno-scientific mobilisation we might recognise as constraints an 'esprit de corps' that favours 'closure' and discipline, and collective mind sets of of 'innovation', 'progress' or 'revolution' that discourage and disqualify dissent.

These two dimensions of focus and mode of learning constitute a typology of *mobilisation learning by resistance*. The resistance=pain analogy suggests that resistance triggers these learning responses of techno-scientific mobilisation, depending on constraints. 'Collective learning' is not just a statistical notion that most people in a group learn, but refers to structural changes in mobilisation (see Jost and Bauer, 2005). Table 5.1 show the four types of learning: avoidance, restructuring, assimilation and accommodation.[9] These types of learning point to ways in which techno-scientific mobilisation responds to the challenges of resistance. The evaluation of the techno-scientific project that arises from controversial public discourse is thus productive, as it allows to learn from resistance and develop sustainable technological trajectories.

## Proposition 6: Resistance Increases Assimilation Pressures

*Assimilation* refers to ways of sense making within existing competences. Assimilation combines routine activity and additional symbolic work. Control is extended not by changing the comportment, but by widening its appeal. The techno-scientific mobilisation effort is shielded and immunised from scrutiny by externalising the problem of resistance as 'public deficit', either as ignorance, irrationality or risk aversion. Deficit concepts of resistance identify information processing issues and deflect from the techno-scientific project as the source of grievance.

In assimilation mode the mobilisation effort keeps control by improving the diagnosis and early warning in order to intervene promptly and better. Interventions arising from 'resistance=deficit' mind sets, abound with expensive campaigns, monetary incentive systems, change of rhetoric and capacity building to turn uncertainty into predictable risk, for example by modelling 'what causes resistance'. In the context of techno-scientific mobilisation one can think of Monsanto's advertising campaign of 1999

*Table 5.1* A Typology of Collective Learning from Resistance

| Focus<br>Mode | Situational attribution (external)<br><br>Symbolic learning<br>Focus 'situation' = about resistance | Action attribution (internal)<br><br>Operation learning<br>Focus 'action' = the movement |
|---|---|---|
| Training<br><br>Regulating<br><br>Mode I<br>Primary | *Assimilation*<br><br>*Diagnostic extension*<br>• Early warning<br>• Monitoring<br><br>*Reduce Deficits*<br>• Information campaigns<br>• External capacity building<br>• Incentive systems<br>• Change of Rhetoric | *Avoidance*<br><br>*Tactical retreats*<br>• Lower expectations;<br>• Postpone<br>• Go elsewhere |
| Thinking<br><br>Directing<br><br>Mode II<br>Secondary | *Structural learning*<br><br>*Resource interventions*<br>• Frame change, mindset<br>• Change of personnel<br>• Wider scoping of concerns<br>• Procedural change | *Accommodation*<br><br>*Substantive alterations*<br>• Output change<br>• A different movement<br>• Strategic adaptation |

to undermine the protest against GM crops and foods in Europe, of monetary compensation schemes to site nuclear waste, or suggestions to avoid the words 'genetic engineering' or 'embryonic' in public debate because of reputation risks arising from negative connotations of the terms. Assimilation knows the remedy and seeks to widen the scope of application. In the pharmaceutical industry this is called 'line extension', to extend the indication of a drug, with or without further testing, to widen its market.[10] Here, additional resources are mobilised for acceptance research, propaganda, bribery schemes and eristic rhetoric.

Assimilation comes naturally to a mind set that considers public opinion being ignorant, emotional or otherwise unsophisticated. Remedies include increased public information and cognitive capacity building programmes for 'probabilistic risk assessment' to foster 'realistic and correct' thinking and to neutralize biases and mere heuristics (Fischoff, 1982; also see Chapter 7).

Assimilatory mobilisation fosters immunises the techno-scientific project against critique. It avoids the crux of the matter, which is the projected future itself. It leads to actions from a limited definition of the problem. Assimilation activities are a call-up for social engineering within a *technocratic* division of labour: scientists discover, engineers innovate and social scientists deliver public acceptance (e.g. Rogers, 1983). Assimilation provides the fixes on 'how to overcome resistance to change' and continues to stock the research literature, offering employment for commentary, consultants, academics and students. The pressing question is how far assimilation can carry before it becomes part of the problem. Campaigns to mobilise conformity can backfire. Assimilation is a risky response to the challenge of resistance.

## Proposition 7: Resistance Leads to Avoidance Learning

*Avoidance learning* from resistance=pain manifests itself in tactical retreats. Consider the lowering of expectations, spatial concentration of activities and partial avoidance in the face of difficulty. Hype and exuberance might mobilise support and raise expectations for a project to sky. 'Energy too cheap to meter', the 'paperless office', 'IT revolution', 'GM crops to beat world hunger' and 'stem cell research cures cancer and Alzheimer's' are just some of the slogans epitomising the high hopes pandered by techno-science. Encountering resistance, these expectations are often lowered. This happened to nuclear power, when 'too cheap to meter' became more modestly 'part of the energy mix'. The biotechnology revolution is quietly called off as promises in the drug pipeline fail to materialise.

Resistance makes certain localities 'unsuitable' for mobilisation. The diffusion, assumed to be globally distributed, becomes geographically concentrated. The techno-scientific mobilisation responds to NIMBY attitudes by avoiding the backyard in question. Finally, some parts of the mobilisation might stall or be postponed while others carry on, as happened with the 'terminator' concept in plant genomics. Public sensitivities and risks of losing markets in controversy over GM crops made it advisable for actors to take an attitude of wait and see with GM wheat and genetic use restriction technology (GURT). Partial avoidance can be a diverting tactic to allow some activities to roll out as planned. On gene technology, some countries stalled on agricultural green biotechnology to allow biomedical red applications to proceed with less public attention (see Chapter 8).

## Proposition 8: Resistance Leads to Reorganisation of the Mobilisation Effort

Structural learning refers to the combination of reflective thinking and external attribution. Here the focus is on new insights within an *adapting framework*. Gestalt theory models this insight learning: a gradual rehearsing of elements, which suddenly reorder into a new coherence that solves the problem, all pieces of a puzzle falling into place. Periods of routine activity, where assimilation reaches it limits with growing contradictions and anomalies, are punctuated by periods of 'paradigm shifts' where mind sets and practices change to make better sense in the context.[11]

Reorienting the mind set of the mobilisation effort in the face of resistance constitutes such a form of structural learning. Sense making of resistance occurs in discourse, and this discourse can change. A previously dominant deficit concept of the resistance could be replaced by a notion that 'resistance is a resource'. This changes the focus of analysis: learn not how to get rid of resistance, but how to make the project sustainable in the circumstances. Such changed perspective arise when people change their minds, but also from new peoplejoining the mobilisation effort. It might involve a generational transition. Restructuring mentalities also includes a shift from cognitive to *communicative rationality* (see Habermas, 2001). In such a transition, the aim shifts from controlling others to reaching a common understanding, from telling people what do to, to negotiating a common future.

The polemics over science communication, urging a shift from an extension-diffusion model to a dialogue model, from concerns over risk perception and science literacy to public engagement, can be seen in this light (see Chapter 7). The wider scope of concerns, the commitment to listening, leads to 'technologies of humility' (Jasanoff, 2003) such as citizen juries, consensus conferences or round tables. In many countries, nuclear power projects are legally obliged to consult widely in order to qualify for a licence to produce electricity. Agricultural biotechnology has to present an impact assessment for a particular GM crop to be liberated. Public consultations became 'due process'; non-compliance can be challenged in court. Such changes alter the process of and the *procedural legitimacy* of decision. New social arrangements, such as technology assessments and deliberative forums, create arenas by which the state moderates technological progress.

It is very plausible that these new institutions and procedures are responses of techno-science to public resistance.

However, there lingers a doubt that much changes remain 'pretend play' that wants to present itself as 'real change'. The mobilisation of a new discourse is one thing, and empirically verifiable through discourse analysis. But like old wine in new bottles, an assimilation effort could lie behind a facade of accommodation, not for reasons of 'conspiracy' of powerful actors, but also for reasons of structural continuity. This is the empirical question of distinguishing the types of collective learning that is in evidence (see Chapter 7).

## Proposition 9: Resistance Increases the Potential for Accommodation Efforts

*Accommodation* refers to substantive alterations to the mobilisation efforts; it involves adaptation of strategy as well as the repertoire of actions. Accommodation refers a shift in practice and the collective mind set that amounts to a fundamental reorientation of the techno-scientific movement. An actor used to mobilise for X in particular ways, henceforth they will mobilise for Y in very different style. The shift from X to Y represents the accommodation mode of learning in both substance and style, and this as a response to the resistance encountered

Notions of periodic upheavals are common in organisation studies (e.g. Bigelow, 1982) and in histories of productive sectors (Tushman and Anderson, 1986) or the world economy (Trebilcock, 2002): routine practice is punctuated by ruptures and reordering of activities. Three questions are important: are these ruptures destroying or preserving established competences; are these ruptures regular or irregular; and are they explained by internal or external dynamics?

Accommodating of strategy and practice must include the re-specification of major parameters: what is required, not required, and why? Old plans are discarded and new ones are drawn up. Thus projects change in substance and their discourse of the future will be different from where it was initially supposed to end up. Here resistance often challenges *outcome legitimation* by shifting the mobilisation for research and development away from outcome to process. Projects cannot be justified only by their outcome; techno-scientific projects tend to produce the very criteria of what makes them a success.

Empirically, the quantitative and qualitative differences between initial plans and later outcomes are good indicators of the degree of accommodation. Large alterations to projects include a shift in project identity over time. As project P alters into project P', by responding to the challenges of resistance, we have to ask for how long does P remain within the realm of similarity to P', and when does P' become different from P. How far can a spade morph until it is no longer useful to call it a spade?

Strategic adaptation incorporates resistance as 'pain'. The mobilisation effort reconsiders the situation in the light of resistance and anticipates what is possible and what impossible under the new circumstances. Resistance changes this context of future mobilisation. In the past, countries might have built nuclear power stations in order to produce plutonium for nuclear bombs; henceforth they build only single-use electricity stations. An example of such strategic adaptation is the 'life science' project in the 1990s. Resistance to GM crops created a new situation where the vision of a new sector, based on genetic engineering and combining food and pharma, was no longer realistic. Strategic adaptation required a separation of farming and pharma, the red from green biotechnology (see Chapter 8).

## Proposition 10: Shifting from Mode I to Mode II Learning Requires a Predisposition

This leaves a key problem open for further research. This typology of responses to resistance suggests a cycle of progress from avoidance to assimilation, to structural change and accommodation (marked by the arrows in Table 5.1). Mode I learning is most likely to be the first response of any movement to resistance. Tactical retreats and improving the capacity of control within the available framework is the first option. What then are the contingencies for shifting to mode II? We need to identify empirically the disposition of movement to mode II learning. Free-floating thinking and observations independent of given mind sets, being able to go beyond the causal models of resistance, assessing the discourse of resistance in their ritual function for blame shifting, seem to be likely pre-conditions for mode II learning. The absence of any of these is likely to constitute what one might call structural *learning difficulties*. We can expect that one of the key blockages to mode II learning arises from strong mind sets, ideological commitments, or promoter signs, which not only discourage dissent but make alternative futures unthinkable. Strong promotor signs such as 'progress', 'revolution' and 'economic imperative' create commitments, but they make blind for resistance and block the move from mode I to mode II learning. Over-committed and dogmatic mobilisation efforts are constrained in their learning; they are bound to avoidance and assimilation processes. Being able to put into perspective powerful promoter signs is therefore a precondition of being able to learn in mode II. We can expect that only mobilisation efforts that paradoxically encourage and thus carry their own dissent are able to break this learning difficulty.

The functional analysis shows that resistance is no deficiency, but an asset of mobilisation. Resistance is an epistemic principle (Schmidt, 1985)—*the reality principle* of an imagined future which opens space for alternatives. This is the working hypothesis of this entire book. The subsequent chapters test the functionality of resistance on the key technologies of the later 20$^{st}$ century: nuclear power, computing and information technology and genetic engineering. However, functional analysis is only a start. It will be necessary to keep an open mind for empirical *dysfunctions of resistance*. The resistance=pain analogy will again be our guide. The main factor of chronic pain is not any grave injury but the mind-body response to the experience of pain. Chronic pain arises from a lack of care and misguided therapies, and mind sets that become part of the problem. We can expect analogous difficulties to arise when techno-scientific mobilisation responds to the challenges of resistance: the response to the challenge of resistance is far more of a problem than the reasons for resistance might ever be.

## NOTES

1. When I started to work on this problem in the mid 1980s, I was repeatedly advised to avoid the term 'resistance', because the term seemed loaded with connotations like 'backward looking', 'ignorant', 'stupid' or 'conservative', which rendered the term useless for analytic purposes. Since student times I felt that the term resistance had an existential dignity which needed to be recovered. The prejudice about the term probably arose from a managerial bias in the analysis of change processes (see afterword on 'resistology'). However, semantics have shifted towards more positive connotations across the social sciences together with an inflationary use of the term (see Bauer, Harré and Jensen, 2013).
2. The pain metaphor has diverse origins and convergences: Lawrence (1954) on managing change in the business context; Seidman (1974) on the existentials of pain, anxiety and resistance in human affairs; Bauer (1991) on monitoring new technology; and Luhmann (1997) on social movements as irritants of an otherwise autistic society; protests are the 'pain' of complex societies, they irritate and induce reflexivity. What is needed is a metaphor that suggests irritation from within rather than from without, and pain makes exactly this point. Pain is a function of the entire organism, not of some specific stimulus, and irritating, wherein lies its very contribution. Note that this idea differs from day-to-day language, which suggests that protest is motivated by pain of individuals, such as in economic crisis and social deprivation. Pain indeed motivates mobilisation, as the global movement for the recognition of 'chronic pain' as a social problem demonstrates. However, our present concern is for 'pain as a model' as to how and why already mobilised movement changes its course.
3. I was unable to trace this apocryphal statement to a defined Napoleonic source.
4. Famously, when on the London Underground a train needs to discontinue its journey for some unexpected technical reason, the famous announcement is: all change, all change, please!
5. The motto of the London School of Economics and Political Science (LSE) 'rerum cognoscere causas' (to know the causes of things) reflects the hopeful confidence of its Fabian founders that knowing the causes of things will be the key to improving social affairs.
6. The reader will notice that my own model of resistance, the resource model based on a pain analogy, is recognisable within this very theory; in other words my theory is self-referential, it contains itself as a special case.
7. This lovely formulation I take from my former student and now colleague, Sue Howard; I remember the source, but not the occasion.

8. Operational learning enhances the way we discriminate and categorise the situation, and how we learn to control our motor behaviour and build smooth and efficient patterns of movements. The competent piano player can read complex sheets of music and his or her fingers slide effortlessly over the keyboard, translating what the eye see into patterns of muscles innervations. Skilled social movement activists know how to organise a protest march and to invite large numbers of people to move in the same direction and make a public impact. Operational learning includes avoidance, learning not to do something because of adverse consequences. Burnt children avoid the hot stove. The pain stimulus has played an important role in research because of its strong link to the withdrawal reaction. Pain is the paradigm of experimental avoidance learning and the modelling of depression. Randomly experienced pain is depressing; it leads to 'learned helplessness' because there is no recognisable link between own action and pain relief. On the other hand, anything that relieves pain is strongly addictive; pain relief is a strong operant reinforcement. The logic of penal repression is based on violence, to shape behaviour by fear of pain punishment, but the result is more likely apathy, depression and hopelessness, rather than constructive activity (see Bandura, 1969).
9. The pair 'assimilation-accommodation' originates in Piaget's theory of cognitive development (Piaget, 1972, 172f). Assimilation refers to the integration of new elements into an existing action schema without changing the activity. Novel elements are forced, classification is pushed and actions remain the same. Children assimilate when doing 'pretend play'. They take a broom, declare it a horse and ride away. Pretend imagination has few limits. The schema integrates without modification, but this reaches its limits as anomalies accumulate, create tensions and trigger a complementary process. Taking a broom for a horse might suddenly appear too silly or too strenuous for long distances. Accommodation refers to the adaptation of activity to new elements by which the practice is modified; the broom is now only used to sweep, a new activity, while for locomotion a real horse might be at hand. The schema integrates and modifies; classification and repertoire of actions widen. This cycle of assimilation and accommodation is a part of any process of framework or paradigm evolution. Assimilation absorbs uncertainty within an existing schema, but only up to a limit. Beyond that limit, accommodation refers to changes to the schema which follow a changing practice.
10. An example is the cancer drug Avastin, worth about $4 billion a year and owned jointly by Genentech and Roche. Currently the drug is indicated only for metastatic cancers. However, research is under way to demonstrate that the drug equally works for pre-metastatic cancers. A positive result will add $1 billion to its value. This procedure is called a 'line extension': rather than developing new substances, research focusses on testing efficacy beyond the original drug target (source: NZZ, 2009, 2 April, no. 77, 13).
11. This similarity between Piaget's theory of cognitive development and Kuhn's theory of the scientific progress is not accidental, but arises from the same intellectual tradition. Kuhn explicitly acknowledges Piaget and the psychological 'Gestalt' theory as inspirations for his conception of 'paradigm shift' (Kuhn, 1962). However, caution is needed as we move from a theory of ontogenetic development and cognitive learning to one of social processes. For Piaget this equilibration process is part of an ordered development of the human mind from infancy to adulthood through stages, and the dialectics of assimilation-accommodation press these stages forward to a given telos: the adult human being. This analogy breaks down in social systems and

networks. There is no fixed order of stages nor is there a telos of maturity to be reached. The positivist vision of a ladder of history out of an era of myth and magic, via dogmatic religion, to the age of science and technology has no reality beyond the rhetorical resonance as a promoter sign in public controversies. An attempt to capture this difference between ontogenetic and societal development is the very motive of the theory of social representation: theorising changing mentalities without a predefined telos. For social processes there is no real ladder of progress to climb (see Duveen, 2001; Jovchelovitch, 2007; Sammut & Bauer, 2011).

# 6  The 'Bytes' of Mainframes, PCs and Social Media

> *Gregory:* "Is there any other point to which you would wish to draw my attention?"
> *Holmes:* "To the curious incident of the dog in the night-time."
> *Gregory:* "The dog did nothing in the night-time."
> *Holmes:* "That was the curious incident."
>
> (Conan Doyle's 'Silver Blaze')

In the series of post-war technologies, computing and information technology (IT) is, after nuclear power, the second expected transformative factor of the economy. In relation to public resistance, however, IT seems like Sherlock Holmes' Silver Blaze case: significantly, we note, the dog did *not* bark. One question is central to this chapter: Why is there so little resistance to computing and IT?

## SOME FEATURES OF THE HISTORY OF COMPUTING

Computing has a long past and a rather short history. The long past traces a perennial quest to substitute 'human thinkers' with a device that quickly and reliably deals with numbers and symbols. The Spaniard Ramon Llull or Lullus (1235–1315) and his 'Ars Magna' is credited with the first attempt to develop a symbolic language (in his case, to generate by combination all the attributes of God) and to create an artificial memory without an anchor in living emotion (Yates, 1992, 175ff). Later pioneers in that line are Leibniz (1646–1716) who experimented with mechanical calculators, and Charles Babbage (1791–1871) and Ada Lovelace (1815–1852), designers for the British Navy. Their 'analytical engine' (1871), programmable on punch cards, implemented the distinction between hardware and software but never went beyond the drawings (see Swade, 2001).

### Smaller, Faster and Cheaper

Many technical achievements are the product of war efforts, and so was the computer. The short history of computers starts in WWII and leads us to a globally networked communication in the new millennium. Until 1945 'computer' meant a person who calculates: henceforth it denoted a machine that does the same by storing and retrieving text, generating and manipulating images, networking with other machines, and steering robots and cruise missiles (Ceruzzi, 2003). The war effort brought the *mainframe*, room-filling machines such as Colossos (UK), Mark and ENIAC (US), and Zusa (Germany). For military purposes, they calculated ballistic trajectories and broke the secret code of the German Enigma machine, which gave a decisive strategic advantage. John von Neumann and Howard Aiken in the US and Alan Turing in Britain became belated war heroes.

The computer world is binary; though this is not written in stone. In the 1950s the Russians tinkered with trinary logic (0, 1, 2)—more elegant, memory saving, and with fewer switches less prone to error than binary logic—but operators failed to convince ignorant superiors (Hunger, 2007).

Binary or trinary, these machines needed an entire department to run a capacity less than a PC of 2000; cooling off the excess heat was a problem. Atomic scientists needed IT to control the chain reactions, but during the 1950s companies like IBM and GE started to build computers to make a name. In 1951 the US census office bought the first commercial computer from Remington-Rand, the UNIVAC, and in the mid 1950s the US government was the major computer user. Large corporations in insurance, steel and chemistry followed suit. In the UK, it was famously Lyons Tea that first computerised its valuations, payrolls and inventories since 1951. In 1952, IBM launched Systems 701 and 650, a line that by the 1970s dominated the business world. These machines continued to be room-filling, air conditioned and operated by a priesthood with card punchers, a control terminal and memory on magnetic tapes. This expensive stuff was often rented rather than bought. Until the 1970s this market was saturated by defence agencies, research universities and large corporations. Hardware was the main cost factor and the software was programmed as required with languages like FORTRAN (Formula Translation) or COBOL (common business oriented language); both were developed in the 1950s.

The replacement of vacuum tubes by transistors and by integrated circuits on small silicon chips made a difference. The transistor was patented in 1951, the integrated circuit in 1959 and together these created powerful devices of ever smaller size. The miniaturisation of processing capacity was a major ingredient of the Cold War missile programme. The silicon chip defined 'Silicon Valley' in California, and the rise of the semi-conductor industry made names like INTEL. Japan and Korea succeeded in building their own industry, where France and Brazil failed (Botelho, 1995). *Moore's Law* described the processing capacity of microchips, which doubled every 18 months (Ceruzzi, 2003, 217), and one wondered whether there were any upper limits (a challenge which nanotechnology would take up). Miniaturised processing became the key ingredient of a general electronics boom. In the 1960s, mini computers arrived with terminals (no longer card punch data entry), real-time *interactive* time-sharing and linked into a network. This made Digital Electronic Equipment (DEC, founded in 1957) a name; its PDP-8 of 1965 sold 50,000 and demonstrated a larger market than ever imagined. But its customer base was still large corporations and scientific laboratories. IBM had entered that product line with system/360 in 1964, and later came the SUN workstations. Cornering global markets landed IBM in anti-trust litigations that lasted through most of the 1970s. IBM was an attractive employer; until the 1980s it had never laid off an employee and was hailed as 'true excellence' (Peters and Waterman, 1982). This was premature, as it turned out when Microsoft entered the game.

## From Hardware to Software

With the mini and later the micro PC, came the software industry. As hardware became cheaper more money went into software, on tailored designs and increasingly into standardised software packages. Software development and testing became increasingly focussed on the user-interface. By 2000 software development was a partially outsourced industry with booming centres in India's Bangalore and other developing economies.

Figure 6.1 shows the progress of computing across the world since 1946, from a handful of mainframe computers in the early 1950s to the one-billionth PC and far beyond in the new millennium. It also shows the cycle

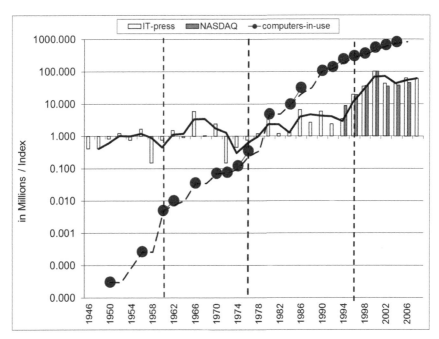

*Figure 6.1* Numbers of computers in use worldwide (dots) and NASDAQ stock index (dark bars); and number of news items on 'computer', 'PC' or 'internet' (light bars) in a single UK press outlet (indexed 100 = 7,300 in 2000). *Source*: Bauer et al., 1995; Cortada, 1993; Winston, 1998; Computer Industry Almanac, 2010 . All figures are on a logarithmic scale.

of waxing and waning computer news reflecting the four phases of this history: 1946 to 1961 (mainframes), 1962 to 1974 (the minis), 1975 to 1995 (micro PC) and after 1996 (mobile, internet and world wide web).

The PC made a big difference. Hand-held calculators had replaced the logarithmic ruler or tables for some time. Electronic enthusiasts in computer clubs supported the arrival of DIY kits. 1974 saw the Altair for $400. Commodore and Apple followed in 1977; in 1981 IBM launched its 'Billion-Dollar Baby'. In January 1983 *Time* magazine put the PC on its title page as 'machine of the year' (instead of person). The PC was both workhorse and a leisure devise. Devices with operation systems DOS, Windows or Mac and external keyboard, printers, scanners, floppy memory discs, mouse, cameras and microphones, etc. came to define the mobile PC environment. The competition split the PC world into 'arty' Mac and 'square' DOS; IBM left it to Microsoft to develop software products for the DOS world.

For a short while computer programming became a youth pursuit.[1] The spread of programming skills gave rise to the hacker culture in the 1980s (see below). However, the rise of Microsoft made these skills obsolete. Standard applications could be learnt without mastering computer language; user-friendly interfaces became the key challenge. Lock-in drove a business model of selling regular updates to customers without any need for it. Evidence is mounting that with updates the efficiency of use perversely declines (Hilty et al., 2005). Equally the spectre rose of a generation able to buy a computer, but 'unfit' to run it properly.

## From Stand-alone to Networking

The third dimension of this history is *networking*. This vision was promoted by DEC and Sun, with the UNIX language using the Ethernet of Xerox in 1973. Sun's slogan was 'the network is the computer'. Networking required workstations capable of time-sharing and link-up in parallel processing. This increased limited capacity by linking up and sharing files in *local area networks* (LANs). Computer 'intelligence' was pooled through file sharing, common back-up memory and exchanges to be known as e-mail. Central back-up was helpful, because individuals always struggled with lost files. But this all took the 'personal' out of computing, and allowed central monitoring (Ceruzzi, 2003, 291ff).

An even bigger step was the global network. First, the ARPANET of 1969 was set up initially to share expensive mainframe resources from the East to West Coast (of the US), and later to study network behaviour and to have a system that was resilient to node failures (e.g. under nuclear attack). E-mail, not part of the original design, emerged as a sideshow. The demand for *remote access* revealed a gap in the infrastructure: the telephone lines were slow in transmitting data. Modems, broadband, fibre optical cabling and wireless transmission responded to this challenge. Finally the *internet* (1982) and *world wide web* (1989 at CERN) appeared for sharing scientific data. The web 'inventor' Tim Berners-Lee wanted to allow access from different computers. WWW, HTTP and HTML became the markers of interconnecting computers and creating a new virtual space. This allowed the rise of the file sharing 'social media' businesses such as Netscape (1994), Yahoo (1994), Google (1998), Napster (1999–2001), Wikipedia (2001), Facebook (2004), YouTube (2005) and Twitter (2008) and some of their now billionaire founders. Google epitomised the opportunities; 'to google' became an activity. The internet offers global distribution of contents and products, a super-highway where information, knowledge and entertainment converge for anyone with the skill and funds to access it, at declining costs.

Computers spawned an industry on the basis of *miniaturisation* and *convergence*. Not only did products converge but entire sectors of the economy. IBM, Bell Telephone, News International and Hollywood merged into the brave new world of powerful 'new media'. Computers, telephone lines, radio and satellites converged and transformed the way the world is informed, observed and entertained by ubiquitous computing. There is no limit to the hyperbole which this mix was able to generate towards the end of the century.

## THE MOBILISATION OF RESISTANCE

The paradox of information technology seems to be the absence of resistance despite many reasons for it. No critical masses emerge that mobilised against computers like they did against nuclear power or biotechnology. Nelkin (1995) argued that the reasons to resist information technology are stronger than those to resist genetic engineering, but public opinion followed a different logic. This conundrum seeks an explanation. Let us start with the many good reasons to worry about and resist information technology.

### Four Issue Cycles of Computing: Platforms to Express Concerns

The mass media index of Figure 6.2 shows four waves of public attention to computing since WWII. These figures, collated for the UK, are likely to be prototypical of similar trends in a world dominated by the US.

In the first wave, which lasted from 1945 until 1960 with a peak in 1956, computer news is a monthly, occasionally a weekly item. The slant is generally positive, and getting more positive as we go along. It is early days; people start to take notice that these machines exist. Wartime secret machines are opened to the public. At seminars at Harvard and Pennsylvania University, Mark 1 and ENIAC were introduced to hundreds of engineers, but news seeped to the general media at a low rate. Cybernetics generalised computing to the social sciences with popular books and conferences (see Wiener, 1950; Heims, 1975 and 1991). Popular science magazines carried

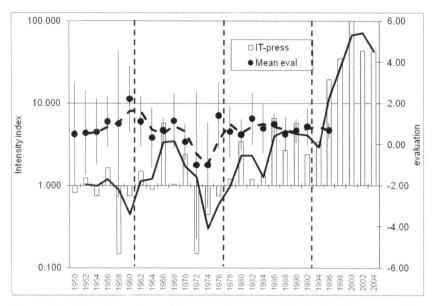

*Figure 6.2* News intensity and slant on 'computers' in the British press, 1946–2004; indexed to 100 = 7,300 article in a single UK news outlet in 2000. Note, computing news is so intense that a log scale is required to represent it adequately. Sources: Bauer et al. (1995) and updates using Nexis/Lexis.

very few articles. The installation of computers in large companies had news value, as well as UNIVAC successfully predicting the US elections of 1952 on the basis of preliminary results of CBS television. The *New York Times* classified the news under 'mathematics', 'calculators' or 'processing machines'. The main forum was the trade press and associations of middle management. At the time it was far from clear whether UNIVAC or IBM 650 would supplant existing office equipment; applications were limited to accounting and scientific research. Industry leaders like General Electric were cautious and saw this line of products as supply and not demand driven. The creation of a glamorous image for the computers was therefore imperative from the very start. Computers were framed as 'giant calculators' and 'scientific management'; in the 1960s the frame of 'information society' appeared (*Scientific American*, 1966). The first phase reflects the seeding of the computer business whose main clients were defence and public administration (see Cortada, 1993, 107ff).

This second wave lasted until the mid-1970s. Computer news was still a monthly, at best weekly, item, but brings an emerging discussion of societal impacts. The Sputnik shock of 1957 instigated the space race and mobilised talent and funding, including into an emergent computer science. Missile technology is the paradigm of cybernetics and control engineering. The news slant is overall on a negative trend into the 1970s, and characterised by the automation debate. The 1960s brought intensified public imagination. The *'computer revolution'* was declared, and Machlup (1962) mapped the rise of information as the fourth sector of the economy, after agriculture, industry and services, accounting for about 30–40% of the labour force. In 1963 the first survey of public opinion in the US (Lee, 1970) presented an ambiguous computer, somewhere between 'beneficial tool' and 'awesome thinking machine'. The *automation debate* raised the spectre of technological unemployment and other concerns reflected in publications with titles like 'the technological threat' (e.g. Douglas, 1971) and visions of an axial transition. And Toffler (1971) diagnosed the 'Future Shock' from the speed of technological change.

The *post-industrial society* (Bell, 1967 and 1973) put information processing as a new societal paradigm: industry and services would be displaced by research and information. Society would split along the lines of knowledge and not property; theoretical knowledge and its producers, the academic scientists and researchers, were the technocratic elite matching the old property elites in an era of affluence.

The third phase lasted until the mid-1990s. Newspapers adopted special computer sections and beats. Coverage increases continuously during that period. The 1970s are a time of global economic crisis. The post-war boom has ended, and the crisis brings home the message of limited resources. In the UK and other places, political uncertainty rules when governments change in fast succession. But this is also the time when a new breed of computers comes to market, the new workstations promise order in this political disorder. These devices capitalised on the microchip, the fall-out of space-race engineering and the nuclear ballistic surveillance programmes.

IBM's dominance with 70% of the computer market worldwide was challenged with anti-trust litigations by the US Ministry of Justice. This became a lingering story until 1982, when the case was found 'without merit' (Ceruzzi, 2003, 170f). Public attention followed the arrival of the personal computer (PC) and new companies like Apple and Microsoft. *Time* magazine depicted the PC as the 'machine of the year' in January 1983. Computer news recovered from a dip, became strongly positive and has remained so ever since. With the PC, computer news had become daily news.

The fourth phase started in the mid-1990s and took news into the stratosphere. In terms of intensity, it rivals only war reportage or terrorism with its high presence. By 2000 a single UK news outlet carries over 7,000 references per year, making internet, IT and computers more than a daily news item, but a news value per se. The new topic 'internet' exploded computing and IT news to 100 times higher than that of the 1980s. My slant indicator ends in 1992, but it can be taken as given that the hype of the 1990s was more than enthusiastic, it was hyper-enthusiastic. It created the expectations of a 'new economy' that is reflected in the stock market valuation of everything with the digital prefix 'e' such as e-learning, e-business, e-government. UK media intensity and the NASDAQ index between 1992 and 2006 are perfectly correlated, testifying to the tandem of coverage and public expectations (r = 0.97, see Figure 6.1 above). In this gigantic symphony of an IT revolution, only the occasional dissonance was audible.

## Framing the Reasons for Resistance

It appears that four major topics could potentially frame the resistance against computing. These frames call the bluff on a bogus revolution, highlight worries about work and privacy, point to the insult of cultivated sensitivities, and identify the dangers to individual health and the moral fabric of society. Let us briefly see how these issues played out over the years.

## Calling the Great Bluff

Any claim to *axial transition* finds a mirror in the rhetoric of reaction: the path to hell is littered with good intentions. The consequences will be perverse, the efforts futile and jeopardising past achievements. This calls for *public enlightenment* on the emptiness of the claims made that drive computing from the outset.

The trope that 'information technology is axial for society' gave rise to an entire genre of sociological literature. Beniger (1986, 4) lists over 50 claims to 'revolution' related to IT between 1950 and 1984, and such claims continue to proliferate, e.g. Castells (1996) on a network society. But, where most see revolution, Beniger sees only continuity: to control society through monitoring, planning and feedback and to reassert bureaucratic rationality by different means. For Winston (1998) the hyperbole of 'revolution' masks the reality of the *'law of the suppression of radical potential'* according to which a moving horizon of expectations covers up the thwarted radical potential of every innovation. This genre of writing illustrates the power of imagination which the computer seems to inspire. To the sceptic this 'smoke screen' needs to be tested against reality, unmasked as ideology and false consciousness. Like the emperor's new clothes in Anderson's fairy tale, computers are not what they are imagined to be.

This creates a warning literature, some of which on the theme of *'speed of change'*. The exponential growth of IT developments challenges the human capacity of sense making. This results in 'future shock' and engenders reluctance; people are naturally resistant to change (e.g. Toffler, 1971). The *perversity argument* has empirical support. Hilty et al. (2005) document experimentally the rebound effect of update cycles of software. While hardware speed increased many times, the productivity of text processing and file management decreased after 2000; 60% more mouse movement is needed by 2003 compared to the same operation in 1997. Updates reduce user friendliness and efficiency.

The futility argument comes as Solow's *productivity paradox*: the computer is visible everywhere except in productivity statistics. Economists study the IT investment of firms and fail to find enhanced productivity, which was the reason for investment. IT productivity depends on skills and organisational structure. During the 1990s productivity gains occurred in the US, but not in Europe, more in large companies than in local smaller ones, and only in some sectors (Van Reenen and Sadun, 2005). Historians hold their breath, and compare to steam, electricity and railways: it can take a long time to see the benefits (Crafts, 2001).

## Quality of Work and Surveillance

The automation debate of the 1960s highlighted the impacts of computers in the production process. The mechanised factory might finally go robotic; in the car and tool making industry numerical control (NC) machines had already replaced the skilful hand of the craftsman. Similar things would happen in the office. Mass unemployment was expected, and the future of work envisaged societies running out of work; jobs should be redistributed and working hours reduced. Others suggested retraining for different jobs. For those remaining in jobs, the *quality of working life* was the concern. Will computers increase the pace of work and social isolation and thus increase job strain and stress, and thus deteriorate health at work? With machine-dictated workflow, stress factors such as loss of control and of social support could lead to cardiac conditions (see Karasek, 1979). The working world was concerned about *deskilling* (see Wood, 1982). Mechanisation expropriated motor skills, automation would take our mental skills and put them into algorithms. Work psychology documented that unless work was redesigned, humans would be left with residual tasks and suffer bad effects to health and motivation (e.g. Hacker, 1986).

The totalitarian potential is explored under the title *surveillance society* (e.g. Lyon, 1994). Orwell's dystopia *1984* envisages a state that monitors citizens and tells them how to think with a language police. 1984 became a test year of IT; the media coverage resonated in 1986 (see Figure 6.3 above). The emergence of the internet gave this vision even more reality. The internet means that one person may monitor everybody else. Record keeping and intelligence gathering reach a new level. What hitherto was exceptional and inefficient becomes routine and efficient. The information available on any individual constantly increases, but does so without much attention. Passport identity used to content itself with recording age, sex, height and eye colour. Now, EFT, electronic fund transfer is the new game of identity monitoring. Credit and loyalty cards are very precise on shopping habits; linked to geographical information this pinpoints the lifestyle of the local street. Advertisers target this 'glass consumer' ever more specifically. On the internet, this is already routine: 'people like you also buy Y'. Novel fuel for privacy issues arise from internet providers and social media such as *Google* and *Facebook*. These web services collect large amounts data of about user activities which are stored as capital for data mining. Users are unable to delete their traces despite reassurances to the contrary. What poses as service provider and social media are de-facto gigantic surveillance operators. Google equally faces challenges over privacy issues when its Earth Street View project shoots pictures of streets and neighbourhoods, thus picturing recognisable people and putting these up on the internet; it raises issues of being photographed and exhibited without consent. The project was legally challenged in France, Japan, Switzerland and Israel.

*Work place monitoring* can be enhanced, and the 'transparent worker' results. Telephone and internet use are easily monitored, raising issues of fairness and privacy protection in using this information (OST, 1987). Counter-control is a motive for resistance. Factory floor and office staff are ingeniously 'soldiering' the system. Surveillance is nothing new in *security and crime control*. The UK now operates close to 2 million CCTV cameras in public places; such intensity of video surveillance exists nowhere else (BBC, 2009). The average citizen is framed 300 times a day from one camera for every 32 inhabitants. The UK is also proactive with 35.9 million genetic 'fingerprints' from 9% of the population (in 2012). The tracing of terrorists on CCTV is used to justify this level of supervision. Traditionally, surveillance followed suspicion. This now potentially reverses: suspicion arises from surveillance. Because everybody is within the frame, everybody is suspicious. Hollywood explores this theme in the thriller *Enemy of the State* (1998; with Gene Hackman and Will Smith): a former CIA surveillance expert helps an innocent citizen to evade the gaze and uncover a political conspiracy: it's not paranoia if they're really after you. The BBC series *Spooks* (2006 and 2007) offer high-tech surveillance drama under terror threats. The panopticon that creates designed self-control is now realistic. Inefficiencies of data overload, theft and error increase the risk of stolen identities and misidentification. Information stored might be false, out of date, or missing, and thus falsely incriminating or excluding; inversely, a stolen identity opens illegitimate access to personal privileges. Such risks grow on the back of policies to dispense more efficiently the conveniences of welfare, crime protection, marketing, voting and money transfers.

## The 'Computer Culture' Touches Humanistic Moral Sensitivities

The early mainframe computers were operated by a 'priesthood' of engineers. The number of computers was small, so was the number of operators. But as machines became smaller, more powerful and networked, a new priesthood of computing emerged. Observers saw an assimilation effect: humans became like the machines. It is a variation to an old story: the creation overpowers its creator. A whistleblower on this theme was a MIT computer scientist. In 1976, drawing on experience as teacher of computer engineers, he saw the new technology strengthening the hands of technocracy. Instrumental rationality, the maximisation of means to fixed ends, dominates over reason and judgement of both means and ends. Not only does the machine bias rationality, but people are led to believe that rationality other than instrumental has no role to play. The computer is the model for thinking, how to interact with others, and how to think oneself. A programmer's joke rang the alarm bells (Weizenbaum, 1966 and 1976). The ELIZA programme mirrored conversations; it would start 'how do you do; and what is your problem?' The conversant then types name and a problem; the computer recognises string X and throws it back: 'tell me more about X' or 'can you think of an example of X'. If after some turn-taking the word 'mother' had not appeared, the system would ask 'and why don't you talk about your mother?' to startling effect. If the conversant gets suspicious and puts a challenge 'you are not really talking?' the machine would ask 'you don't like me, do you?' . . . etc. The reception of this program was alarming. Some students preferred the machine to talking to a real person; intelligent people *attributed empathy* to a stupid machine that only follows if-then rules and mirrors what is entered. Therapists welcomed ELIZA as a step towards efficient diagnosis and substituting expensive therapy by cheaper machine interaction. Weizenbaum feared for human interaction and humanity.

Not only were his 'hacker' students easily duped by ELIZA, but they were also an elite with very narrow interests. He charted the 'hacker culture' and its emergent personality type: a *compulsive, socially inadequate, amoral, ambiguity-intolerant and power driven control freak*. At MIT, this was in the main a male culture substituting opaque human relations with the certainties of computer interaction. Control meant to pursue the hack, breaking into the protected systems of powerful corporations, military or civil agencies. If this character type was spearheading the information revolution, this might not be a nice place to live. Another MIT researcher studied this subculture. Turkle (1984) observed how computer scientists developed notions of 'self and other' derived from computer language. The body became 'hardware' and the mind was 'software'. Habits were 'defaults'. Irritations became 'the need to clear a buffer'; emotions were 'program errors' flagging the need for 'debugging'. Common sense, often psychoanalytic, terms were replaced by computer jargon; the unconscious became the 'machine code'. The private world of hackers was full of unusual metaphors.

Hackers compensate for an unrewarding world. The computer allows one to express repressed desires, and to catch missed opportunities. Here, some observers were startled over contours of a *'new Gnosis'* (e.g. Boehme, 1996; Dinello, 2005, 24ff). The internet and its virtual reality had gained religious overtones. The science fiction novel *Neuromancer* (Gibson, 1984) coined the words 'cyberspace' and 'matrix' for the paradise of a perfect brain-computer interface that is lost and regained by going through squalor, humiliation and pain. Hackers and Gnostics endorse a strong dualism. The body imprisons the spirit; being out casted, squalor and pain liberate those who know the path. Hackers celebrate the triumph of software over hardware, mind over body. Cyberspace takes on attributes of God: immaterial, ubiquitous and permanent. If virtual reality liberates, the 'real' reality can be neglected—purity there, pollution here. Liberation is for the few. In self-controlled cyberspace, the limits of the body are overcome by technology: God-like, omnipresent at the centre of a universe through purification and the apocalyptic destruction of a rotten world. This dream should clearly worry those who love the real world.

The *programme of artificial intelligence (AI)* touched moral sensitivities. The artificial intelligentsia (according to Weizenbaum) overlapped with hacker culture. Turing (1950) put the question: can machines think? His answer is a test protocol: If C enters a conversation with a person A and a machine B, and C cannot make out the machine, the device passes the 'Turing test of intelligence'. This challenge inspired a quest for making machines do things which require intelligence if done by humans (Minsky's definition). Starting with mathematical reasoning, AI later gave up the generic route, and specialised on *expert systems* for specific problems (chess, medical diagnosis, credit ratings, buying or selling stocks) and on *robots*. People like Simon, Newell, Minsky, McCarthy, Winston and Feigenbaum became household names who put the earlier 'cognitive psychology' (Neisser, 1967; Miller, Galanter and Pribram, 1960) on a new footing. *Cognitive science* mixed computer science, psychology and language philosophy (see Johnson-Laird, 1988). The strong programme modelled the process between sensory input and motor output digital processors layered in biological, semantic and intentional pathways (Pylyshyn, 1986). AI was popularised in Hofstadter's fugue (1979) weaving together the themes of formal logic, Bach's music, *Alice in Wonderland* and Escher's paintings. The controversy flared in the 'affair Dreyfus'. Hubert and Stuart Dreyfus, the one a philosopher, the other an engineer, considered AI a fallacy, a waste of money, even dangerous. AI was taken in by the *computer metaphor*. An 'as-if' model and protocol analysis became 'how the mind works', and turned into 'how we ought to think'. AI mistook a model for reality. A protocolled input-output device with storage facility became the new paradigm of cognitive science. Tool became theory and design turned ontology, thus prompting the question: 'Design' or 'Dasein' (German for 'being, existence')?. The 'broad AI' models of mind fuelled polemics (e.g. Boden, 1988). What started as modelling showed how the mind does not work: humans are not von Neumann machines after all (Dreyfus and Dreyfus, 1986).

The abstract concretises in 'narrow AI' *expert systems*. Machines operate well on know-that, knowledge containing 'x is y', ordered by rules. However, experts rely on know-how, their experience, skills and tacit knowledge, is difficult to verbalise. Ironically, the mundane is unreachable; a robot that cooks good pasta or rides a bicycle seems utopia. For more specialist tasks, novice doctors or pilots might train on them. But to the proficient expert, these systems are insufficient. The expert does not only follow rules, but also judges when to dismiss them, and that is no matter of meta-rules but of pre-codified experience and precedent. Chess machines are competent, but the champions in flesh and blood tend to prevail (Dreyfus and Dreyfus, 1986; Collins, 1990). However, with billions of dollars in military-industrial funding, the AI movement had enough momentum to launch expert systems in battle field operations, flight control, industrial production, banking, courts and hospitals with the potential to create more damage than anything else. A system might mistake the rising moon for a ballistic missile, and this pattern recognition triggers the counter-attack; computers dispense court decisions without considering circumstances, returning us to barbaric justice. Stock market crashes are aggravated by expert systems (buy, hold or sell), which apparently avoid the irrationalities of human brokers. What operates well in normal markets creates havoc in panic conditions. Concerns over AI resurfaced under the heading of catastrophic risks arising from new technology, for which anthropogenic climate change is a prototype. The Cambridge Centre for the Study of Existential Risks (CSER, founded 2012) puts AI and robotics into the same category. 'Intelligence explosion' and 'singularity' points to devices that learn beyond human control, one more reason to be concerned about AI. Robotic warfare is already reality in counter-insurgency warfare. Unmanned flying machines, so-called 'drones', remotely controlled via satellite communication thousands of miles away, spy over hostile territories and deliver deadly ordnance. Killing of suspects without trial has become practice and changes the rules of warfare, because 'unmanned systems' cannot distinguish between combatants and civilian bystanders (Sharkey, 2008; Unmanned Systems, 2011). Unmanned systems are fast pushing into civilian use, adding ubiquitous and anonymous surveillance capacity; drones come as 'animal cyborgs', i.e. roaming beetles and rodents mounted with micro sensors and antennas (source: *Observer*, 17 Feb 2013; *Time*, 11 Feb 2013). Here as elsewhere, science fiction movies such as *Terminator* (1984) are the 'diegetic prototype' (Kirby, 2011, 195).

## Dangers to Health and Moral Panics

The computer and the internet give rise to many moral panics. Approaching the new year 2000, the IT world was caught by millenary visions of doom, the Y2K effect or 'millennium bag'. Computers were built with a timer for the 20$^{th}$ century; On 31 Dec, 1999 midnight computers would jump 99 + 1 = 00, i.e. back to 1900, creating mayhem. The ensuing anxiety created a billion dollar business for consultants who were doom mongers and saviour at the same time. In the end there was a big millennium party, and it remains unclear whether Y2K was a case of successful prevention or of lucrative doom.

Widespread are worries about *children on the internet*. Parents worry about a new world which was not part of their childhood. Are kids gaining from gaming and roaming the WWW compared to other pastimes? Internet use might lead to withdrawal and social isolation. Or is it simply part of a trend towards private leisure amid anxieties over public places. Is this a matter of censorship or parental mediation (see Livingstone, 2002 and 2006)? The internet and TV share the same socialisation syndrome: children growing up in their shadow might withdraw from civic life into bowling alone (see Kraut et al. 199). Concerns emerged about *computer dependency and addiction*, suggesting a psycho-pathological deviance from normality. What for the hacker is a way of life is for the *computer addict* an obsession difficult to shed (hence the drug analogy) (see Excursion 6.2).

Social life on the internet exposes children to false friends. The spectre of paedophilia is every carer's concern. The internet is indeed a large market place for *pornography* and *paedophilia*. The crackdowns on illegal sex rings become a test case for policing the cyberspace. Notwithstanding these efforts, sex is a major driver of internet activity; sex-related keywords regularly top the search engine statistics. Then there is *cyber-crime*, such as the illicit acquisition of credit card details, the obtaining of bank details through counterfeit websites, or the wholesale theft of identities. These worries have kept people from fully trusting the brave new world of internet transactions. To address these issues of online trust design features, the logging of references and interaction histories was the big issue during the boom of the late 1990s.

Many *health issues* arose as computers became standard equipment at work and at home. The screens before LCD emitted electromagnetic radiation, and the video display terminal (VDT) debate of the 1980s raised awareness of the dangers of prolonged exposure. Concerns resonated with the daily press when birth defects in an accounting department were associated with VDT work (Foster, 1986). Experimental and epidemiological studies attempted to calm public unease. Work psychologists recommended precaution and set time limits for VDT interactions. Prolonged sitting on a computer screen led to another issue. *Repetitive strain injury* (RSI), sore eyes and shoulder and back tensions became an epidemic among high frequency typists. What had afflicted the odd violinist, prolonged repetitive fine motor movements in a constrained posture, became endemic among journalists, typists and academics fixated on their computers (Van Tulder et al, 2007).

The VDT issue was resolved by LCD technology. However, the concern over non-ionising electromagnetic radiation lingers on, flaring again with wireless technology. While landlines stagnated, mobile phones leapfrogged technology particularly in the developing world. People imagine '*electrosmog*' and 'boiling brains' emanating from ubiquitous antennas which potentially turn neighbourhoods into gigantic microwave ovens. News about brain tumours, the loss of male fertility from pocket phones and of leukaemia around overland pylons and antennas resonate with worried imaginations. Issues of exposure monitoring, definition of thresholds, cover-up of evidence and the guidelines for antenna locations are controversially discussed (see NRPB, 2004). These fears follow on older concerns over X-ray, radon, radioactivity, microwaves and television (Burgess, 2003).

128  *Atoms, Bytes and Genes*

Clearly there was and is no shortage of worries about computers and IT that could mobilise a public against this technology. The bogus revolution, the surveillance society, worries over unemployment and job quality, the control-freak computer culture, AI's false claim on the human mind and the finally moral panics over health, crime and children's safety offer a long list of potentially serious worries. But none seem to result in large-scale collective action and public protest actions. I continue to ask: why not?

### Potential Actors of Resistance

Consumer organisations had an established voice on product safety and transparent markets. Issues arising from computers potentially fell into their remit. 'Electrosmog' links up with the safety of microwave ovens. In the 1970s, people doubted the wisdom of introducing a potent source of electromagnetic radiation into the household. The issue was framed into safe handling—how not to burn yourself, and not to dry your cat in it—while the snobs engage a culinary distinction: one does not cook with microwave. Equally, the security of credit and debit cards, e-commerce and of internet banking is of concern to consumers subject to potential fraud. Cybercrime and 'identity theft' are issues where consumer organisations indeed raise awareness and lobby for protection (see Consumer International, 2014). The collection, storage and trading of private information about customer transactions is a concern to consumers. The sophisticated monitoring of habits, fashions and tastes and the delivery of tailored advertising amounts to intrusion of the private sphere.

The privacy issue overlaps with the remit of *civil liberty NGOs*. These groups originate in the 1960s with struggles against discrimination on race, religion, gender or sexual orientation, fighting for and defending the right of minorities and individuals against the intolerance of majorities. Protection of privacy, freedom of speech, the fight against censorship and surveillance are agenda points. A watchdog organisation called Privacy International (1990) publishes a seven colour ordering of countries by the degree to which privacy rights are safeguarded against abuse (from 'consistently protected' to 'endemic surveillance'). Some countries have an information commissioner (or similar) who monitors the state of the surveillance. There is mounting concern that for example the UK is a 'surveillance state' with a system of ubiquitous CCTV in public places. Some countries experienced revolts over the *census*. Most governments collect statistics on every inhabitant of the country. The census was an early application of computing in the 1950s. German plans for a national census were blocked in courts in 1980 and 1983 and postponed to 1987. Public protest was mobilised because beyond traditional items on age, sex and household, the scheme included 'personal questions' on holiday destinations. At the time Germany was caught in a dilemma of avoiding terrorism (Rote Armee Fraktion, RAF) and an authoritarian state. The census was framed as a tool of the latter. Similar concerns cancelled the census in Holland of 1981. The Dutch held vivid memories of Nazi occupation when their first move was to use the census to identify the Jewish members in their community in order to deport them.

Privacy is a frame that can mobilise many people. Opinion polls in 1984 had shown that 44% of Japanese worried about their privacy in the context of IT. In the US privacy is a respected value and concerns over its protection increased in the 1980s; more threats were expected. Before the internet, people did not seem prepared to trade privacy infringements in return for security and benefits (Katz and Tassone, 1990). By 2002 51% of Britain is worried, the older more so than the younger, 58% do not trust governments to protect personal information, 66% worry about personal information on the internet, but 72% want to trade privacy for security, men much more so than women (ICM poll for the *Guardian* July 2002).[2] In this context, policies to introduce *identity cards* pose an opportunity to mobilise against computer surveillance on matters of privacy protection. In some countries, there is a proud tradition of not having to show an identity upon demand by a person of authority. Britain and Australia share this civic pride. Schemes to introduce identity cards on the back of large computer innovations in Australia and Britain failed.

The hacker is an ambivalent figure: hero and villain, explorer and intruder at the same time; they spearhead the IT revolution and reveal its dark side (see Jordan and Taylor, 2004). Their origin lies in 'phone phreakery' of the 1970s, a technical 'underground' that tapped computerised telephone systems in search for free calls. They anticipated the convergence of computing and telecommunication. The villain status arises from successfully hacking into secure systems then selling their skills to organised crime and the Russians. The cyberpunk hero arises from their rebellious success in embarrassing the powers that be, entering the 'secured' systems of AT&T, IBM, DEC, the CIA, the Pentagon, and NASA, revealing their secrets and testing their 'security'. They are also famed for releasing 'viruses' into the internet. These actions demonstrate everybody's vulnerability. Hackers who are caught are either jailed or employed to fix the security holes. In the early 1990s, the law cracked down on notorious cases of hacking involving espionage for the Russians. New laws were tested in US and German courts (see Hafner and Markoff, 1993). An international hacker movement, Chaos computer club (1984), organises annual festivals and advocates decentralised, free and open access to information and a belief in the transformative power of computers (Levy, 1984). Internet activism of those who care about free and uncontrolled access to the internet is on the rise. With over two billion users worldwide (2010), the internet has become a large transport infrastructure like the open seas, which has the potential to mobilise people, if they ever perceive their space at risk from undue privatisation and control. Social media such as Facebook raise privacy issues (Jones and Soltren, 2005). A user platform, *'Europe versus Facebook'* initiated by a group of law students, monitors the privacy issues arising from commercial operations and has challenged the provider on a number of counts since 2011. They mobilised against abusive practices on five principles: transparency, opt-in not opt-out, decide for yourself, data minimisation and open social networks.

Whether privacy becomes a collective action frame with mass appeal is difficult to gauge; historical privatisations of common property of land, air and sea went through without large public challenges. The Pirate Party has gone global and gained political representation in Sweden (in 2006), and this indicates a new phase in the counter-mobilisation effort; commentators compare it to the environmental movement spawning green parties in the 1970s (source: *Economist*, 5 January, 2013, 14–16; see Chapter 4). Campaigning against private and state control , they defend the internet commons and 'net neutrality' and equal treatment of all traffic on the internet. There might be potential to bring many issues into coalition under the 'privacy and piracy' frame: the freedom of news making, the use of data, surveillance, business practices and open access to information. Much will depend on how the billions of enthusiastic users around the globe feel about being charged for access (copyright) and being snooped into their private lives (data protection) [3].

*Excursion 6.1* A Critical Mass of One 'Unabomber'

*The most spectacular form of active resistance against computers was the violence of a single person, the* Unabomber. *The Unabomber sent lethal letter bombs to computer companies, advertising agencies, airlines and scientists around the US. Active over 17 years he killed 3 persons and maimed 23. The FBI mounted the biggest manhunt in US criminal history. He blackmailed leading newspapers to publish his 'manifesto' in which he justified his deeds. His acts were to raise awareness among an 'oversocialised public' able neither to see nor feel the 'oppression and psychological pain' caused by the system. It should open eyes and spark 'revolution rather than reform' of the scientific-technological society. Current society marches blindly into a future 'where good outcomes cannot be separated from bad ones'. In mid-August 1995, the New York Times printed his declaration on the front page. Reading the piece and recognising style and content, a person identified the estranged brother-in-law who had written similarly back in the 1960s. Thus, the Unabomber was finally traced. In April 1996, Theodore J. Kaczynski, a 53-year old mathematician from Harvard and Berkeley was apprehended in a lone cabin in Montana. 'Odyssey of a Mad Genius', Time magazine put it (15 April 1996). Many were wondering whether he had operated alone; the Economist (13 April 1996) wondered about 'the Luddites' lost leader?' However, without roots in 1968, loathing both left and right, the Unabomber was an isolated loner, a one-man revolt against modernism launched from the American wilderness. In January 1998, he was sentenced for live without possibility of parole (see Chase, 2003).*

*The Unabomber found an imitator in the UK: in January and February 2007 Miles Cooper, a 27-year old school caretaker, had sent a series of injury-causing letter bombs to forensic laboratories and traffic control agencies to protest 'against an overbearing and over intrusive surveillance obsessed society' (source: Daily Telegraph, 25 Sept 2007, 11). What was initially believed to be an animal liberation action turned out again to be the isolated action of an individual.*

## Making Resistance Visible

There is no shortage of worries about computing and IT, and there are actors who are prepared to mobilise on issues of work quality and privacy. An important condition for successful public protest is to make visible the wider discontents about computing and IT. To what extent was this possible or not?

## Attitudes to Computing and the Internet

It seems that the 'law of opinion polls' applies particularly well to computers and IT: without controversy, no opinion research. Compared to nuclear power or biotechnology, opinion research on IT is perfunctory, and continuously shifting the issues.

The earliest poll seems Lee's (1970) report for IBM of US attitudes of 1963. Lee finds that questions grouped into two sets: those referring to a 'beneficial tool' and those referring to an 'awesome thinking machine' taking over vital human functions. The latter view was more likely among women and the less educated, and correlated with alienation and low ambiguity tolerance. A qualitative study in Britain of the 1960s showed the 'wait-and-see' attitude, 'nightmarish' visions of things to come and much joke and caricature (Sheldrake, 1971). Office clerks and factory workers were apprehensive, fearing the downgrading of jobs and redundancy. Three attitudes emerge: the 'clay pigeons' believe that nothing can be done, they just sit tight and wait; the 'false optimists' expect things to happen to others but not themselves; finally, the 'gravy train' view spots a bandwagon to jump on. And apparently the International Society for the Fight against Data Processing Machines (ISFADPM, 1963) funded a book on *How to Sabotage the Beast*, which seemed largely a spoof (Matusow, 1968).

Polls multiplied in the 1980s, testing Orwell's ominous year 1984. Polls in Germany, Belgium, Sweden, Japan and US showed that into the 1980s attitudes to computers polarised (see Bauer, 1993). Across Europe, except in Italy, a quarter of the population was *pessimistic*, considering the societal costs of computing larger than the benefits. In the early 1980s, 80% of British expected a deskilling effect of computers, a figure that declined to 10% in 2005. In 1992, 30% expected more interesting work from computers; in 2005 this increased to 55% with similar figures reported for Belgium. On the other hand, in 1999, 50% of Brits reported stress and frustration using computers (MORI, 1999). Through the 1980s, around 70% of British and Germans saw the computer as a job killer; and in Germany also as a major stress. Job killer expectation, always lower in the US, Canada or Japan, declined much faster in Britain than in Germany (see Figure 6.3). Eurobarometer data showed that between 1991 to 2010, Europeans were largely and consistently optimistic about computing and IT, as they are about solar power, much more so than they were about nuclear power, genetic engineering or nanotechnology (see Gaskell et al, 2011).

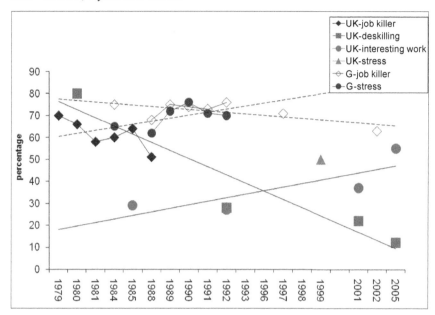

*Figure 6.3* Computer means job killer, deskilling, stressor or more interesting work in UK (continuous lines) and Germany (dashed lines). *Source*: Williams and Mills, 1986; various Eurobarometer surveys and MORI polls.

It appears, that public attitude to computing and work, its risk of bringing unemployment, deskilling and stress, showed discontent and potential for larger public mobilisation in the early 1980s, but in recent years these sceptical attitudes have largely dissipated.

## Non-users and Refuseniks

Resistance to computing might be found in the people who have no part in it. Access to computers and the internet became an index of societal development. However, differential rates of diffusion can be investigated as markers of resistance in the system; we might consider those who had access, and then rescinded.

The diffusion of IT has given rise to the international panics over the 'digital divide', defined as a lag in access across region, race, age, gender, education or income. The digital divide is a version of the 'knowledge gap hypothesis' (see Bonfadelli, 2005): news of innovations reaches people with higher education faster; and this can create disadvantages. Consider health information: the internet offers comprehensive coverage of health issues, but you need access otherwise the digital divide becomes an information gap (Rogers, 2004).

Regional divides are well documented. In Mississippi less than 40% of households had access to the internet in 2003, while in Alaska this figure was 70% (Day et al.,, 2005). In 2006 66% had access in the southeast UK, 50% in Scotland and Northern Ireland. The growing gap between urban and rural areas was analysed (Schleife, 2007). Europe varied from below 1% in Albania to over 60% in Iceland; Asia from 1% in Cambodia to 55% in Korea and Singapore. In 2002, Australasia led with 33%, followed by North and South America 24%, Europe's 20%, Asia's 5% and Africa's 1% of internet and computer access (ITU, 2003). Note that in the 1950, 90% of all computers were located in the US, but only 27% by 2004; in 1998 78% of all internet access occurred in the US, by 2006 this was only 18%.

Within any country, there are high correlations between income, education and computer/internet access. In 2003, 80% of the least educated had no access to the internet, compared to 80% of those with higher education who had (Day et al., 2005). Income differentials show similar disparities. In the UK, 93% of affluent household had access, while 57% of low income households did not. The gender gap is closing. In 1984 in the US, 20% fewer women were using computers at home than men; by 2003 the gap was 2% (US Census, 2003). In 2007 UK this gap was 5%; in 2003 it was 10% (Dutton and Helsper, 2007). In Germany the gap was stable at 13%, but not among the young. Older women were holding back. In sum, internet access is a matter of being young, urban, educated, and affluent. Communities with many single households, low unemployment, qualified jobs and less immigrant population are better networked. To reduce the digital divide it was suggested that training was offered to older citizens, immigrants and the unemployed, and to provide access in libraries, cafes and schools. But is a networked society necessarily a better society?

Because of a social imperative to use the internet, the *non-user* received attention. Non-use was not only a matter of lack of resources, but researchers discovered the *motivated refusal*. Why are very affluent places such as France, Germany, Japan and Switzerland lagging? Maybe there is a difference between lagging due to resources, and motivated non-use. Researchers classified and estimated the refusing types: TechNos, inter-nots, refusers, drop-outs and non-liners, some of whom have computer access but do not use the internet. In Germany all non-users, 34% by 2007, were called *'refuseniks'* (see Initiative D21, 2007). In the UK and US non-users are typified: *'proxy users'* use somebody else's computer; *'indifferents'* have access but do not bother; *'passives'* have no access and do not mind; *'refuseniks'* actively reject either net or computers. Suburbia in particular shows motivated resistance, despite being close to facility. Of 46% non-users in suburbia, 59% had no intention to join, a much higher rate than in the city or the hinterland. Non-liners think that the internet is unnecessary, dangerous, pornographic, just entertainment or confusing, and do not expect to miss out on anything. Costs are a lesser concern (Lenhart, 2000). Of these 46% non-users, 17% are 'drop-outs', who no longer use internet due to technical complications; 20% are 'net-evaders'; 8% have access but proudly claim no interest; 24% are 'nevers' who have no access and never had; of all these

56% had no intention to link up (Lenhart, 2003). Thus, some 8% of the US population, mainly in suburbia, refused the internet in 2002 even if they could have access, when 58% of US households were connected.

In the UK in 2007, 32% of the population were not using computers; 28% were not using the internet, of which 90% will not seek access in the near future, an increase from half in 2004. Of the 5% who had let go of computer use, a third were unlikely to resume. Twenty five per cent of non-users, or 7% of the population, had access but did not use it; and 40% ex-users, or 2% of the population, were not interested anymore. This suggests a 'refusenik' rate of 9% in the UK of 2007, when 61% of all households were connected and computerised (Dutton and Helsper, 2007). In 2003 this figure was 7% refuseniks and 18% indifferent, who had access but did not bother (source: *Economist*, 20 September 2003, 43), when 48% of households were connected. This seems to suggest that while the internet further penetrates households, about 10% of the population are motivated 'non-liners' in the US or the UK. A group that might increasingly attract attention (see Wyatt, 2003).

## Why is There No Critical Mass?

Clearly there is no shortage of worries about computers and the IT revolution: the bogus revolution, the threats to work and privacy, the computer culture and the risks to health and morality. In the 1980s, there were a solid 25% of Europeans who had serious reservations about the benefits of computing. The digital divide is in part due to lack of resources, but also due to motivated refusal. In the new millennium, some people vote with their feet and exit the cyberspace, particularly in suburbia. There are some 10% non-liners who actively refused or rescinded their internet access, something not envisaged by diffusion researchers. In many affluent countries like France, Germany, Switzerland and Japan people seem to be more reserved in their enthusiasm for the internet. All seems to point to a sizable proportion of the population that does not feel totally at ease with IT, and for whom the internet is no bandwagon.

But this potential seems hard to mobilise into public protest. IT witnesses isolated flare-ups of discontent, mainly over work issues, surveillance and protection of privacy. But this opposition is neither cumulative nor persistent, nor does there seem to be an over-arching meta-frame in sight that combines a range of issues and could bring about a broad coalition of discontents and protest.

Some of the issues are divisive among the activists. Those concerned with civil liberty focus on the protection of privacy, the hacker movement on open access. User-friendly interfaces are a marker of quality of working life for trade unions and every computer user in the land, while the hacker elite detests standardisation as 'Microsoft fascism'. Humanistic intellectuals who worry about computer culture find no ear among the hackers.

Thus, the counter-voices of the 'information revolution' are fragmented. Civil liberty groups, consumer organisations, trade unions, intellectuals and hackers have little common ground, except the occasional issue such as a local census in Germany and Netherlands, or ID cards in Australia and the UK.

Furthermore, for many burning issues in modern societies, the internet is instrumental in the mobilisation of a public sphere (Myers, 1994): anti-globalisation, anti-nuclear, environmental and biotechnology activists make effective use of the internet as operational and administrative tools. Thus, none of these activist groups has a strong enough motive to undermine the IT revolution. Opting out from computers and the internet would undermine their own operations. Any qualm there is over IT remains therefore minimal.

Finally, the most vocal polemicists in IT are 'hyper-technological'. Hackers do not work against computers, but deplore the under-utilisation of its potential; they deplore the 'law of the suppression of radical potential' (Winston, 1998). They are impatient with the pace and direction of progress. Many later are co-opted by the system as pioneers, innovators and security experts. IT seems to be a history of discontented activists forging the next step of the development.

## TECHNO-SCIENTIFIC RESPONSES TO RESISTANCE

The lack of critical mass and mobilisation against computing seems to suggest that there was no need to respond to any challenges on the part of IT protagonists. However, absence of mass mobilisation does not mean an absence of resistance at the local level. A closer look at this history shows a number of small-scale revolts and responses, but also large-scale ideological mobilisations by IT protagonists which most likely pre-empted any seriously critical engagement with the developments. Let us consider in turn the mobilisation of public awareness for IT, delays to local projects, alterations to system designs and evidence for institutional learning.

### Mobilisation of Public Awareness of IT (PAI)

Until the mid-1970s, computing was very costly. In 1950, one counted 15 computers in the US and a handful in Britain, France and in the USSR; in 1955 there were 240, 1960 there were 10,000 and 1966 there 35,200 units, mainly in the US. Britain in 1960 operated 50 units (McCarthy, 1966; Cortada, 1993, 120). A doubling to 85,000 units in the US was projected for 1975. The global market for computers was saturated and IBM dominated, supplemented by the residual 'bunch' of NCR, UNISYS, AT&T, NEC, CRAY, BULL, etc. For example in Switzerland, the market saturated at about 30% of business units or about 2,500 machines in the

1970s (see Bauer, 1993). Figure 6.1 above showed this market plateau and the parallel decline in news attention in the mid 1960s. The attention cycle for computers waxed four times since the 1940s. This put the technology into the public mind, but raising expectations requires an imaginary. From the start, the computer business was more about supply than demand. Apple's Steven Jobs epitomised this logic with his famous: 'people don't know what they want until you show it to them'.[4]

As shown in Figure 6.4, the PC took computers into the household; what initially was a capital investment became a consumer item. In 1984 about 10% of households in the US and UK used a PC, 15–20 million units. In 2002 there were 600 million PCs worldwide, distributed to less than 1% of people in African countries and to over 50% of inhabitants in the US, Europe, Australia and some Asian countries (ITU, 2002). The global stock of computers was fast reaching the one billion mark by 2005 and has since gone far beyond (see Computer Industry Almanac, 2010).

The rates of diffusion vary. For mainframes and micros, the unit of diffusion was the business unit. Computing was a capital good. The early analysts counted on businesses that use computers, maybe employees with computer access. Since the PC, diffusion is calculated to the household or the individual user. PC access became an indicator of technological prowess of the nation and a source of policy anxieties over the 'digital divide'. The US led the world, but its share of computers declined from 90% units in 1956 to 27% in 2004. An IT gap between US and Britain widened in the 1990s, but after 2000 the UK caught up to overtake the US. This catch-up was driven by the internet, with unprecedented diffusion rates in the last 10 years; for many, internet use was the entry into personal computing (Rogers, 2004). Figure 6.4 shows that in the US 50% of households reached internet access in 2001 and in the UK in 2004. Smaller countries like Korea, Singapore, Iceland, Sweden or Finland reach for internet access even faster. The PC needed 15–20 years to reach 50% of households; the internet got there in less than 10 years.

However, Hannemyr (2003) has shown that many of these 'tropes of speed', in particular historical comparisons with earlier devices such as telephone, radio and television are poorly grounded hype that became 'urban legend'. The story of the fast diffusion of IT is part of the effort to grab attention and to make it happen. The diffusion approach makes a taboo of the key question: will a networked society be a better society?

## Framing the 'Computer Revolutions'

Kling and Iacono (1988) observed that demand for computers was not only created directly by the computer industry, but indirectly by different strands of mobilising expectations. The computer movement was based on

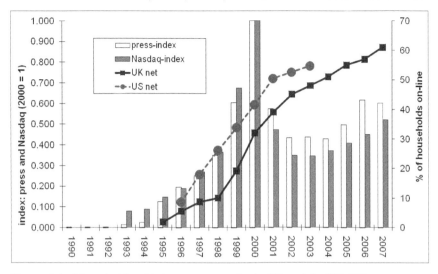

*Figure 6.4* Households with internet access in the US and the UK, UK press coverage of computing and the internet (index 1=7300 articles in single news outlet), and NASDAQ composite stock index 1990–2007. *Sources*: NexisLexis; British Social Attitude Survey (1988); ONS (2006); Shepard (2007); Day et al., 2005). UK press coverage and NASDAQ correlate very highly: r = 0.96 between 1993 and 2007.

five constituencies that made it an object of desire: urban information systems, artificial intelligence, computer-based education, office automation and the personal computer. Each strand defines a social problem for which they offer a solution and mobilise images and material resources to bring this about. These solutions are not only practical, but are aspirations for a future society.

*Urban information systems* engage the police, libraries, tax collection, local administration and urban planning. Local government reform takes computers to put administration into professional hands away from political appointees. Urban planners, librarians, police, tax officers and social service providers bring in computers with a common vision: a symbol of improved public services, transport flow, efficient tax collection and crime control. The latest developments include CCTV and biometrics for identification and surveillance. Computers are the panacea to foster competence, rationality, efficiency and transparency, and enhance the autonomy of local government.

The *AI* movement held up the 'thinking and feeling machine' as their quest. What later retreated into specialist engineering was in the 1970s and 1980s a popular story of in mass media coverage. Machines that solve problems intelligently, that move as robots, speak and converse, interact socially and express feeling captured the science fiction imagination. The Turing Test (Turing, 1950) defined a machine as 'intelligent' if a blind observer was unable to distinguish human from machine. AI started in the secrecy of the ARPA defence context; psychology and cognitive science excited ambitious students, and popular science the wider audience. The computer provided the root metaphor for the revolution that replaced the dogmas of behaviourism.

The idea of every pupil and classroom with a PC mobilised for an *education revolution*. Education must provide a computer environment for computer-assisted instructions, simulation and demonstration, and as gateway to the world wide web's global educational resource. Schools engage the technology for three reasons. *Computers enhance learning* by stimulating creativity and interaction. Interacting students are no longer taught mathematics, they become mathematicians. Computers *equip students for the labour market*. 'Computer literacy' adds to reading, writing and numeracy. For the computer business, schools and universities are places to 'capture souls', to form future customers. Governments encourage this with tax allowances for educational sponsorship where schools gain equipment for free or at rebate prices.[5] Finally, IT became *a symbol of progressive education*. High-tech in the classroom is the marker of the 'good school'. Parents are keen to see modern IT, although they may have little understanding of how it is used. Computers associate with 'excellent education' among teachers, educationalists and policy makers. Academics like Papert (1980) reached guru status on such matters. In the UK, the BBC promoted its own best seller home computer as early as 1981 and became part of the country's computer literacy campaign. The computer would renew the spirit of the protestant reformation, reinvigorating the idea of education with a new form of literacy.

*Office automation* rallied people into the brave new world of computing as early as the 1950s. The vision of scientific management seeks to improve productivity in the office as well as in the factory. This meant initially large data-entry pools populated by key-punch operators to feed mainframe computers; later, processing terminals replaced the card punchers. The desktop PC allowed for a different vision of office work. Every desk becoming a networked workstation made the centralised writing pool an anachronism only surviving in outsourced call centres and data processing operations in developing countries. Scenarios glossed the new office world: the author-operator, one author plus one operator, many authors plus one operator, many authors plus many operators. These scenarios had different implications for deskilling or upgrading of office jobs (OST, 1985). The networked desks, with mail, calendar, text processing and file handling raised the expectations of a 'paperless office'. Paperless office automation was promoted by professional organisations, in trade magazines and in the marketing of computer software and services.[6] But paper consumption increased (Sellen and Harper, 2003). Word processing created its own dynamics of multiple drafting, cut-and-paste multipliers and generally more words.

With features of a grass-roots social movement, the *PC arrived*. Young, usually male, computer amateurs built their own devices, soldering and fiddling parts in sheds and garages. These hobbyists had roots in radio amateurism, HIFI electronics and the 1960's flower-power individualism, and sought self-actualisation, fun and non-conformity, and to challenge the powers that be. 'Computers for everybody' supported many Davids against Goliaths like IBM. The combination of functionality and fun, work and leisure, at ever lower prices was appealing far beyond the nerdy user. Computer clubs sprouted to discuss issues and organise support around particular hardware and software. New magazines like *Byte* spread like bushfire, established ones created regular PC columns and at trade fairs the enthusiasts met the industry. Open software and standards reduced the need for self-programming, and what had looked briefly like a new culture technique retreated quickly to specialist occupations. Programming with MS-DOS or BASIC was soon irrelevant. IBM-Microsoft and Mac became 'installations' and unavoidable references: path-dependency set in (Allerbeck and Hoag, 1989). What started with grass-roots and utopian aspirations was co-opted by fast players like Apple and Microsoft who cut off their own roots. The movement made into a hip object of desire what was initially a thing 'only secretaries' would use.

These imaginaries of PC, office automation and education reform exploded in the cyber-hype of hysteric proportions when the internet arrived in the late 1990s. It transformed devices into a mobile, universal, multi-functional, multi-media environment for work and entertainment. The computerisation of society envisaging every citizen networked into cyberspace defines an axial transition of society. It opens the prospect for bloodless 'revolution', an historical opportunity not to be missed (Castells, 1996).

For this representation of the future, *computers are the key* to overcome reactionary education, to liberate productive forces by removing tedium from the factory floor as well as the office space, to close the gap between work and leisure. *Only the latest technology will do*. What is needed is a continuous quest for smaller, faster and more powerful computing. The quest for the latest gadget becomes an end in itself against a background of an every retreating horizon. *More computers are better than fewer*. Access needs to be universal; everybody is to become an early adaptor; the faster, the better. Quantity before quality; quantity will translate into quality of work and life. 'Real life is on-line' and not having access is a deprivation. *Computing is a win-all game* without conflicts of interest; fostering co-operation, rationality and general happiness in the factory, the office and the classroom. Losers are temporary and few compared to the winners. However, *unco-operative people* hamper progress. Contrarians are sectarian only and will not affect the game. They are just poorly trained. The *acceptance problem* emerges in the

1970s and preoccupies much research (see Williams and Mills, 1986). Undisciplined and otherwise deficient users were holding back the revolution. The resistant user becomes the culprit of technological failure. IT came to see the need to re-configure its users (Woolgar, 1991).

Finally, the computer movement relied on a *hall of heroes* who personify the ideals. A generation of young entrepreneurs founded what later become global operators and made gigantic fortunes. Much glamour of computing comes with the fables of Ken Olsen of DEC, Steve Jobs and Steve Wozniak of Apple, Bill Gates of Microsoft and Larry Page and Sergey Brin of Google. These young wizards became billionaire philanthropists in their 40s and inspire the young. Many IT wizards were university drop-outs who put themselves otherwise into the lime-light. 'If they made it, I can make it; and computing is the path . . .', seems to be the message. Bill Gates himself revealed the latest version of Windows, as did Steve Jobs for Apple Mac, iPod and iPad, until his death in 2011. These figureheads personify the computer ideology.

The various strands of computer movements and their ideological tenants have exerted a powerful grip on our imagination. However, as a representation of the future, their ideological function must be recognised. Finlay (1987) and Winston (1998) went to great lengths to reveal this brainwashing, so that we can recognise the ideology that inspires our actions, but blind our eyes. The enthusiasm becomes part of the problem: hallucinating, we are insensitive to any downsides of the technology. The problem always presents itself as the solution.

However, neither the parochialism nor the isolation of anti-IT protests condemned the widespread computer malaise to technological determinism. There are a number of subtle impacts of anti-computer sentiment which can identified.

## Delaying the Computer Revolution

The agitated debates on digital divides over computer and internet access testified to considerable inertia in the population. Some of this inertia is due to lack of resources particularly in economically deprived contexts, but for others staying away is a conscious option for resistance. This brought an upper ceiling to the penetration of computer and internet use, which is an impact of resistance.

The 1950s and 1960s saw similar inertia in business computing. Computer entrepreneurs faced the indifference of the business world and needed to convince them of the need and the cost-effectiveness of these devices. Through the 1960s, the high expectations of a global spread of computers and production automation were not met, and uptake of home computers in the 1980s was rather modest compared to other electronic devices. There were considerable delays at all stages of computing due to 'supervening social necessities'; revolutionary new designs were hijacked into existing uses (see Winston, 1998). The reality of computing lagged much behind its revolutionary rhetoric.

*Excursion 6.2   Computer Addiction and Phobias*

*It is widely assumed that time spent in front of computer screens in a population is normally distributed. This leaves two tails to worry about. On the right tail are the heavy users who are obsessive, but spearhead the revolution. According to Shotton (1989) 'addicts' are mainly male, unmarried, or married and childless, putting their marriage at risk. They have a mother-dominated upbringing, are shy and unhappy at school, but academically accomplished. They are resilient to social influence and dissociate emotional and intellectual life, and think mechanically. Whether computers cause addiction or whether they are a new outlet for people already obsessive remains unresolved. But like other obsessive behaviours such gaming and gambling, it raises the issue of access restrictions and bans, and of moral panic: addicts are the dysfunctional new elite. These hackers are the cyberpunks, the computer underground who fight the hegemonic forces of the industry who standardise life and control access. The numbers of hackers is not known, not least because their definition remains unclear. However, faithful to the ethos of the early day computing, they spearhead the 'open access movement'. All information on the internet should be open and free of charge; and this is demonstrated with spectacular 'hacks' into well-secured computer systems. Computer clubs like Chaos are the voice of this crowd who fights for open access to music, images and data. When the law cracks down, new ingenious ways are invented to free access without 'breach of copyright'. The movement finds sympathy among libraries, scientists and funders of research. Some initiatives have started to reverse the costing of production: the author pays to make access free. This reversal of principle is controversial not least because it makes life difficult for journalists, authors, musicians and film makers by undermining the traditional business model of user-pays without securing the alternative income stream.*

*The other tail of the normal distribution defines those who do not use, who avoid the digital world and are computer phobic. In the 1980s, psychological research diagnosed and offered treatment of this deficiency. The 'clinical eye' on resistance is conceptually and empirically problematic (see Bauer, 1995). Epidemiological estimates suggested that 3% of US professionals were cyberphobic; other sources claimed a rate of 25–33% of most occupations (Brosnan, 1998; Rosen and Maguire, 1990). College dropout rates were blamed on it. 'Computer rage', 'techno-stress', 'technology related anxiety' or 'futuretense' are synonyms for this malaise with computer interaction. Seventy-five per cent of British employees witnessed colleagues 'letting it out on machines or the IT department' (abusing, kicking, bullying, pulling the plug) and 50% experienced frustration and stress with computers (MORI, 1999). These figures suggest critical masses that sympathise with action against computers. Trade unions defended jobs and the quality of working life, but did not mobilise against computers. Technological strikes like in those of the British Printing Industry in the 1970s seemed a thing of the past (Martin, 1995).*

## Alterations to Large IT Projects

The census revolts in Germany and Holland in the 1980s were partially successful. The census was postponed or abolished. This became another episode in the history of the census which is closely linked to data processing machines, but recently worries about declining participation and other 'big data'. Population censuses move away from enumeration and replaced the 'knock at all doors' with a system of repeated sampling. The rising sensitivity of the wider public to issues of data protection and privacy and thus the risks of non-coverage seem to have brought about new ideas of how to collect national statistics without the census. The social statisticians responded to the challenge and changed their ways.

Local protests managed to halt a software project of Lotus in the early 1990s. A group called Computer Professionals for Social Responsibility (1981) originated among staff at Xerox PARC in California. They specialised in 'educating the public and policy makers' on the impact of computers on society, including participatory design, the internet commons and open access. These activists meet annually and were dignified with the Norbert Wiener Prize. Their campaigns became known as 'computer populism', one of which convinced Lotus in 1991 to discontinue the project Marketplace: Household. It was intended to construct a consumer database on 120 million US households using social security numbers as universal identifiers. The protection of identities and their privacy was insufficient (see Lyon, 1994, 174).

Australia revolted in 1985 against an identity card scheme that was supposed to reduce tax and welfare fraud (see Lyons, 1994, 174). A similar project in the UK was launched in the late 1990s. ID cards linked to a National Identity Register including biometric data (see Home Office, 2006). A group of *academics* challenged the feasibility of achieving objectives of safeguarding against identity theft and avoiding fraud within the budget (see Whitley et al, 2006). Polls showed, depending on item wording, a majority of 55 to 65% in favour of ID cards in the UK; women tended to be more supportive than men, with small differences across age, regions and social class (ICM polls 2004–2006). The UK government was determined to carry the project through with public support after the July 2005 terror events in London. However, the scheme was cancelled in 2010 by a conservative-liberal coalition government; the liberal partners had long campaigned against the scheme (Martin, 2011). Another surveillance revolt was staged in New York. The Taxi Workers Alliance, organising a quarter of 40,000 taxi drivers, called a strike against plans to install *GPS* and credit card facilities in theircars. Cabbies wanted to avoid central monitoring of movements, of credit card payments and of their discretion in cash transactions (source: NZZ, 7 September 2007, 7).

Many successes of resistance to computing are indeed local, do not make the news headlines and are thus invisible in the grand scheme of things. Alterations to IT projects in response to local resistance enter into the many case studies of organisational or institutional change.

*Excursion 6.3*  Altering the Decision Making Criteria

---

*The organisation 'bfu' is a Swiss public service agency dedicated to the prevention of traffic and household accidents with 40 full-time positions in the 1980s, of which half were women. In the 1980s central computerisation of its operations was introduced for accounting, administration and statistical research. The case study mapped the implementation process from the first system in 1982 to the second solution after 1986. The first system optimised hardware costs. It had operational problems and triggered considerable 'resistance to computing' among the office staff. Users complained about increased stress, sore eyes and backache. There was considerable anger towards the IT team about lack of training and consideration. Some refused to do additional work arising from system failure, others refused VDT word processing, consciously underused its memory functions and ridiculed the system's pretences to productivity and reliability, and triumphed whenever it broke down. Women distanced themselves from the 'male toy' called 'Bruno'. Staff protested officially and compiled long lists of defects and problems with the system. The IT team oscillated between defending the system, deploring the incompetences of employees and reconsidering their design. By 1986 the system had increased from 4 VDT units with 20 users to 12 work stations with 60 users, and a new processor was in place. The new strategy of 1985 came up with an 'extended relevance structure' to make a decision. The decision criteria increased from 26 to 36 and shifted from purely functional to human criteria for both hardware and software, as well as the 'orgware', i.e. the ideas about supporting work organisation. In reaction to the challenge of staff resistance, the criteria for procuring an IT system increased in number and widened in scope. The resistance of office staff altered the thinking of the IT project team of what constitutes a 'viable system' (Bauer, 1986).*

---

*Excursion 6.4*  Resistance and the Splits in the Project Team

---

*A second case study investigated the computerisation of banking operations in a Swiss Bank, the co-operative Raiffeisen group. In the mid 1980s, the banking group, the sixth largest in Switzerland by business volume, faced the task of providing data processing for a PC network to over 1,000 branches across Switzerland. Some of the branches were too small to operate a local system. Thus an IT project was set up to adapt existing bank clearing software for the needs of different branches. An in-house team worked in collaboration with the software provider. The project involved moving from an NCR to an IBM system, both of which had local allies in the project team. Raiffeisen lagged behind the rest of the Swiss banking on IT but not in operating efficiency. The IT team was aware of resistance among its branch managers, in the early phase and later as the system rolled out after 1989. Raiffeisen is a co-operative banking movement with a traditional emphasis on branch autonomy and local self-determination. The case study revealed that the project team had considerable difficulties in understanding the concerns of the branches, which generated, through various channels, over 500 points of criticism of the system. The branches known to be resistant and performing well were most critical. It was the split in the IT team, which gave this criticism an open ear. The IT team was split into 'technocrats' who implement the 'perfect system', and others, mainly the former NCR alliance who had lost an earlier 'battle', who documented, and thus heeded and supported the criticisms arising in the branches. A follow-up two years later revealed that around 40% of criticised defects of the system were altered pertaining to data input, data processing and output format. About 20% of the points were considered 'operational misunderstandings', and 15% were system constraints nothing could be done about. The case demonstrated how resistant attitudes of branch managers were constructive; and the better performing ones produced more criticism. Resistance is necessary, but not sufficient; it required the conflict in the project team togive criticism the 'hearing of voice' and to bring the team to heed the points raised (Bauer, 1993).*

## Institutional Learning and Conceptual Change

Hyperbole and visioning are not areas where resistance taught the IT protagonists a cautionary tale. While the early history of computing might have suggested more modesty about future claims making, the later stories of the the IBM 'billion- dollar baby', and internet and social media hype had rather the opposite effect: to teach the virtues of exaggeration and hype in the making of new technology. The lesson of IT seems to be: open all floodgates of enthusiasm and imagination to flush out any frustration that might follow. For the newspeak of the IT revolution, its unreality is never a question.

Some of the issues of computing and IT led to responses accommodating public concerns. The concerns over non-ionising radiation of VDT units and RSI led to epidemiological studies confirming the risks. This resulted in organisational solutions, such as limiting exposure time to several hours per day, and the regimes of rest taking.

A conceptual innovation of the insurance industry responds to the challenges of public concerns. Insurances manage risks. And *'phantom risks'* are now recognised in this industry (see Swiss Re, 1996). Phantom risks are imagined dangers, so not demonstrated by scientific evidence. Electrosmog emanating from mobile phones and mobile phone antennas is the case in point. Nevertheless, they mark what citizen-consumers perceive relevant for precaution: people behave as if these dangers are real. So for example, the housing market responds to the siting of a mobile phone base station with a depreciation of properties. Such 'risks not yet demonstrated' must be considered for insurance purposes, because once they are public opinion, it must be understood that the courts will consider them litigable in the future, as courts are obliged to consider public opinion, and this could be costly at some point in the future. For insurance purposes, a risk is real when costly consequences arise with a probability greater than zero. The phantom risk innovates this way of operating.

Resistance to illicit surveillance activities by private and public agencies has led to *new institutions and procedures*. For example, in 2001 the UK set up the Information Commissioner Office with the remit to report regularly on freedom of information and the progress of the surveillance society.

Into the early 1990s the PC was for most users still a version of the typewriter, and as such destined for the office staff, mainly women. IT was an innovation carried by middle-ranking technical staff, but it was destined to change only the jobs on the shop floor and in the clerical office. Office automation focussed on clerical staff and until the later 1990s it was difficult to find a computer on any desk higher in the ladder. It is quite feasible to argue that the reluctance of the higher management to adopt the desktop computer led to the discovery of the 'user' and participatory design. *Management's resistance* to computers was a wide-spread worry in the 1970s and 1980s and a major block to business efficiency as apparently between 20 and 30% of US and UK managers were 'techno stressed', costing $4 billion a year and contributing to the recession of the 1980s (Howard, 1986). In trying to convince managers to use PCs, user-interface design became a focus of attention. Indeed, the design of human-computer interaction became key to resolve IT acceptance problems during the 1980s (Frese, 1987). Software was now a major cost factor, while hardware costs had declined. Investment moved into software testing and the development of design concepts. Software testing initially sought to avoid programming errors (Gelperin and Hetzel, 1988). But within organisations, IT professionals lost their status: the PC was everyone's business (Markus and Bjorn-Andersen, 1987). Then the focus moved to the operator: how to make the system 'fool proof' so that the normal mistakes of variously skilled operators do not result in major consequences. The focus on the manager users doing more than word processing and number crunching brought about the *user-centred design movement*, combining software design, psychology and 'user participation'. Much of these efforts to develop and test interfaces with users came from Xerox PARC, the Palo Alto Research Centre in Southern California, in the early 1980s. Xerox, a photocopy business, was alarmed by the prospect of the 'paperless office', and funded research into office activities beyond photocopying. Concepts like 'user-friendliness' aimed to create a computer environment that should encourage, stimulate and support the use of the multi-functional PC. The move from command language and key functions to pull-down menus and 'icons' on touch screens were major innovations of human-computer interfaces. User-centred design brought human memory, perception and motor learning to the centre of product design. It was recognised that software and hardware configured a 'user', and it was advisable to do this purposefully and with the awareness of human psychological constraints (Dagwell and Weber, 1983; Card, Moran & Newell, 1983; Shneiderman, 1987; Sutcliff, 1988). Ergonomics advised on body posture, grip and physiology to build comfortable hardware; cognitive psychology advised on software design and the user-interface. The narrow model of the 'information processor' was extended to include the social and aesthetic needs of real users. As PCs moved up the hierarchy, pure functionality was superseded by *simplicity, versatility and pleasure* (Norman, 1998, 67). The computer became an object of desire, an accessory of lifestyle and status. The competing worlds of PC and Mac played this out when the internet brought new design challenges. Here functionality, usability, simplicity, versatility and fun were not enough. As content providers compete for voluntary attention, the new targets are *persuasive design* and emotional attachment. 'Captology' (Fogg, 1998) or design for activity-based emotions (Norman, 2004) systematise the new ways of overcoming the recalcitrance of website users and to engineer their attention and trust. The internet led to the recognition of the *non-user* when news about an apparent decline among users aged 18–29 years rang alarm bells in the late 1990s. This non-user group cannot be explained with economic or cognitive deficiency. Thus, the resistance brought the re-appreciation of the motivated non-user in the wider context of technology development. Non-users matter in the social construction of technology as the 'not yet-relevant social group' (Oudshoorn and Pinch, 2003), the potential of untapped markets.

In summary, computing and IT morphed from a few mainframe computers to billions of multi-functional communication devices in the 21st century. Despite ample reasons to worry about hype and false promises, productivity gaps, deskilling and killing of jobs, increased stress, electro-smog, private and public surveillance, AI and a neo-gnostic elitism, computer addiction, pornography, privacy and social isolation, and the de-humanisation of human interaction, none of these issues created a mass mass that put computers and IT into question. IT has seen nothing like what nuclear power or genetic engineering had to face in terms of global challenges. Resistance there was, but scattered, unsynchronised and on particular local issues. The scattered sceptics of the IT revolution had little to show in terms of mass mobilisation. The dog of public resistance did not bark at computers and IT; the diegestic prototyping of the 'IT revolution' was too effective.

The computer arose in the war effort and its protagonists had to define a civil use for it. All through its history it has been more supply than demand driven, and fuelling imagination and hype with an ever-shifting horizon of an IT revolution. Hype and images of an extraordinary future disqualified sceptical voices from the very beginning as 'latterday luddites'. This does not however, testify to the absence of resistance, which, as we have seen, flared locally and sporadically. It does however testify to the power of imagination and enthusiasm a) to distract public attention from the thorny issues, and b) to raise the threshold for framing resistance action, and c) to pre-empt the meta-framing of a counter-mobilisation on something like health, culture or pricacy. It also testifies d) to the fact that resistance can challenge techno-scientific projects under the radar of public attention and without critical masses of public opinion. The experiences of micro-resistance to IT projects were a rich source of conceptual innovations grappling with a recalcitrant user. We owe it the notion of 'user-centred or participatory design'. The discovery of the 'implied user' amounts to a social psychology and a rhetoric of design, seeking to persuade the non-user as a not-yet user to join the game. And the notion of 'phantom risks' is a powerful new idea with relevance for other emergent technologies.

## NOTES

1. I recall my language proficiency dilemma as an ambitious psychology student at University of Bern in the early 1980s: learning PASCAL or Spanish. I opted for Spanish and never regretted it.
2. The press recently reported the rise and change in privacy actions of British courts. Privacy actions are rising, and they are no longer predominantly protecting high-profile celebrities from mass media attention, but a more varied public from abuse and widespread misuse of personal data. The courts are getting more concerned with 'Big Brother' databases (source; *Independent*, 22 July, 2013, 10).
3. In early June 2013, 30-year-old Edward J. Snowden, an employee of a US computer consultancy firm, blew the whistle on secret programmes of total surveillance of cyberspace communication. The US NSA programme PRISM and the British equivalent at GCHQ TEMPORA had signed up the collaboration of all major Internet platform providers. Snowden chose Hong Kong as safe haven to make his revelations to the world media and Wikipedia represents his interests. Snowden now lives in Russia, where he has been granted temporary asylum, to avoid a US criminal court. The particular scandal of these revelations lies in the fact that these secret surveillance activities have been ongoing since 2007, that secret services of one country use the services of other countries to overcome legal constraints of spying on their own citizens and major Internet platforms providers are fully implicated. The new euphemism for this kind of surveillance activity is 'big data' and 'meta-data', the making use of transaction traces for other purposes (see Wikipedia 'Edward Snowden'; accessed 11 April 2014).
4. Steven Jobs dropped this famous line in an interview for Business Week On-Line, published on 12 May 1998. Correspondent Andy Reinhardt asked him: did you do consumer research on the iMac when you were developing it?, Answer: No . . . ..
5. Ironically, my own university, the LSE, has apparently never received a free computer from any of its providers; some people were outraged at that missed opportunity when other institutions continuously modernised their IT services with donations.
6. I got my first paid post-graduate job with NCR Switzerland where I joined a small team that was bundling standard software and some organisational ideas into a product package called 'NCR office automation'. That was back in summer 1987.

# 7   Public Opinion and Its Discontents

*'You ask me, Sir, whether I lead or follow public opinion?*
*The truth is I meet it on the way.'*

(Edmund Burke)[1]

The discussion of nuclear power has shown the growing relevance of public opinion for science and technology. I have highlighted the long-term changes in public opinion and the key events that fostered the change. This history alerted scientists, engineers, activists and politicians to the fact that public opinion can be both lost and gained. Finally, the nuclear debate shows the social sciences in innovative spirit: new 'names' were coined for a recalcitrant public. Scientists and engineers eagerly adopted what social scientists elaborated: the public understanding of science (PUS) resonated among scientists, risk perception (RIP) more among engineers. Both notions deal with a recalcitrant public at times doubtful over nuclear power, genetic engineering or information technology.

## THE PUBLIC SPHERE, PUBLIC OPINION AND ITS MEASUREMENT

In order to understand 'public opinion', it is important to clarify three related notions: the public sphere, public opinion, and opinion measurement. They are often confused. Public opinion is not the same as the public sphere, and opinion polls do not fully capture public opinion. Like a family snapshot is not identical to the family process, and this particular family partakes in the larger reality of 'family' as a historical institution. The three notions are however nested.

### The Modern Public Sphere

Eighteenth century Europe saw the fostering of a distinction between public and private spheres of life. The public is understood in contrast to the private spheres of kinship relations and of economic exchanges between private individuals. 'Private' means not involved in the affairs of state. Language is sometimes confusing: in the UK private 'public schools' educate the public elite, while public 'state schools' educate for a private working life. The private sphere formed in the 'third estate', among commoners of independent economic means who were hitherto excluded from public affairs. Princes and nobility governed the state as a private matter. But, Kings or Queens had no privacy.

Thus, the *'bourgeois public sphere'* is a feature of modernity (Habermas, 1989). Private people meet in public to discuss matters of common interest, to engage in polite conversation and subject the dominion to the rule of law. Based on the fictitious identity of owning property and being 'human', individuals were idealised as shielded from social influence and buffered against passions (see Taylor, 2007, 185ff). The ideals of a public sphere and of individual rights are counter-factuals that underpin the discontents of the ancient regime. The 'mandate of heaven' no longer suffices. Government is to be supervised not only by the will of God, but by the rule of law and by public reasoning to provide legitimacy. Government ought to listen to something outside formal proceeding (Luhmann, 2000, 274ff). Public spaces in piazzas, salons, literary societies, theatres, promenades, arcades, parks, coffee houses and pubs, the printing press with papers and books, and more recently the worldwide web, are the forums of a public sphere, where civil society expounds topics and interests, challenges the political order, during the electoral cycle or outside of it. The public sphere is part of the long process of enfranchisement of the 'commons' that culminates in the practices of democracy and universal suffrage: free and fair elections are the only legitimacy to govern. Mass media are an important part of this process. Book printing, newspapers, pamphlets and magazines, satirical and serious writing, and writers and journalists and their protection define the freedom of speech.

Public opinion thus forms when private people gather in public to debate, inform and educate, and to reach a common understanding on what matters. *Public opinion* is this process, safeguarded by rules that differ from those of the market and of kinship.

This liberal ideal of a public watch-dog for authorities is however one among several notions of 'public opinion'. The medieval 'vox populi, vox dei' (Latin for 'God speaks through the people') suggests that if the people were to acclaim a new saint, the Church was advised to listen. In the modern analogy, public opinion is source of inspiration for governments in need of it (Key, 1961; Luhmann, 2000, 286). Other traditions are more patronising. The public is the barbarous multitude, mad, bad and uncultivated. This justifies a selective inclusion of opinion to the club of grandees. Public opinion is demonised in the mass psychology of the late 19th century. Gabriel Tarde (1901) and Gustave LeBon (1895), fearful of revolution and unrest, saw those who gathered in the street as de-individuated masses, incapable of reason, given to hysteria, somnambulant in a hypnotic state and thus prey to demagogic leadership. Mass opinion correlates with totalitarian rule, the dictatorship of the majority controlled by a populist strong hand, as in the disasters that brought fascists and communists in the first half of the 20th century to power (see Noelle-Neumann, 1984). The public is disqualified; it must be controlled. However, only a free public can be inspirational. The nature of public opinion remains controversial.

## Public Opinion: Attention, Rules, Frames and Process

The public opinion process focusses attention and selects themes with a rhythm of change, and what emerges as public opinion is schematic. Hilgartner and Bosk (1988) capture this with the public arena model. Public arenas have limited attention span. Problems compete for public attention; some make it, but not necessarily the most pressing ones. Competition over issues places a premium on drama to increase the likelihood to be selected. Areas are politically biased, and changes in culture can change the bias; selected issues resonate with cultural symbolism, and repetition and redundancy saturate a theme into oblivion. New events must reinvigorate a theme to keep the attention otherwise it is displaced. Social issues therefore have a cycle of rising and falling from public grace. Different arenas suppress or reinforce each other; some themes have a niche life without the prospect of becoming mainstream. Some problems become institutionalised and form communities of operators who sponsor big themes such as crime, war and peace, economy, civil rights or science and technology.

In public opinion modern society observes and describes itself, and flags up important issues for the agenda. Opinion (doxa) contrasts traditionally with knowledge (episteme) (Waldenfels, 1982); the latter is true or false, while the former is varied and distributed. People have different opinions and are to be persuaded; it is risky to ignore opinions, which are schematic and scripted. The talk of *'crisis'* focusses attention from the past to present and future options. The talk of *'reform'* makes ideas practical and mobilises political identities for change. Confidence is gained by forgetting why previous reforms failed. The talk of *'revolution'* makes this point more radical. The discourse of *'abnormality'* points to deviance to prevent further deviations. Finally, the talk of *'cause-effect'* points to trajectories which, considering preferences and values, are to be altered or established. Without public opinion there is no democracy, but public opinion does not advise on right or wrong; it specifies the frame through which this might be determined. Public opinion thus audits, inspires and legitimises political action of the minority that rules the majority; it is a way of living with a narrow top (see Luhmann, 2000, 299ff).

Living with public opinion requires appreciating its *Eigen-logic* (Neidhardt, 1993). The public is a space where people claim freedom of expression. But one enters the public sphere in a role. One might be audience, attending and listening, free to leave. The mass media play a key role as mediators, highly in demand with an acute sense of whim which allows them to win and keep audience attention. The media trades attention for scoops. They tend towards excitement and entertainment rather than argumentation. Finally, speakers air their views not to reach consensus but to make dissent transparent. Speakers must consider news values (Hansen, 1994). News selects the novel and alarming, the extraordinary and catastrophic, and opinions are polarised: expertise is balanced with counter-expertise, irrespective of authority of evidence.

Besides its functionality, Habermas stressed the historical ideal-type. A public sphere that produces reasonable outcomes is based on four presuppositions of power-free dialogue: inclusion, equality, authenticity and absence of coercion. Nobody capable of making a relevant contribution should be excluded, and everyone has a right to speak. Speakers are obliged to mean what they say, abstain from deceiving others and deluding themselves, and nothing but the better argument shall prevail (Habermas, 2001, 45). These presuppositions are counter-factual, not empirical descriptions. They preserve the ideal-type, derived from Aristotle's rhetoric, the Greek polis and polite society of 18$^{th}$ century. Rationality is here no calculus to maximise means against fixed ends, but a dialogue that shapes means as well as ends. This rationality is grounded in universally binding rules of communicative action. Claims making oriented toward a common understanding face a triple check: true to the world, right in the community and sincere and truthful in relation to the speaker. So one must ask three questions in public: is nuclear power safe and cost effective(true facts), do the safety implications challenge our way of life (right) and is the real issue energy or weapon capability? The three musketeers of logos (Arthos), pathos (Porthos) or ethos (Aramis) ride together or not: one for all, all for one. The purpose of reaching a common understanding constrains the rhetoric; no perlocution and to keep all three musketeers fencing is the benchmark for any existing public sphere.

The public sphere having emerged from feudal times risks 'refeudalisation'. When strategic public relations dominate, attention grabbing display of pomp (pathos) and personality cult (ethos) overpower the arguments (logos). The public becomes private and vice-versa: the very distinction erodes. If power or kinship regulate access, and money stifles the argument, then 'bullshit' prevails over truth seeking (see Frankfurt, 2005). The modern public sphere remains resilient (see Habermas, 1990), but is at risk when markets fail the newspaper, television or the internet; mass media monopoly and censorship end a functioning public sphere.

This *counter-factual standard* remains a useful test of the public opinion process. This utopian twist allows us to be concerned and raise alarm when conversations degenerate into a display of power. 'Guns' are no argument. Whatever is visible as public opinion, the key question is: how did it come about? This question remains valid for both old and new media. Is Facebook a perfect public sphere or an exclusive club of segmented conviviality? Does increased visualisation manufacture consent or inform opinions?

## Public Opinion Measurement

In public opinion, society observes itself; this self-observation can be observed as representative (with unlimited regress) and as a matter of the social sciences. The multiple meanings of 'representation' confuse matters though (Derrida, 2003, 68ff). 'Representation' means to *stand in for somebody*: the legal meaning. Society is *represented by bodies* who speak for the people. These bodies are elected presidents or prime ministers but also parliamentarians and civil society actors represent particular social issues and constituencies. They *make visible the invisible* by rendering *present the absent or neglected* issues of society. These bodies need to be legitimate, having gained their role by proper procedure rather than by birth, by claim or by violence or threat thereof. Thirdly, public opinion *represents by proxy*, an index correlated with due process. An index of public opinion is mediated by a *calculus of representation*. This introduces the statistical notion of measurement: the equal probability of every person to be selected, the unbiased sample of opinions. This resonates with the idea of 'one person, one vote'. But who counts as a person? What about animals and things (see Latour, 2005)?

The social sciences observe public opinion through many concepts rooted in common places. In the extreme, the public is the *'mad, hysterical, effeminate crowd'* of LeBon (Ginneken, 1992). Ever since, rational individuals tend to distance themselves from uncivil crowds. What fascinated Tarde in the late 19th century was the *synchronisation of opinion* through mass circulation media—action at a distance. The fact that public attention in the late 1890s was focussed on the miscarriage of justice in the Dreyfus scandal rather than on France's colonial war in Africa needed an explanation. What caught Tarde's eye was the collective attention and the political pressure arising from this (Tarde, 1901).

Since these speculations of the 19th century, the social sciences developed various ways of gauging public opinion: opinion polls, focus groups, media monitoring, document analysis, stock values and betting on certain outcomes, and more deliberative exercises. The 'gold standard' of public opinion remains the democratic 'vote', but proxy measures are useful to make public opinion visible at any time. Actors need to know, and want others to know, whether a course of action carries support or not. Opinion measures are taken very specifically and timed at will to move on the climate of opinion: observation is often intervention.

Ever since Mr. Gallup predicted correctly Roosevelt's presidential victory of 1933 on the basis of a small but *representative poll* when *Readers Digest* got it wrong on the basis of much larger numbers, the Gallup poll, i.e. questionnaire based interviews plus statistical sampling, assumed a 'gold standard' of opinion measurement. What predicts a vote must be valid, and the key is sampling quality not sample size. Polling created a billion dollar industry which anxiously protects its methodology in every voting cycle and transfers credibility to anything else they do. Failure to pick the winner, as in the US in 1947 and UK in 1992, shakes the industry (see Katz, 1949; Jowell et al., 1993). At elections politicians are taken to account, but pollsters too. Clearly, systematic polling is superior to haphazard attempts of making public opinion visible. Internet sites, radio and magazines present straw polls, where audiences respond quickly to questions, but create hopelessly biased data. Such self-selection samples of data are useful at most to grab attention.

Questionnaire design has become an artful industry, juggling question types, pre-testing, response formats, item ordering, incentives to respondents and forms of interviewing (see Schumann and Presser, 1996). Equally, randomised survey sampling involved the choices of a population frame, stratification and clustering, sample size and selection of interviewees (Kish, 1965). All efforts are part of the 'total design' to avoid bias and to balance error margins with the costs of collecting data. *Standardisation* is a necessary ingredient of comparisons. In order to constitute a signal, measures need a benchmark. Compared to 60%, 75% is 'larger than' within a margin of error (of +/- 5%). And results need framing. A glass of water can be described as 'half empty' or 'half full'. Both are correct, but the rhetorical effects are different: half empty suggests a bad deal, give me more; half full suggests a deal to be happy with. Numbers do not speak for themselves; they need benchmarks and framed interpretation.

*Excursion 7.1*  German 'Technikfeindlickeit' and the Uses of Opinion Polls

*Some survey items achieve notoriety. Allensbach reported that the Germans, in particular the younger folk, had left 'Fortschritt durch Technik' (Audi's famous 'progress through technology') behind and become 'technikfeindlich' (against technology). On the question 'do you belief that on balance technology is rather a blessing or a curse for humanity', 72% of Germans, 83% of teenagers, saw technology as a blessing in 1966. This had declined to 30% by 1981, 23% among the young. Ambivalence, considering it part-curse and part-blessing, had increased from 25% to 57%, or from 8% to 58% among the young (see Figure 7.1). No such change can be observed in the US, where recognising the benefits of technology seem second nature for 75% of the population.*

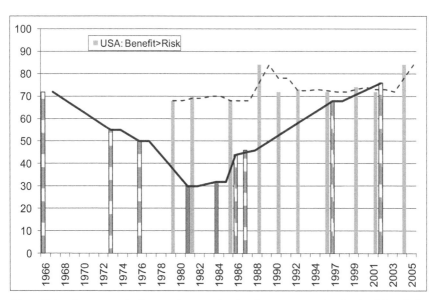

*Figure 7.1*  German attitudes to science and technology construed as 'Technikfeindlichkeit' according to Allensbach polls (percentage responding 'on balance technology is a blessing or a course') compared to US responding 'on balance modern technology has more benefits and risks'. *Sources*: Hennen, 1994 and 2002; NSF 2006.

*This was the stuff of scandal, and fine-grained analysis could not stem the moral panic. The reputation of Germany as a high tech country was in jeopardy. Others saw an opinionated public ignoring facts, ushered on by biased media reportage. The generation of 68 politicised technology from editorial positions in the national press and broadcast media (see Kepplinger, 1992 and 1995). Others saw methodological issues: there was little evidence of 'Technikfeindlichkeit'. Different response modes were used: the middle option neither/nor (1966) produces more 'blessings' than a middle option (1981) part-blessing-part-curse (Jaufmann et al., 1989, 51ff). A similar item positioned Germany in 1981 midfield among Western countries. The Danes and French were more ambivalent than the Germans (see Jaufmann and Kistler, 1986). The Swedes, Irish and Dutch were more sceptical and more polarised; only the US, Canada, UK and Australia were more positive. The German Technology Assessment Bureau (TAB) published three reports (Hennen, 1994, 1997 and 2002) showing that Germans were neither anti-science nor anti-technology but doubtful with regard to nuclear power and genetic engineering. To no avail, the moral panic continued, at least into the 1990s, fed by political expediency. Some actors profit from pointing to a malaise of the nation. By 2014, the new term is 'Technikaufgeschlossenheit' (open to technology.)*

*Figure 7.1 shows that the 'scandalous' figures of 1981 in the end were episodic. The reunification of East-West in 1990 brought a discourse of modernisation which put to rest the panic. Also, with Marxist schooling and less post-materialist aspiration, the 'Ossis' were less sceptical than the 'Wessis'. By 2002 Germans were as positive as they were during the post-war miracle. The moral of the story: a) public opinion needs to be measured over time; b) the opinion measures represent and thus stimulate debate; c) some results receive heightened attention because they are politically useful. Public opinion measures thus influence debate.*

## The 'Social Psychological Liquidation' of Public Opinion and Reflexive Turn

One of the concerns of polling is the reification of opinion. The success of random sampling as the 'gold standard' of social research—maybe the one real achievement of the social sciences in the 20th century (see Converse, 1987)—supports the mindset of 'operationism': public opinion is made to be what the polls measure. Everyday conversations, but also seasoned social scientists, equate *public opinion with opinion measurement*. This conflation of measure and process, of temperature and fever, conflates the signifier with the signified, which is 'fetishism': attributing powers that measures do not have. A successful polling industry and practice amounts to what Habermas (1989, 236ff) called the *'social psychological liquidation of public opinion'*. The concepts of opinion and attitude are reduced to individual information processing, tied to a mind-computer analogy (see Chapter 6); this remains controversial in social psychology as part of a paradigm competition (see Farr, 1996). The use of polling in Cold War propaganda had made opinions into effect measures of communication efforts of private eyes and secret services. The 'battle for the hearts and mind' is modelled on an artillery guidance system: the media are delivery systems, the messages are ordnance, press officers the gunners, and opinion polls provide the target intelligence. Privatised observation of public opinion ignores their formation in free conversation; it becomes 'nothing but' an aggregate of individual responses. It requires new ideas such as 'social representations' (Moscovici, 1961; Bauer and Gaskell, 1999) to reconnect opinions, attitudes and stereotypes to communication milieus in order to disprove the liquidation thesis.

In the late 1980s *focus groups* came to rival opinion polling, in the same way polls had rivalled media monitoring before. Like before, winning elections dignified the 'new method'. The campaigns of Reagan 1980 and Clinton 1992, and Blair's UK victory of 1996, were guided by gauging the language of decisive electoral segments. The focus group was developed in WWII to efficiently conduct interviews with many soldiers (Merton and Kendall, 1946); it survived in product testing and audience research. By the 1990s it resurfaced as a key tool of social research. The focus group was claimed to empower the subjects, though good intentions were confused with methodological procedure (Bauer et al, 2000). Qualitative enquiry proved more powerful to 'colonise a life world' than polling measures could ever be.

The *reflexive turn* of observing the observers of public opinion shows how metrics are constructed. Public opinion is made visible by constructs in a literal sense; measures designed for a purpose. If we treat the questions as the data, we see that items imply a stereotype of the public. Items measuring the attitudes to sciences can address the 'public' as *students* in need of education, *citizens* with allegiances, concerns and engagement, *believers* endorsing orthodoxy and deference to authority or *consumers* aware of products with more or less satisfaction. Each stereotype suggests different questions, albeit with overlap. Survey items frame the object of research. Giami (1991 and 1996) has demonstrated how human sexuality was variously framed in Kinsey's studies of the 1950s as pleasure, in contraception studies of the 1960s as choice, and in the context of AIDS in the 1990s as health risk. The framing of citizens as 'deficient' has led to polemics against measuring 'science literacy' per se as if the 'deficient public' were a necessary stereotype of the method (see Wynne, 1995; Bauer, Miller and Allum, 2007). Research has revealed that constructs of the 'public' come with any format of public engagement (see Brown and Schulz, 2010).

Two meta-frames of public opinion of science and technology deserve a bit more attention: the 'public understanding of science' and 'risk perceptions'. These are functionally equivalent ways of representing the 'public' in public to tackle recalcitrance in the face of techno-scientific developments. They also represent the aspirations, opportunities and efforts of the social sciences to be useful.

## FRAME A: SCIENCE LITERACY AND UNDERSTANDING OF SCIENCE

The 'public understanding of science' means different things to different people. For some, it is the rallying cry to co-ordinate activities of reaching out to a wider audience for science. For others it stands for research on how communities relate to science and technology, and for others it is the sack they hit when they mean the donkey, a conflict of substance misplaced into one over methods. Recent research on science and its publics comes in three phases (see Bauer, Miller and Allum, 2007). Each phase diagnoses a problem and opens up different solutions; this is what frames do (see Table 7.1). Historically speaking, we thus find ways of framing public opinion adopted by mainly scientists and their representatives in many countries. The UK's Royal Society has taken a leading role in this since the mid 1980s, for reasons not yet elucidated.

### From Literacy to Public Engagement

In the 1970s concerns centred on *scientific literacy*: what do people need to know about science? Not knowing enough, the public has a cognitive deficit which prejudices science policy and politics; many questions of energy, health and the environment require familiarity with science. Debates waged on how to measure literacy, as matters of fact (literacy), understanding experiments and probability (process) or appreciating the workings of institutions such as laboratories, communities, funding sources and peer review processes (savvy). Literacy encouraged measuring a cognitive deficit and invigorating education: the public needs to go to school with a different curriculum.

*Table 7.1* Periods, Problems and Propositions Relating to the Public of Science

| Period | Problem Attribution | Propositions for Research |
|---|---|---|
| Science Literacy | Public deficit Knowledge | Literacy measures Education |
| 1960s–1980s | | |
| Public Understanding | Public deficit Attitudes | Know-attitude Attitude change |
| 1985–1990s | | Education Image marketing |
| Science & Society | Trust deficit Expert deficit | Participation Deliberation |
| 1990s–present | Notions of public Crisis of confidence | 'Angels' Mediators Impact evaluation |

*Source*: Adapted from Bauer, Allum and Miller, 2007.

In the 1980s the concerns moved on. The Royal Society's report 'The Public Understanding of Science' (Royal Society, 1985) still found a public deficit, but this time attitudinal: people do not appreciate science enough. More by wishful thinking than sound science, scientists linked cognitive and attitudinal deficits with the assumption that cognition drives affection: the more you know, the more you love it.

One of few merits of the 'deficit concept' was to stimulate research into public perceptions of science on a global scale (see Bauer, Shukla and Allum, 2012). Eurobarometer has carried questions on scientific knowledge, interest and attitudes since 1977, though consistently since 1989. This database allows the assessment of long-term changes in literacy, attitudes and interest in science across Europe. Figure 7.2 shows that average literacy in Europe has increased by about one-third of a standard deviation from 1989 to 2010. Literacy is measured in quiz format. The number of correct answers gives the literacy score. Interest in new discoveries increased by one-tenth of a standard deviation between 1989 and 1992, but declined by the same

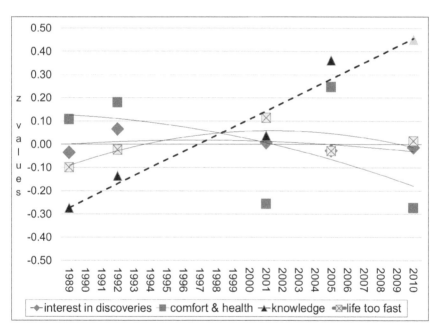

*Figure 7.2* Knowledge, interest and attitude facets in EU-12 countries from 1989 to 2010; knowledge is measured with a 13-item quiz; (figures for 2010 is projected from trend); attitude 'science makes our lives easier more, more healthy, and more comfortable' (comfort and health; agree with) and 'science and technology make our lives change too fast' (life too fast; disagree with); interest is based on an average score from 'not at all' or 'somewhat interested' (= 0) and 'very interested' in new scientific discoveries (= 1) (DK = 0). All indices z-scores (m = 0; sd = 1, n > 60000). *Sources*: integrated data Eurobarometer 1989, 1992, 2001, 2005, and 2010: see Bauer, Shukla & Kakkar, 2009).

amount by 2005. Positive attitudes to sciences, measured by agreement to 'science and technology make our lives easier, more healthy, and more comfortable' increased by about one-sixth of a standard deviation between 1989 and 2005, but shows erratic change. Over the last 20 years, the evidence in Europe is: on aggregate, scientific knowledge increases, positive attitudes fluctuate and interest in new scientific discoveries decreases.

Much hope was invested into a positive knowledge-attitude relationship, with mixed results (see Allum et al., 2008; Einsiedel, 1993). The technical literature on attitudes was largely ignored: knowledge is no driver of positive attitudes, but of attitude quality. Information-rich attitudes crystallise and are thus more stable (see Eagly and Chaiken, 1993). The well informed might hold strongly to very different attitude positions on an issue, depending on their espoused values.

Another stream of thought ignored cognitive deficits and adopted propaganda. The tactics of emotional appeal used in advertising and image making show how to design perceptions, essentially, public relations.

After 2000, the discussion followed another influential British report (House of Lords, 2000). *Science and Society*, the new term, emerged from the critique of the 'public deficit' idea.[2] The public deficit idea was part of the problem. Many scientists and policy actors held caricature views of public opinion. And a public that is taken for 'stupid and irrelevant' is unlikely to respond trustingly to the authorities (see Wynne, 1992). The Lords diagnosed a crisis of confidence in science and suggested more public engagement to regain public trust. 'Technologies of humility' (Jasanoff's nice term of 2003) became the way forward: citizens juries, consensus conferences, scoping of public concerns, hearings, extended debates, deliberative opinion polling, study circles, future search conferences and scenario workshops: the labels are proliferating (e.g. Einsiedel, 2001; Medlock et al, 2007). Much effort goes into such events, which are too cumbersome for academics to organise. A market of 'private angels' emerges who mediate between science and the public. Private consultants, often academic spin-offs, stage these public engagements. As a consequence, funding shifted from research to event making and taking the 'performative turn'. The role of research though remains threefold: to justify events as 'participatory democracy', to develop typologies and new events with a 'unique selling proposition' (see an early typology: Arnstein, 1969), and to evaluate events (see Delli Carpini et al., 2004). There remains the risk of presenting the old as something new: the public often remains at the receiving end. 'Dialogue' is the talk, 'monaude', or I speak, you listen, is the walk.[3]

## Public Understanding of Science Activism

The promotion of public understanding of science comprises old and new actors, many motives, and an expanding repertoire of activities (see Gregory and Miller, 1998).

Old actors are science museums, such as the British Science Association (BSA) and the American Association for the Advancement of Science (AAAS) and its sister organisations in many countries, collecting and exhibiting objects. Increasingly, learned societies such as the Royal Society of London or Edinburgh or the New York or Chicago Academies of Sciences pursue outreach activities, although their priority is the lobbying of government policy.

New actors of public understanding of science are science centres with a focus on interactive exhibits. Since the 1980s in the UK, several academic chairs for the public understanding of science were set up in the UK alone at Imperial College, in Oxford and at Bristol University. Funding agencies, public and private, make increasing efforts to exhibit their research and to oblige grant holders to perform 'public engagement activities'.

An old-new actor is the mass media system in print and broadcast that covers science. In the 1980s and 1990s, many mass media outlets built up their routine in weekly pages, sections, features and programs and expanded the science beat with specialised staff. Among the new actors, internet service providers and social media have become a prominent forum for scientific information and debate. The provision of free content on the internet has led to a crisis of the traditional business model of newspaper, they lose their advertisers to the internet, and this affects the science beat. When newspapers close or reduce their staff, the science beat is first to go. This leads to a paradoxical situation in the new millennium: science communication expands, while science journalism is in crisis, though not so globally (Bauer, 2013).

The analysis of mass media content offers a measure of media activity and a proxy measure of public opinion at the same time. Newspapers comment on issues and on each other; they are and make public opinion. Dual in nature, they are medium and actor-vector of opinion. The methodology of *content analysis* was systematised during the Cold War propaganda effort (see Krippendorff, 2005; Osgood, 2006).

The British public opinion of science can be gauged from a longitudinal analysis of British post-war press (Bauer et al., 1995). Over 6,000 science news stories, published between 1946 and 1992, were systematically sampled and coded. The database was later extended. This science monitor considered three dimensions of news: attention, slant and framing. Figure 7.3 shows public attention to science with a characteristic phasing: rising public salience of science from 1946 until 1966; then a decline into the mid-70s; since recovering, and very strongly so across the turn of the century. It appears that science news in the 21$^{st}$ century has reached historically unprecedented levels (Bauer, 2012).

The slant of science news follows a similar phasing: more positive into the 1960s, more negative into the 1970s and more positive again since. However, the direct correlation is not very strong; slant and level of attention diverge. In the early 1960s, slant turns when attention is still rising;

160  *Atoms, Bytes and Genes*

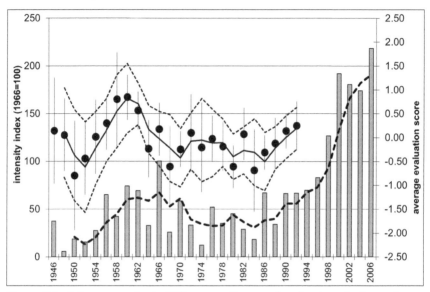

*Figure 7.3* Science news in the UK press: salience and evaluation of science; 'intensity' reports the number of articles (1966 = 100); 'average evaluation' reports evaluation, moving average and the confidence interval (+/- 1 SD): *Source*: Bauer et al., 1995. the figures of long-term intensity and evaluation of science news are currently being updated by project MACAS, mapping the cultural authority of science: see website http://macas-project.com/

similarly in the 1970s, slant is recovering when attention is declining. But since the 1980s, slant seems to improve in parallel to rising attention. On the basis of this data I was able to show the extent to which science news has become 'medical news': biomedical stories have displaced physics and engineering news; the popular papers have led this trend since the mid-70s (Bauer, 1998). Similar long-term trends can be ascertained for Bulgaria (see Bauer et al., 2006) and Italy (see Bucchi and Mazzolini, 2007).

The repertoire of event making for science has become larger and very innovative. New ideas proliferate in science weeks, festivals, cafés scientifique, pop science, street theatre, science-on-the-beach and the sponsoring of art installations. New is also the 'angel sector', which, initially a university spin-off, handles well this growing repertoire, though is increasingly absorbed by larger marketing and PR firms who thus expand their portfolio of services. Environmental and consumer organisations, and also patient organisations that support sufferers of particular diseases, have emerged as activists and platforms for scientific information. Finally, in a knowledge economy with a large high-tech sector, the differences between science communication and product marketing and reputation management are blurring (see Gregory and Bauer, 2003; Bauer, 2008).

Increased mobilisation at the boundary of science and society requires competent staff. Many universities, first in the US in the 1980s, then in the UK, later in France and Germany and other places, offer specialist degree programs in science communication, mainly for natural scientists on a career change into the communication world (see Gregory and Miller, 1998). With these programs come journals that encourage scholarly writings on the topic. The journal *Public Understanding of Science* was founded in 1992. These programmes expand de facto into public relations; students learn to organise media events, to cultivate media relations and to sustain a media presence for their employers and future clients (see Bauer and Bucchi, 2007).

Scientific outreach goes back to the 18th century, as an emerging historiography of popular science reveals (see Jacques and Raichvarg, 1991). Science outreach seems to come in waves: more activity between 1850 and 1880, again in the early 1900s to the mid 1920s, 1950 to the early 1960s, and after 1985 to the present (see Bauer, 1998 and 2012).

There is little research that maps the growth and diversification of the PUS mobilisation (e.g. Bauer and Jensen, 2011; Miller et al, 2002; Shinn and Whitley, 1985). There is a clear need to develop comparative tabulations of these efforts, develop a sound typology of activities, and report on the manpower involved, the motives espoused and on the financial commitments. Such information might revitalise an old UNESCO idea of creating global indicators of scientific culture, the SRAs or 'science related activities' (see Godin, 2005 and 2012; Godin and Gingras, 2000).[4]

## Two or More Cultures of PUS

The picture of public understanding of science that emerges from longitudinal and cross-sectional comparisons is more complex than many people think.

The picture on knowledge is fairly clear: economically more prosperous countries in Europe are generally also more knowledgeable of science; this holds over time and across countries. The trends on attitudes and interest are not straightforward, and depend much on the questions asked and the issues highlighted.

I have suggested considering this problem under the two-culture of PUS hypothesis. Not only might we have a two culture split between a literary and a scientific elite, but also the public understanding of science takes two or more recognisable forms (Bauer, 1995). The science-society issue is one of *two cultures of popular science as a function of economic development*. The question arises of how to characterise and frame this variety of public understanding of science. Let us consider two ways of doing this: first on the distribution and correlations of knowledge, attitudes and interest, and second on particular questions that tap in our understanding of the nature of science.

The deficit concept of public understanding of science—the more you know, the more you love it—is probably correct, but only within the bounds of a developing industrial society: interests drive knowledge, and literacy predicts positive attitudes to science, which reflects the high expectations populations might bring to science and technology to change their lives for the better, and to offer opportunities for the young. Scientific knowledge is unequally distributed, even with a low and high mode, some know a lot and most know little; knowledge is thus highly socially segmented. While in post-industrial, more high-tech and prosperous contexts, knowledge is normally distributed and social segmentation is lower; knowledge does not relate to attitudes; highly knowledgeable citizens hold negative as well positive attitudes, as do less informed members of the public. Interest in science is declining because much of it is taken for granted: what is normal does not need our interest and attention (see Durant et al., 2000). This hypothesis is consistent with the empirical observations of an inverted U-shaped relation between knowledge levels and positive attitudes across European countries and Indian states: across India, knowledge correlates with prosperity and higher scientific literacy. Across Europe, prosperity correlates with higher literacy levels, attitudes are negatively correlated: the welfare expectations arising from science decline with higher literacy (Shukla and Bauer, 2012).

A different way of framing this cultural variety is through items that tap into our understanding of the nature of science. Agreement with statements such as 'Science and technology can sort out any problem', 'new inventions will always be found to counteract any harmful consequences of scientific and technological development', 'one day science will be able to give a complete picture of how nature and the universe work', 'there should be no limits to what science is allowed to investigate' define an unrealistic image of science and adherence to a modern myth of science.

Figure 7.4 shows how level of knowledge, strongly related to prosperity of a country, is negatively related with adherence to a mythological image of science. Across Europe, it appears that economic modernisation does away with the myth of science. Less prosperous countries are more likely to adhere than more prosperous countries do. The modernist myth of science as omnipotent, without limits, self-correcting and approximating a true picture of world, has become a dubious ideology. In 2005 countries with higher levels of knowledge are more likely to reject the modern scientific myth (Bauer, 2006). Closer inspection shows that for 2010 the negative correlation of adherence to myth and level of scientific knowledge continues to hold across Europe; however, the gradient is less; the slope in 2005 is -0.18, for 2010 it is -0.05. This could means that while in some countries, at the lower end of the knowledge spectrum, adherence to a myth of science declines, in others, mainly at the upper end of the knowledge spectrum, the same myth asserts itself. We thus find a *re-enchantment of science* into the new millennium in places where the event making for science is most highly intense. It could well be amount to 'medialisation of science', an over-adaptation to the lure of public attention (see Weingart, 2012). The enlightenment effect of science might reach limits in enlightening its own operations.

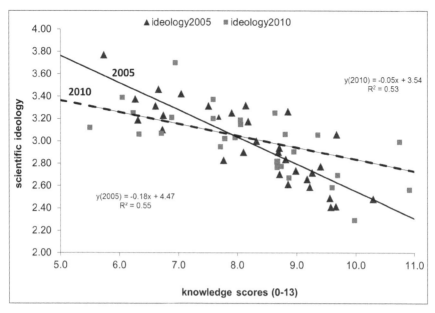

*Figure 7.4* Relationship between knowledge of science (score on a quiz of 13 items; M = 8.23; SD = 2.95; n > 15000; 2005 and 2010; dots correspond to country average scores) and 'scientific ideology' expressed in agreeing with 'science and technology can sort out any problem', 'new inventions will always be found to counteract any harmful consequences of scientific and technological development', 'one day science will be able to give a complete picture of how nature and the universe work', and 'there should be no limits to what science is allowed to investigate' (4 items, m = 2.97; SD = 0.82; 1–5, n > 15000). *Sources*: EC 2005 and EC 2010: EU27 + Iceland, Norway, Switzerland, Croatia and Turkey).

## FRAME B: RISK PERCEPTION

While literacy, public understanding or attitudes are the terms scientists prefer when dealing with public opinion, engineers prefer the term 'risk perception'. The 'vita contemplativa' looks for attitudes, where the 'homo faber' sees risks. Actions can fail (risk = probability x damage), and engineers are men and women of action. So the idea comes easily that we should relate to the world in 'risk' terms as we are all responsible designers of our lives. However, people differ in 'perception' of the same action. For some, a meteor hitting the earth is highly likely and they adjust their lives accordingly, while for others this has never even crossed their minds. Some people do not eat pork because it is risky (for health or for salvation), while for others it is a culinary delight.

Risk assessment and risk management are now common language. Investments are risky, as are nuclear power stations, but so are sports, food and relationships; sex is risky, and so is, it seems, getting out of bed. Life can fail and you can be blamed. The question is less whether this is correct than what changes if we frame our way of life in these terms (see Rothstein, Huber and Gaskell, 2006).

Risk is the 'scientific' form of premonition. Forget menace, danger, fate, or fortune. We calculate our future and thus control the not-yet. Astrology offered, but never achieved it. Responsible actors must consider failure, compare the odds of particular options and take precautions against litigation. It is now impossible not to think in this way. Risk management becomes blame avoidance: we design procedures to reduce risks and we follow them, so as not to be blamed if things go wrong (and they will). Risk management is *'blame avoidance reengineering'*: more effort is put into avoiding blame than prevent failures. Because of Murphy's Law 'if things can go wrong they will' the best you can do is avoid the blame (Hood and Rothstein, 2001). The actor faces risks, where observers see danger. The meaning of 'risk' shifted from something to embrace to something to allocate blame for. Risks are the 'sins' of modernity; without sins, there are only miscalculations.

## The Risk Society

The idea of the 'risk society' (Beck, 1992) has spread the notion of risk. Beck suggests an axial transition, with unspecified timing, into a society where the distribution struggle is no longer over wealth, but over risks: who bears the 'collateral damage' of large technological projects? Modern industrial society undermines its own foundations. Old 'godly' hazards such as earthquakes, floods and storms are recognisable as delayed consequences of human activities of 'fracking', deforestation and $CO_2$ emissions. But whose responsibility is this? Chernobyl and climate change are consequences of scientific-technological progress initiated long before any side-effects are visible. Two things are new: first, the undesirable effects of technology do not stop at the border of the country; nuclear fall-out from Ukrainian Chernobyl travelled to the shores of Lake Constance and to Cumbria in the UK. Second, traditional ways of copingno longer serve: the welfare state and its insurance systems are stretched to the limits, unable to absorb the costs. Who, then, will carry the bill? In Beck's scenario dread and danger will replace wealth as the fundamental social cleavage. Hence, risks are political: the tribe decides what to fear and how to police responsibilities (see Douglas, 1992).

A historical account of the risk society is offered by Luhmann (1998). Societies have three modes of absorbing contingency in human affairs. The first is *familiarity* with the universe based on myth and religion. Familiarity guarantees certainties in a taken-for-granted world. The unfamiliar makes us anxious, so we try to extend the familiar. The second mode is *confidence* in institutions that have no alternative. Without alternatives, the actor can only have confidence that the system works and is in good hands. We tend to confide in the police as we do not want vigilantes. Finally interactions with others involve choices and trust, and *taking risks* by trusting others. A course of actions can fail because others are not predictable. We take higher risks because we know more or we trust others. Trust and knowledge are functionally equivalent: they allow us to act. With little knowledge, we must trust more; but with knowledge trust is less needed. Modernity is burdened by risk and trust as the dominant modes of looking into the future. Confidence and familiarity are relegated to social niches of state monopolies and religious life.

## Risk Assessment and Risk Perception

The origin of risk assessment is gambling: how to bet your stakes when the outcomes are uncertain. This logic found applications in managing investments in ships sent to shores patrolled by pirates or in insurance schemes for afflictions like floods, crop failure, fire, illness or premature death, and more recently to calculate failures of technological systems, such as nuclear power stations or industrial plants. Insurances are ways of collectively absorbing the big failures of an individual by everybody making a small contribution.

This story of risk assessment is one of technical success and public failure (see Carlisle, 1997). Risk is technically defined as expected costs $p(c)$, i.e. the probability of a costly event $c(E)$ minus its benefits $b(E)$: $R = p(c) = p(E)*[c(E) - b(E)]$. This simple formula suggests equivalences between high probability events of low costs, and low probability events with high costs. The formula opens up a rationale for calculating a chain of events by coming up with a resulting probability. Risk assessment can identify weak spots in a system and open up courses of action: either to fix the problem or to avoid blame.

Observations on traffic behaviour shows that people do not respond to regulation as expected: *risk compensation* and *moral hazard* suggest that absorbing one risk leads to risk taking elsewhere. People seem to operate to a stable risk portfolio. Making working life safe makes us pursue risky leisure; safer cars make us drive more recklessly (see Adams, 1985 and 1995); bailing out one bad bank makes others neglect the necessary precautions.

One risk of 'risk management' is its take on public opinion. Over nuclear power, researchers noted that public fears did not match the expert assessments. Henceforth, expert–lay differences were investigated eagerly in analogy to psychophysics. It was assumed that there were *objective risks* and *subjective risk perceptions,* like physical sound pressure and psychological loudness; and these are in a lawful relation. It was found that the public overestimates low risks like 'nuclear reactors going critical', and underestimates high risks like 'smoking causes cancer'. The public deviates from experts on 'novel' and on 'dread' risks with grave consequences (see Slovic, 1987; Fischoff et al., 1978). The expert was objective, while the public was deficient: this is a matter for risk *communication*.

Baruch Fischoff (1998), a man of the first hour, reviewed the field with self-irony, in six stages. First, risk was seen as a matter of *getting the numbers right*. Technically correct assessment was needed: specifying hazards, measuring exposure and classifying the resultant risk level. But as public opinion deviated from these calculations, it was difficult to motivate policy solely on these assessments. Secondly, risk communication came in to *tell the 'true' numbers*: openness and transparency legitimised policy preferences. It became clear that openness can increase as well as attenuate public fears, and create ambivalence over the facts (Otway and Wynne, 1989). Thirdly, the matter became *explaining what the numbers meant*. Irrational humans bias information intake and are statistical illiterate, dealing poorly with probabilities. Others saw irrationality in the cognitive architecture: common sense uses heuristics rather than probability calculus, follows the law of small numbers rather than that of large numbers, using available information rather than systematic data collection. But to de-bias people by providing information and statistical training was unsuccessful. Fourthly, risk managers tried to *show to people that they have accepted similar risks in the past*. Contradictions between past and present risk taking creates cognitive dissonance, and that motivates people to change present behaviour. Eating ice cream is as risky (food poisoning) as a nuclear meltdown (i.e. the same burden of death and illness per year), and people have been happily eating ice cream, so might now also go for nuclear energy. This did not work very well either. Fifthly, it was then suggested that *deals might be offered*. Nuclear waste disposal sites 'bribe' communities by offering compensation. Events are framed as benefit rather than cost. This seems to work better, but it is still no magic bullet for aligning public opinion and expert assessment. Lastly, *people are made partners to the project*, with participation and community involvement. Here risk communication and science communication converge in the deploying technologies of humility. Risks are negotiated with all the stakeholders. What was initially an objective calculus turned out to be politics by other means: objectivity had lost its authority.

The field of risk perception is ultimately a dubious theory of dual cognitions (see Wagner and Hayes, 2005), and it thus attracted several lines of criticism.

The idea of 'perceptions' does not explain why people opine the way they do. Fears are unequally distributed in society, and situated. Nuclear power, 'Waldsterben', smoking, cloning, terrorism, extreme sports, vaccinations, pork meat and GM crops are worrisome for some while unproblematic for others. How to explain this variation; and how to deal with it? Framing the problem as one of irrational common sense deviating from expert rationality makes for an exercise, but is out of sync with reality.

The duality of expert and lay rationality is challenged by psychology. Developments of a theory of heuristics demonstrate that 'fast and frugal' judgments, traditionally known as 'rules of thumb', intuitions, gut feelings or yuck factors, are not residues of uncivilised barbarism, but evolutionary achievements. They are the human form of pragmatic rationality, and to be found among experts as well as lay people. Gigerenzer (2007) tells of a Nobel Prize winning economist who manages his own investment portfolio with the rule of thumb 'equal capital allocation' and not with his own prize winning complex algorithms. Or, the expert on decision making, faced with an option between two university positions, is unable to follow his own recommendations on how to solve such dilemmas. Gut feelings are not emotional stupidities but tested ways of reducing complexity. The choice is not one between rational and emotional, but rationality for different occasions.

Sjoberg (2003 and 2006) reviewed the research and took issue. The perceiver model seeks to explain perceptions of vaccination, nuclear power, GM crops, alcohol and smoking by choice, trust, dread and novelty of the issue. However, these factors explain very little variance. Replications of results show the semantics of 'risk' rather than a valid model. It is conceptually confused to assume that fear drives perception; more likely we fear what we see, hear or smell. Sjoberg finds that the main meaning of risk perception is 'tampering with nature', an illicit transgression of the natural order of things. The psychometric approach is successful in producing research and defining policy levers not because of evidence, but because of its simplicity that resonates with existing prejudice. Claiming that risk perceptions deviate from expert assessment because they are based on affect is consistent with widespread notions of public opinion as 'emotional' and therefore second rate. While this technocratic attitude might prevails among experts, the global insurance industry lead a way out. They came up with the concept of *'phantom risk'*, for example over the ionizing radiation of mobile phone systems, which despite the absence of scientific evidence, could incur costly litigations or affect the housing market. Risks do not have to be scientifically evidenced to have economic impact (Swiss Re, 1996).

Culture theory takes issue with the psychometric model from a different position (Douglas, 1992; Wildavsky and Drake, 1982). Risks are not individual's affair. Modelled as aggregates of private judgments, we ignore some people, take others into account, consult and advise, persuade and mobilise common beliefs and values. The model is biased by a *cult of heroic individuals* who perceive risks purely and not polluted by politics and other emotions. But risks reflect a common way of life; they are ordered for attention; most are totally ignored. Vaughan (1996) showed how the 'causes' of the *Challenger* explosion of January 1986 were traceable to an expert consensus on a non-risk ('the O-ring is secured by redundancy'). Within NASA mentality at the time, nobody could have reached a different conclusion. A 'risk perception' different from expert assessment is not the case. The correlation between O-ring failure and low temperatures was only obvious in hindsight, even to experts. Complex systems inevitably fail; we imperfectly and selectively monitor signals against much noise (Perrow, 1984). We ignore the unknown unknowns which cause trouble. Risk hierarchies are thus tribal markers. Not eating pork is an identity matter for a Jew or Muslim, but not for a Christian or Buddhist. Equally, a kamikaze pilot and the mountaineer do not calculate risks, but take them because of who they are and the way they live their lives. It is important to understand the overall mindset. This act is not explained by its voluntariness, obedience to authorities, familiarity with dangers, or low dread of consequences, but rather by the samurai ethos of the kamikaze, and the lure of the flow-experience in the case of the mountaineer. After all, risk is only one among many ways of premonition (Duclos, 1994); the hegemony of risk talk needs to be challenged against alternatives. Might it be that a culture of litigation, rather than increased safety, is the main outcome?

In this chapter I have made three points so far. First, I distinguished between the public sphere, public opinion and public opinion measurement. Conflating these notions blinds us from understanding public opinion as both process and input into modern techno-science. Secondly, I reviewed two concepts by which social sciences tried to be helpful and innovative: by framing *recalcitrant public opinion* either as public understanding of science, more for the contemplative scientist, and as risk perception, more for the design minded homo faber. Both meta-frames in that sense are implicit and *operational theories of public resistance*. Both traditions converge in the advice to build *community*, to establish procedures by which a common understanding can be achieved. And this requires recognising that public opinion is neither hysterical, nor malignant, nor ignorant nor otherwise deficient, but a mediator without any fait accompli. Edmund Burke was right: one should neither seek to control nor follow public opinion, but encounter itrespectfully. In the remainder of this chapter, I will explore the risks of trying to make public opinion an ally of one's dreams.

## SOCIAL MOBILISATION AND ITS RISKS

Techno-scientific developments increasingly shape up as if they were social movements. Nuclear power led us into the atomic age promising 'energy too cheap to meter'. Information technology 'revolutionises' the way we make war and peace, communicate, work, memorise and relate to self and others in the information society. Biotechnology experiments with 'nature' and disposes of our moral foundations. Just around the corner are nanoscience and synthetic biology, setting out to re-engineer matter and life from scratch, not yet, but maybe soon, leading us into a 'synthetic nanoworld' of endless enhancement of human capabilities, and all for the better.

## Why and How is Techno-science a Social Movement?

A social movement is a *network of people* identifiable by *references* in opinions and beliefs that *express preferences* for social change (see; Ahlemeyer, 1989; McCarthy and Zald, 1987, 20; Tilly, 1978, 9). In our case, the common reference is 'new technology'. Social movements capitalise on a *low level of formal co-ordination*. Social movement organisations have cadres, staff and workers like a small firm, but pull in much voluntary contributions. Such organizations fall between fads and fashions, riots, crowds, and milieus with no formal membership and political parties and clubs with defined membership. A club operates with an already mobilised and fee paying membership; social movements mainly have potential members (Useem and Zald, 1987, 275). When the club organises a recruitment drive, it becomes a bit more movement-like. Movement organisations maximise mobilisation by all available means including show times, civil disobedience, disruption and even violence; this action repertoire is a defining variable. A movement seeks to mobilise public opinion and make it an ally to its cause.

Science and technology seen in this light is a movement raising the world from a moral vision (on technological visions: Dierkes, Hoffmann and Marz, 1996). It mobilises people into pulling the same rope, though not necessary in the same direction.

One might argue that science has become a societal institution, and is no longer a social movement. Researchers in industry, government or universities are members of *academies* such as the Royal Society of London, or of *learned societies* of physics, chemistry, astronomy, biology, history or psychology. These peer organisations are based on *formal membership*, clubs often by invitation or by subscription, and renewed by annual fees, election of committees and participation at 'tribal' gatherings. However, in a historical perspective these institutions were once dynamic; change is relative to a time frame. After all, there was a 'scientific revolution' that mobilised resources and liberated knowledge from the shackles of authority, traditions and religion; at least that is how the story goes.

However, science as a social institution variably exerts authority over social affairs; it influences policy, secures the allocation of resources with variable success. The 'fifth branch' of governance adds to the executive, legislative, judicative and the mass media (see Jasanoff, 1990). For sure the history of science is complex and contorted. For our purpose it is important to recognise that the authority of science has been challenged and public opinion is therefore a target of intensified outreach either over public understanding or over risk perceptions.

The challenge is twofold: on the one hand rising environmental concerns maintain that science and technology is part of the problem, safety and liability are consumer concerns when things are designed for obsolescence, and disasters such as Thalidomide, TMI, Chernobyl, Fukushina, Challenger, Bophal, Seveso and Basel undermined the belief in scientific and technological progress. And when the US Congress did not approve the Supercollider Project in 1993, the physics community went into a state of high alert. On the other hand, the authority of science is challenged by utilitarian expectations in politics and business. More private patronage shall lead to a more productive and more useful science (see Mirovski and Sent, 2005; Krimsky, 2003; Etzkovitz and Leydesdorff, 2000).

Social movement research predicts that institutions resort to mobilisation when their influence on policy is jeopardised. Useem and Zald (1987, 275ff) showed how the nuclear power complex, in the 1960s still an integral part of energy and defence policy, started to seek public support in the 1970s when anti-nuclear protest eroded their grip on policy. What emerges is an industry of *quasi-social movements* that seek grass-root support with evolving action repertoires. Techno-scientific agencies are learning to mobilise effectively ever since, but need to face the unintended consequences of good intentions.

## Thesis: Resistance is Functional, Because It Points to the Risks of Mobilisation

For an institution, to refocus on external communication is not without risks. For example, professionalisation brought the green movement into a dilemma between effectiveness and credibility. A slick and successful campaign can undermine public support. Equally, science communication has faced sceptics before. Traditional scientific elites consider the extension of knowledge a wasted effort.[5] Beyond quaint elitism, scepticism might be reasonably based on empirical observations of the unintended consequences (Merton, 1936). Communication is necessary to extend scientific research to peers. However the *professionalisation of science communication beyond peers* has implications that might come to haunt the designers (Bauer, 2008).

### Risk 1: Getting Bad Press That is Out of Control

Social mobilisation and outreach might create more bad than good news, either because it is badly done, or because mass media work in a way that is out of scientists' control. Seeking attention might include surprises when the gaze of the observer is not controlled.

Media coverage accentuates opposition because it thrives on controversy. It confers public status for those in the limelight and can legitimise their effort outside the sphere of science. As the only bad publicity is no news, visibility seeking might enhance the status of the wrong actors, and not those that deserve prominence (from the point of view of core science).

Finally, publicity for a particular actor opens the field for opponents to enter. Without that presence there would be no pretext. With publicity comes the challenge. Visibility is oxygen, and it can suffocate. Social mobilisation invites counter-mobilisation.

### Risk 2: Determining Leadership Through Prominence Rather Than Reputation

Weingart (1998) pointed out that science traditionally determined its leadership by the *reputation* gained in peer-reviewed publishing. Increasingly, science coverage in the mass media opens a second route for the scientific career whose credential is based on *prominence* through publicity and visibility. Social mobilisation poses the dilemma between reputation and prominence to regulate scientific careers. Shortcutting the peer-review process and feeding results to the mass media before formal publication is increasingly likely as is science by press conference. There is clearly the possibility that prominence could become a functional equivalent of reputation to determine the hierarchy of prestige in science. Alternatively, prominence might be a pathway to enhance a scientific reputation. Apparently, coverage of one's research in the *New York Times* in one year will increase one's citation counts over 10 years by 72% (Phillips et al., 1991). In that case, the effective press officer is an essential requirement for any ambitious researcher, not to seek prominence, but to build reputation (if measured by citations) in the long run. Prominence could then also become a predictor, and as such as proxy for later scientific reputation.

### Risk 3: Hype and Frustrated Expectations

Seeking public attention might invite exaggerated claims making. Hyping research results has become routine in the form of the 'elevator brief' or the Pecha-Kucha (20 slides, 20 seconds each) summary of the content and likely impact of research. Scientific journals produce press releases that attract the attention of the media to the upcoming production. In 1997 a now famous paper on 'adult nucleic transfer' was press released as 'cloning' and 'Dolly the sheep' by *Nature* to huge effect in terms of subsequent news flow (see Chapter 8).

Pre-publication hypes like the one over cold fusion can be risky for researchers and detrimental for the community. The sensation stalled much of the science at the time and directed researchers to reconstruct a false claim, badly documented by the media; a claim that could be easily challenged in the peer-review process (Weingart, 1998).

Catastrophe scenarios of a meteorite colliding with the earth attract public attention. However, such claims might lose credibility once they are seen to be self-interested, i.e. the astronomy community seeking to create a new funding opportunity for mapping the stars by sponsoring a scare story (Mellor, 2010).

The emerging drive for publicity might also be at the root of particular cases of scientific fraud. Individual researchers might entangle themselves in a media personality invested with high expectations. Failure to produce results might push this individual to resort to fraudulent evidence. It remains unclear whether fraud cases are on the increase. However, in medicine, where the stakes seem higher, fraud seems more likely (Franzen, Roedder and Weingart, 2007; Giles, 2007; Grayson, 1995).

Hype creates an exaggerated discourse of expectations. Cures to pressing problems such as global warming or crippling diseases are imminent. Frustrations set in when these 'promises' are not fulfilled that undermine the authority of science. What happens when prophecy fails is an age-old question (Festinger, Riecken and Schachter, 1956). Theology dealt with it under the name 'Parousia delayed' (i.e. the second coming of Christ postponed) which created institutions of the Church to manage these expectations. Similar issues are lately revived as the sociology of expectations (Borup et al, 2006) to explore this risky feature of new technology management.

### Risk 4: Over-powering the Mass Media with Professional PR

The anxiety over reputation leads to the professionalisation of science communication. Research institutions and universities, research centres and even research projects, increasingly employ specialist public relations officers. This is often a lucrative career move for science journalists who know the beat but whose livelihood has become precarious. As a consequence the balance of power between journalists and PR agencies is shifting, and independent science reportage is in jeopardy. The watchdog function of science journalism which emerged in the 1970s is threatened (see Franklin, 2007; Goepfert, 2007; Brumfield, 2009).

### Risk 5: Inadvertent Repercussions for the Conduct of Scientific Research

Attention seeking might have repercussions for the production of scientific knowledge itself. The closer coupling of science and the media jeopardises science's historical achievements of creating certified knowledge through the 'organised scepticism' of peer-review. The competition for public visibility traps scientists in the logic of 'news values'. As public relations becomes a strategic issue for scientific research, the struggle for attention leads to adaptations not only at the point of presentation of results (i.e. information design), but upstream, at the point of formulating the research questions (epistemic influence). Projects are planned and designed with a view to raise public attention in order to generate impact and funding. Symbolic research for attention rather than epistemic progress is likely to increase. Peter Weingart (2012) talked of the 'lure of the media'.

### Risk 6: Making Public Commitment and Creating Learning Difficulties

A project that makes public commitments by raising expectations creates a lock-in on itself. Public commitments bind actors to the promise under risk of losing face. A project ties efforts and attention to a promised solution to a problem. This solution might become part of the problem though.

Sloterdijk (2005) evocatively charts the modern history of project making as the age *of disinhibition*, of unfettered initiative and destruction, of rushing and speeding, of expansion, of license and disregard for collateral damage. A privilege of impunity arises from the suspension of all inhibiting constraints.[6] Such enterprises amount to a *'nautical ecstasy'*: the typical philosophy of captains such as Columbus, Magellan, Cook, on sea, and Cortez, Aguirre or Pizarro on land. In pursuit of El Dorado, the land of unlimited riches, misconceptions of the world are recognised only in hindsight, but we admire the adventures, and the poets sing. In the expeditionary corps, mass psychosis offers immunisation against distraction and doubts; the need for neither exit nor voice arises. As the stories of these expeditions show, to avoid mutiny required a constant effort of symbolic mobilisation.

Social mobilisation creates *collective forms of autism*, highly intelligent pursuits, but blind to the concerns other than the immediate project. All feedback only enhances the effort through selective attention. Failed prophecies are met with increased mobilisation and shifting horizons: i.e. we need to work even harder (see Festinger, Riecken and Schachter, 1956). The nautical ecstasy assembles cognitive and motivational immunisation against *learning*; it becomes impossible even to want to and to be able to consider alternatives. The mission for a particular future blinds towards alternative futures.

Tool making is an anthropological constant compensating for a lack of genetic inheritance and the resulting anxieties (see Gehlen, 1953; Blumenberg, 1990). History charts the magnification of these means to enhance the range and control of human activity. Science as an ideal emancipates itself from the practical concerns and seeks purpose-free knowledge. But it remains tied up with in *techno-scientific mobilisation* to overcome actual grievances. Though children of the Enlightenment, *science and democracy* are not twins, which leaves us with the problem of *avoiding a scientism* in pursuit of the common good. The certification of knowledge remains an elite matter of the republic of science, and probably will remain so, because functionally required. Democracy enfranchises the people to translate public preferences into legitimate decisions. Thus a functionally 'elitist' science needs to learn from the *resistance of the public* in the mode of an *'authority that learns'*. Techno-science faces the recalcitrance of common sense, as this is playing out over nuclear power, information technology and genetic engineering and other emergent technologies. Prometheus is rebound, and the primacy of initiative has returned to the *primacy of resistance*. The reader might now want to reread the earlier chapter 5 on resistance.

## NOTES

1. This saying is attributed to Edmund Burke; however, I was not able to trace the exact source.
2. I will not rehearse the confused polemic regarding the 'deficit concept' in public understanding of science; see our comment on this at times misguided debate in Bauer, Allum and Miller, 2007.
3. The phrase 'talking dialogue and walking monaude' I take from Steve Miller (personal communication).
4. The Royal Society of London sponsored a workshop in November 2007 'International Indicators of Science and the Public' dedicated to the creation of measures of 'science culture' with a global reach. The meeting was attended by 35 researchers from North and South America, Europe, Africa, Asia, and Australia (see Bauer, Shukla and Allum, 2012).
5. An anecdote: on the launch of the journal *Public Understanding of Science* in January 1992, the founding editor, John Durant, received a beautifully handwritten note stating that 'gentlemen do not publish, they engage in conversation'.
6. Sloterdijk's language develops this mindset of conquest in vivid images and nearly untranslatable word cascades: e.g. the disinhibited auto-persuasion of action; speeding into the world like a bullet out of a canon; projectile existentialism gaining identity in the trajectory; from reflection to projection, etc.

# 8 Genes, Biotechnology and Genomics

Genetic engineering and biotechnology bring this overview of three postwar strategic technologies to a close. The genetic engineering 'revolution' is very much ongoing. People have been working, witnessing, living and disputing the genetic engineering of microbes, plants, animals and humans, whole or in parts, for the last 40 years. It remains history in the making. In this chapter I address three questions:

- How did the biotechnology movement come about?
- When, how and over what issues did public resistance arise?
- What was the response of the biotechnology community to this resistance?

The history of public debates over genetic engineering shows the formation of the *biotechnology movement* including promoters and opponents. Two watershed events, GM crops arriving in Europe in 1996, and Dolly the sheep in 1997, created a global debate, from remnants of local controversies that were much older. Shifts in framing and public opinion created a new strategic situation to which many protagonists of biotechnology had to adapt. Biotechnology is a good example of how resistance matters and makes a difference.

## LONG PAST AND SHORT HISTORY OF GENETIC ENGINEERING

The history of genetics is a specialist matter. For our purposes it will suffice to set few important markers. Since the 1970s biology has increasingly replaced physics as the lead science, both at the lab bench and in public attention. But concern over heredity is as old as society, and older than the scientific take on it. Ideas of characteristics passed from generation to generation through 'blood-line' informed kinship and legal notions of inheritance, improved farm crops and animals by selective breeding. The history of replacing 'blood' by 'DNA' (deoxyribonucleic acid) as the locus of inheritance is considered a triumph of 20th century science (see Appendix 2).

The story centrally involves British naturalist Charles Darwin and Austrian monk Gregor Mendel and many others, who demonstrated natural selection and the laws of inherited characteristics. The locus of inheritance was later identified in chromosomes, genes, DNA, the double-helix structure and the linear model DNA—>RNA—>Protein. The 'four-letter code' of A(denine), T(hymine), C(ytosine) and G(uanine) had become amenable to deliberate splicing and recombination by the early 1970s (recombinant DNA or rDNA). What for the atom was the demonstration of fission, for genetics was the recombination of genes, key moments in history.

This history of biology is marked by controversies: mechanism or vitality; reductionism or holism; evolution or design. Mendel's laws of inheritance were ignored for years; the eugenics movement tried to solve social problems by genetic selection and finally selective extermination (see Kevles, 1985); Lyssenko's quest for inheritance of acquired characteristics was a dead end, but pleased Comrade Stalin; Rosalind Franklin's contribution to the double-helix model was side-lined. However, public attention to all this is scant, until the double-helix became an icon of our time (see Chadarevian, 2003).

## Semantic Issues: Biotechnology, Old or New

This history is often a battle over metaphors. A term can affect public attention, funding streams, and legal regulations. Excursion 8.1 lists terms which have served as markers for particular angles on the development of genetic engineering (see Bud, 1991). Indeed, only in the late 1970s was biotechnology associated with genetic engineering, applying technology to live forms to further humanity. In 1976 the European Federation of Biotechnology (EFB) was founded, led by the older German Association of Chemical Industry (DECHEMA), to unite under a new banner the biochemists, chemical engineers, and microbiologists. In 1979 an American stockbroker, banking on rDNA business opportunities, took out a trademark on 'biotechnology' to protect its meaning as 'industry based on genetics' (Bud, 1991, 441).

*Excursion 8.1* Naming the New Development (Bud, 1991)

- *Microbiology (late 19th century)*
- *Eugenics (>late 19th century)*
- *Cytogenetics (>1880s)*
- *Biotechnik (>1910)*
- *Molecular biology (>1938)*
- *Human genetics, a new name for 'eugenics' (>1954)*
- *New eugenics, the privatised eugenics (>1962)*
- *Genetic engineering (>1965)*
- *Genetic modification (>1965)*
- *Recombinant DNA, rDNA (>1973)*
- *Biotechnology (>1980)*
- *Genetic enhancement (>1985)*
- *New and Modern versus old and traditional biotechnology (>1990)*
- *Life sciences (>1990)*
- *Genomics and 'omics research' (>2001)*

For genetic engineering, it mattered whether it was a break with history or continuity. Words mobilise and demobilise. Liberal warriors against over-regulation prefer continuity: genetic modification has been done for thousands of years; existing laws of patent, licencing and liability cover the matter. Continuity justifies assimilation. On the other hand, proud innovators stress the revolutionary novelty of genetic engineering. This attracts attention, funding and investors. But those concerned with dangers and uncertainty will raise alarm and call for containment in rules and regulations. Biotechnology, like other mobilisation efforts, tended towards hyperbole to seek attention at the cost of public alarm. This rhetorical dilemma of novelty and alarm became the preoccupation of the 'sociology of expectations': mapping the risks of making techno-scientific promises (see Borup et al., 2006).

The dilemma could easily be observed in the rhetoric of protagonists, who in the legal arena stressed continuity with breeding, brewing and baking, while for business investment stressing the revolutionary aspects of 'biotechnology'. Novelty is the news value; continuity is not. In 1984, the OTA tried to settle the ambiguity (OTA, 1991): old or traditional biotechnology refers to interventions at the level of the cell such as fermentation and is the remit of industrial microbiology; new or modern biotechnology shall refer to interventions at the level of the gene, to alter or transfer genes from one cell to the other and is the remit of molecular biology using recombinant DNA techniques. The OECD continued this struggle over statistics. As the biotech sector became an attractive place to be, it became necessary to distinguish 'true' from 'spurious' biotech activities (see OECD, 2005). Different definitions lead to different estimates of the size of the biotech sector. The academic journal *Advances of Genetic Engineering & Biotechnology*, launched in 2012, defined by listed its topics as: transgenic organisms/genetically modified organisms (GMOs); stem cell research; recombinant DNA technologies; gene therapy and autoimmune diseases; human genome; and medical ethics, thus listing most controversial topics that arose in modern biotechnology.

## MOBILISING FOR AND AGAINST GENETIC ENGINEERING

A new technology is a gigantic resource mobilisation of symbolic and material capital, and biotechnology is a good case in question. There are two targets to social mobilisation. On one hand there is the state, its policies and laws. Actors compete directly for capture on national and international policy and regulations to control the new development. On the other hand actors compete for public attention. The public is an indirect lever upon government expected to be 'responsive' to public opinion. In the biotechnology debate, regulation was known as the first hurdle, and public opinion as the second. I will not deal with the regulatory frameworks, their national and international dimensions. This complex matter is well documented elsewhere (see Cantley, 1995; Galloux, Gaumont and Stevens, 1998; Jasanoff, 2005). I will focus on the second hurdle, public opinion and its dynamics, which, in the short history of biotechnology, was for many protagonists something of a 'surprising discovery'.

## The Scientific-Industrial Complex

By 2006, the global biotech industry comprised 4,275 companies, of which 17% were quoted on the stock market, 1,465 operated out of Europe, 1,116 out of the US, 601 out of Asia and 363 out of Canada.[1] The sector generated revenues of $74 billion with research and development (R&D) in the region of $29 billion (Lawrence, 2006). A quarter of a million researchers were employed in biotech R&D in 21 countries, and 6,000 patent applications were filed per year in Europe alone (Beuzekom and Arundel, 2006). Biotech R&D was 87% on health (therapeutics and diagnostics), 4% on agriculture and food and 9% on other applications such as environmental remediation (OECD, 2006). For this triple R&D the colour code of 'red', 'green' and 'white' biotechnology has become common parlance. Though, the biotech sector concentrates overwhelmingly on the global health business.

Most biotech firms were small and R&D intensive, having started in a counter-culture similar to that of the IT sector (Vettel, 2006). Founded by academics, Genentech (founded 1976) and Biogen (founded 1978) became leading biotech firms by engineering insulin for the large market of diabetics; others followed their lead to genetically engineered therapeutics. The US Supreme Court decided, with a narrow margin, to grant patents on life forms that were human inventions, but admittedly more to akin to chemical compounds than to honeybees (*Charkabarty vs. Patent Office 1980*; see Jasanoff, 2005, 48ff). The early 1980s saw the growth of university spin-off companies in the US and elsewhere, which changed the relationship between research and commerce. The first issue of *Nature Biotechnology* (by-line 'the science and business of biotechnology') in 1983 and the annual *Ernst & Young Reports on Global Biotech*, since 1986, testify to this take-off period, as do biotech investment funds and the stock indices which track the biotech field since the 1990s.

Private patronage of techno-scientific research is not new. The state-patronage of post-WWII, as part of the Cold War effort, is the historical exception (Dorn, 1991). Nonetheless the normalisation of business-university partnerships, supported by state funding, tax regimes and patent systems (e.g. Bayh-Dole Act, 1980), penetrated the world of research to such a degree that observers feared corruption by profit motives and the erosion of disinterested competences (see Krimksy, 2004, 177ff). University-to-business technology transfer became global policy (e.g. OECD, 1992). *Genentech* was the first to go public among unprecedented enthusiasms in 1980 (Rifkin, 1998, 41ff). Many others went public in the 1990s and created a biotech boom adding to the millennium bubble (Haber, 1996). Modelled on Silicon Valley, governments encouraged regions to become 'science parks' next to university laboratories. Thus Europe presented its BioAlps, BioValley, BioRhine, MediconValley, Oxbridge-London or Stockholm-Uppsala etc. By 2006, biotechnology had become a new global sector. The industry lobbied governments and projected its public voice. EuropaBio did in Brussels what the Biotech Industry Association (BIO) did in Washington: monitor government and public opinion to ensure the sector's interests are served.

As old industries of chemicals, foods and pharmaceuticals saw new opportunities, start-up companies strapped for cash collaborated or merged with established firms. For example the old Swiss firm Roche became a majority shareholder of Genentech in 1990; thus the business-research networks came to characterise the modern biotechnology movement (see Powell, Koput and Smith-Doerr, 1996).

*Excursion 8.2* Monsanto's Life Science and the North Atlantic GeneRush

*For a while the biotech movement was fuelled by the 'life science' vision of which the US company Monsanto was the leading voice calling for famers, chemists, butchers and bakers to unite. In the early 1990s, traditional agro-chemical companies specialising in fertilisers and herbicides envisaged a new industry that brings together food production from farm to fork to create 'novel foods' with added, even pharmaceutical benefits. Chemical, food and pharmaceutical companies strategically positioned themselves in this vision where production of crops, animals, foods and medicines would all merge based on rDNA and cloning technology (pharming). Monsanto, Novartis and others started to buy seed and plant research companies. Patents were sought for genetically engineered seeds. In 1997, Monsanto acquired Calgene, who two years earlier had marketed the first GM food product: the Flavr Savr tomato with longer shelf-life (Martineau, 2001).*

*Monsanto's initial biotech success was built on genetically engineered soya seeds 'Roundup Ready', into which a bacterial gene was inserted. Approved in the mid 1990s, this seed survives the broad-spectrum, glyphosate based 'Roundup' herbicide sold since 1973. Seed-cum-herbicide is the new business model for soya, sorghum, canola, alfalfa, cotton and wheat. Farmers of staple crops are encouraged to buy both seeds and chemicals from the same source, which thus creates loyalty as well as royalty. Similar plant designs appeared for insect resistance (e.g. BT cotton) and enhanced nutritional values (e.g. golden rice). The claim was that genetic engineering is the key to food security.*

*Another milestone of genetic entrepreneurialism is located in Iceland. In 1996, Kari Stefansson, a geneticist, founded deCODE GENETICS with the help of US venture capital. Iceland has a homogenous population of 250,000, an obsession with family trees dating back 1,000 years and a national health database covering the entire population from birth to death (see Specter, 1999). These records and a newly created genetic database were to be joined into a biobank business model that traces complex associations between illnesses and genetic factors, heart issues and schizophrenia being among the first targets. deCODE presents itself as* 'empowering prevention' and markets personalised genome scanning and genetic tests for common diseases (see www.deCode.com; accessed 15 August 2008).

By 2008 deCODE employed over 400 employees, achieved $40 million of sales and conducted R&D of $54 million (Lawrence and Lahteenmaki, 2008). It promised to transform the Icelandic economy, traditionally dominated by fishery, into a knowledge society based on genetic research. However, granting a monopoly over the country's databases was controversial. Icelandic society needed to secure a stake in the profits, and debated the issues of consent to use personal information for commercial purposes. A licence was granted for 12 years initially, and included an annual fee with a fixed and a variable element. Icelanders can opt-out and withhold their data, but the default position is presumed consent. About 10% of the population opted out after heated public debates (see Thorgeirsdottir, 2004). The firm struggled with the stock market crash of 2000, when its stock value evaporated from $1.8 billion to a fraction of that. Nonetheless, Iceland started rolling the veritable biobank bandwagon competing for capital, volunteers and researchers with favourable conditions and high quality data: genetic biobanks were in discussions for England, Scotland, Singapore, Finland, Estonia and Tonga in the Pacific (see Arnason, Nordal and Arnason, 2004).

## State Sponsoring: The Human Genome Project

Partly motivated by concerns about what might be left undone by the market, during the 1980s a project for genetics was envisaged modelled on the man-on-the-moon or Manhattan project. The large co-ordinated research effort under state patronage became known as the Human Genome Project (HGP). The idea was to identify the entire human genome and sequence the billions of basic letters of A(denine), G(uanine), C(ytosine) and T(hymine). Particular segments of this sequence link to visible organic traits. To identify these links is to open the book of life. Far from uncontroversial, the politics of funding and methodology—the US Department of Energy (DoE) advocated 'sequencing', whilst the National Institutes of Health (NIH) favoured 'mapping' the genes—is well documented (Cook-Deegan, 1994). In an attempt to deal with massive amounts of data and to speed up the project through division of labour, the Human Genome Organisation (HUGO) was set up in the US, UK and Japan. Funding came on-stream in 1990 in the midst of criticism of 'big science'. Many feared that biological research would be the monopoly of genetics, and the involvement of the DoE, home of atomic secrets, raised suspicions as a scheme of soon unemployed bomb makers. There was also resentment among physicists. Their Superconducting Super Collider Hadron Project was cancelled by the US Congress in 1993. These concerns were partly addressed by separating HGP funding from normal NIH funds, to avoid competition with other research, and by earmarking 5% of the budget for the study of its ethical, legal and social consequences (Watson, 2004, 172f) known as ELSI. With funding of over $150 million, ELSI created the bioethics boom, the biggest employment scheme for philosophers since Plato's academy.

Once under way much progress was made to speed up gene mapping and sequencing and by developing algorithms to sift through data and identify meaningful segments. Informatics and biology merged into Bioinformatics. But the public-private dilemma could not be avoided. In the late 1990s the HGP was a race between private Celera and the public Sanger Institute in Cambridge (UK), funded by the charitable Wellcome Trust and dedicated to public genetic information. Craig Venter personified the entrepreneurial spirit of the HGP. His venture company sequenced and patented large numbers of sequences, the functional significance of which were not yet defined. Celera Genomics was a factory with several hundred automatic sequencing machines, using public data for private purposes. Notoriously, Venter sequenced his own personal genome in 2007.

On 26 June 2000 a media event was staged where President Clinton in Washington and Prime Minister Blair in London jointly proclaimed the completion of the HGP. Knowing the 'language of God', Clinton declared, human kind is gaining immense new powers to heal. But, the main surprise was the count of human genes, 72,415, much smaller than expected. British magazine *Prospect* gave a copy of the human genome on CD with its October edition (October 2000). The project had ended earlier than planned, but sequencing and mapping were far from over. Sequencing time was reduced from 13 years for the HGP to just 4.5 months for Watson's own genes (Wadham, 2008). Efforts multiplied to link disease symptoms to biomarkers in a veritable genome mining rush. Prospectors rushed into the Human Diversity Project and the HapMap Project I and II (IHMC, 2007; HGSVWG, 2007), and into the 1,000 Genomes Project (launched January 2008) to secure robust statistical associations between biomarkers, and other risk factors for common diseases. Large biobanks of genetic health and lifestyle information of thousands of people are sifted for 'susceptibilities'. The concept of 'susceptibility for X' is key to a new preventive medicine that is based a diagnosing pre-symptomatic disposition with a certain probability of developing symptoms sometime in the future (Bowcock, 2007).

Global biotechnology emerged under the mixed patronage of corporate and state sponsors who invested both money and high expectations. But an important third constituency in this game is civil society in many countries around the world. Civil society has produced a heterogeneous set of voices that reference biotechnology. Some groups are newly formed on the specific issues of biotechnology; others existed before and diversify their portfolio of engagement. Indeed, it is one of the features of technological controversies that structures built up in past conflicts become relevant in new controversies. The increased sensitivity provided by these activist groups is one form of collective learning that arises from controversies. The biotechnology controversy is a good example of this state of affairs.

On the side of the protagonists of biotechnology, we often find patient groups who organise sufferers of particular illnesses and their families. The internet helps to find each other, often across national borders. Through internet platforms patient groups provide information, broker research results, and offer support that complement medical care. On genetic diseases, some of which are very rare, organisations such as the UK's Cystic Fibrosis Trust or the Muscular Dystrophy Campaign operate also as fundraisers for research. HIV/AIDS activists pioneered this form of mobilisation with great impact on directing research efforts through acquired expertise and intensive lobbying (see Epstein, 1996). For example, in the UK, where the approval of drugs for the NHS is decided centrally through National Institute of Clinical Excellence (NICE), patient groups often champion new drugs, challenge negative decisions and project their 'voice of hope' on particular issues of genetic testing, therapy and stem cell research. Not least because of their power to put pressure on drug licencing, patient groups caught the attention of the pharmaceutical industry, which supports such activism that links the hopes of sufferers invested in science and the new bio-business (see Novas, 2006). Others draw attention to the concepts of business-friendly international NGO (BINGO), government operated NGO (GONGO) and international NGO (INGO) to highlight the boundaries between corporate PR, state propaganda and civil society activism.

## Actors of Counter-mobilisation

Environmental groups such as Greenpeace and Friends of the Earth are interested in the implications of biotechnology. The release of GMOs into the environment, initially through field trials and later through large-scale planting, carries risks for biodiversity. With major regulatory frameworks in place for the release of GMOs both in the US and the EU by 1990, defining the procedures and type of assessment that had to be undertaken, it was not clear how far biotechnology was an issue that allowed for successful public campaigning. Apparently, Greenpeace had agonised in the early 1990s over taking on the issue of GM crops and food. The argument could be made that genetic engineering reduces the amount of herbicide and insecticide needed to raise the crop and thus the burden of pollution. The arrival of GM soya in autumn 1996 in Europe would make all the difference in opportunity. How to define risks became a major global battleground in the 1990s. Is this a matter of sound science and closed committees, or a matter of wide public perception and involvement? Should the criteria be limited to, for example, the nutritional equivalence of GM and traditional crops, or should it consider a wider range of concerns, such as biodiversity and agricultural policy?

*Consumer groups* became a biotech reference by actor during the 1990s. Consumer groups took up the issues of food safety, the potential health effects of GM food for those who consume it, and of consumer choice and thus food labelling: how shall food stuff containing GM ingredients be labelled (if at all) to be able to make an informed choice? Consumers should be able to make up their own mind.

Consumer concerns got a high profile in the late 1990s, when the coalition of environmental and consumer pressure convinced major European food retailers such as Carrefour in France, Sainsbury or Tesco in the UK, to avoid or clearly label GM ingredients in their products. This policy was in marked contrast to what was happening at the other side of the Atlantic, where the roll-out of GM crops and food was in full motion. The 'transatlantic puzzle' over biotechnology asked: why is it a non-issue in the US, but a problem in Europe?

*Organic and Small Farmers Associations* took issue with biotechnology. Small and often also organic farmers project a vision of food production that is different from industrial agriculture and heavy reliance on agrochemicals. Organic farming practises crop management with a minimum use of pesticides and free-range animal husbandry. Genetic engineering is seen as unsustainable industrial farming and rejected in favour of low-tech solutions based on biodiversity, crop rotation and timing to control pests.

*Anti-globalisation* is the social movement that resists the neo-liberal consensus on how to conduct economic affairs after the Cold War. Global groups such as ATTAC (since 1998 'the world is not for sale') oppose the liberalisation of markets that undermine democratic control by stealth. The French activist Jose Bove personifies this struggle representing the small farmer *Federation Paysanne*. Genetically modified crops and food are part of a larger struggle against industrial food culture of 'malbouffe' epitomised by McDonald's fast-food restaurants. Agricultural biotechnology, and its entanglement with agri-chemicals like Monsanto, became a main reference of concerns. Genetic engineering of plants and animals creates dependency and continues colonisation by other means.

*Churches and religious groups* make reference to biotechnology in the wider sense of 'engineering life'. They take position in ethical discussions over decisions about giving, taking and extending life, over therapeutic means and dilemmas that arise from scarce therapeutic resources. The bioethics 'protecting life' predates modern biotechnology and is not confined to religious traditions. The religious concern over biotechnology has to do with the hubris element of 'playing God'. Creating and designing life forms, commodifying life and enhancing life according to a cult of perfection rests uneasy with religious sentiment as this is seen as the prerogative of 'higher' forces. Churches are active in official commissions, staging debates, and contributing to the public discourse. The Church of Scotland's Science, Religion and Technology or the Vatican's Pontificia Academia pro vita are prime examples of this activity. Religious sensitivities are at stake when biotechnology touches on human reproduction (IVF), birth control and abortion, prenatal diagnosis, and embryo research as in the case of stem cell developments. The motive of creating the 'perfect human being' through genetic enhancement also concerns these actors. The distinction between healing and eugenic interventions is important here.

Beside these established structures of civil society, there are numerous smaller or larger *groups of activists* forming their identity exclusively around biotechnology issues. Organisations like GeneWatch in the UK or the Council for Responsible Genetics in the US operate as brokers of information on

184  *Atoms, Bytes and Genes*

new developments. Others launch specific campaigns nationally or internationally. Thus civil society has made its voice heard to varying degrees on different applications of biotechnology. Genetically modified crops and food provoked a broad coalition among environmental groups, consumer interests and anti-globalisation activism particularly in Europe. This gave the resistance to biotechnology a colourful public presence in the late 1990s on matters of sustainability, consumer choice and globalisation. Civil society's response to the biotechnology of reproduction, diagnosis and healing of disease is not so clear-cut and identifiable. This is because 'saving lives', a key argument for medical genetic engineering, is very difficult to challenge with credibility.

## THE PUBLIC FRAMING OF BIOTECHNOLOGY

Biotechnology was small news until the second half of the 1990s. Prior to that, in most countries, gene news was a monthly item at most. This sporadic news flow goes through four phases: prior to 1973, 1973–1981, 1982–1997 and after 1997.

Firstly, prior to 1973, very little of modern genetics surfaces in public debate and media coverage. Important for modern genetics was the discovery of the gene structure in the double-helix model of A, T, C and G (Crick and Watson, 1953). A break-through this was scientifically, but wider public attention had to wait until Watson presented a controversial account in late 1960s (see Chadarevian, 2003; Watson, 1968). Gene news was rare and occasional on topics such as eugenics, as at the conference 'Man and His Future' (1962) at the Ciba Foundation in London (Kevles, 1985). Until the early 1980s, the nature/nurture debate over intelligence resonated in public, but without critical mass (see Linke, 2007).

*Excursion 8.3*  Key Public Debates on Genetic Engineering

- *1948–1954: radiation effects on mutation rate (genetics funded by physics)*
- *1962–1965: new eugenics after Ciba conference; genetic engineering*
- *1965 XYY chromosomal syndrome: genes and criminality*
- *1969–1980: IQ nature/nurture debate including Socio-biology debate*
- *1973–1975: Asilomar and other places; moratorium on rDNA research*
- *1978–present: IVF, reproductive technology, amniocentesis for Down's*
- *1985–1995: regulation of biotechnology field trials*
- *1990–present: emergence of evolutionary psychology*
- *1992–2000: HUGO human genome project, human diversity genome project*
- *1994–2002: GM food debate (climax early 1999)*
- *1997–2009: human cloning, stem cell debate*
- *2000: first draft of human genome leads to omics: proteomics, pharmacagenomics*

Secondly, the mid-1970s was marked by two significant events with media coverage that became axial for the development of genetic engineering: the Berg letter (Berg et al., 1974) and the Asilomar conference of February 1975. Scientists at the frontier of micro-biology and genetic research called for a moratorium. Recently developed techniques, patented by Cohen-Boyer in 1974, allowed the movement of genetic material from one cell to another, splicing bits and pieces together into what became known as recombinant DNA (rDNA). Scientists were worried that hazards might arise, as novel organisms could escape and create havoc in the environment. At Asilomar in California, 140 scientists from 16 countries and invited journalists were impressed by potential legal liabilities and discussed how to live with the dangers: they classified experiments by risk level and devised precautionary measures. This vanguard of research called for self-restraint, a novelty in the history of science. The story so far was of a Promethean struggle of breaking the shackles of tradition and authority blocking the path to progress. But growing awareness of the commercial potential of new rDNA techniques, spectacularly demonstrated in an engineered bacterium that produced human insulin in 1979, reversed the moratorium to new drama of 'DNA, Dollars and Drugs'. This mobilised a biotechnology movement (Watson, 2003, 108ff; Yoxen, 1983, 45ff) and also increasing press coverage.

Thirdly, from the early to mid 1980s, gene news became a weekly event, reflecting the entrepreneurial spirit of the 'new age' of Genentech and Biogen and spun-off micro-biology research labs. Both the EU and the OECD launched biotech as the hopeful future investment (Bud, 1991). In the US, a regulatory framework for biotechnology was put into place by 1986, overseeing the release of GMOs into the environment, and similar regulations came to the EU in 1991 (see Durant, Bauer and Gaskell, 1998). By the mid 1990s, genes were daily news in the UK (Figure 8.2). By 1996, the number of items about biotech in the news (for one newspaper in each country) was 878 in the UK, in France 297, 174 in Germany, 204 in the Netherlands, 230 in Austria, 210 in Switzerland, but much less in Greece, 10, indicating its salience in those countries. Debates had started in many countries, identifying biotechnology as a technology for the future, discussing controversial issues and putting a research policy in place. This news stream had become steady, but a far cry from the level of Gulf War news of 1991. Gene stories had a human interest angle on the 'gene for x' format. Over the years, we read of the gay gene (1993), of the gene for breast cancer (1995), schizophrenia (2001), gregariousness (2007) and laziness (2008). Some more serious than others, they engaged the public imagination for the geneticisation of health and human behaviour. At the turn of the century, genes were news value. What had happened? 1996 and 1997 saw two key events

186  *Atoms, Bytes and Genes*

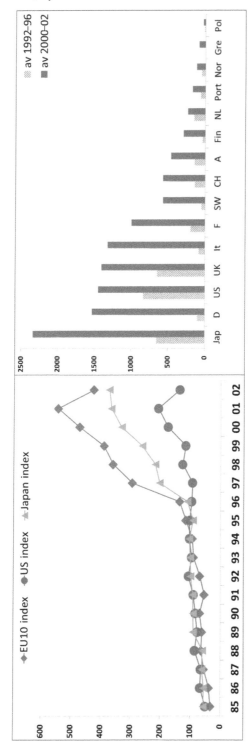

*Figure 8.1* News coverage of biotechnology in Europe, the US and Japan. Number of elite newspaper articles per year indexed to the average 1992–1996 = 100. Graphic on the right shows the increase in coverage from before 1996 to the peak years of the cycle, based on a single news outlet. Search keywords were DNA, IVF, biotech*, gene*, stem cell* or cloning. *Sources*: Bauer et al., 2004; Gaskell and Bauer, 2001.

which catalysed the debate over biotechnology and structured it as two different issues: GM crops and Dolly the cloned sheep.

## A Global Issue Cycle

After 1996 the news coverage of biotechnology exploded to unprecedented levels, peaking in the early 2000s; in Europe coverage increased five times, in Japan four times, but less so in the US, where coverage about doubled. The level and the increase of coverage varied, but in these countries gene news had become a daily item by 2000. Figure 8.1 shows the changing attention to 'gene news' across EU, the US and Japan. Until 1996, there was the constant trickle of news with the occasional spike. But after 1996, things were clearly going in different directions. Most remarkable in this story is the simultaneous attention to 'genes' across the globe. Hitherto, gene news reflected local times and events; after 1996 gene news was synchronised globally.

### The Watershed Events of 1996/97: Making the Global News

In autumn 1996 a US shipment of soybeans arrived in Rotterdam to feed cattle and as an ingredient of processed food for human consumption. This particular shipment contained approximately 2% of GM soybeans, the first harvest of Monsanto's Round-up Ready variety. It was all perfectly legal; the GM variety had been approved in April; however, it did not meet with public approval. Europe was still negotiating how to label this stuff, and how to licence it (see Lassen et al., 2002). In this regulatory uncertainty, Greenpeace and others saw an opportunity. They staged protests and raised the alarm over GM crops and food. GM crops had already met controversy. As the Flavr Savr tomato stayed on the US shelves in 1994, Jeremy Rifkin had launched the 'tomato war' to favour organic produce (see Martineau, 2001); in Britain, Astra-Zeneca found itself peddling 'Frankenfood', as its GM tomato paste was called.

In March 1996, with proof that BSE (Bovine Spongiform Encephalopathy) had moved to humans in the form of vCJD (variant Creutzfeldt-Jakob Disease), the UK government had lost public confidence in handling food risks. In the UK the scene was set for a debate over GM food. Greenpeace was internally split, but took the risk, and Tesco, Sainsbury's and Carrefour, major food retailers, vowed to avoid GM ingredients for their products or label them. GM food became the focus of attention across Europe (see Gaskell and Bauer, 2001). It was a historical first, a synchronous attention to an issue in all member states (Seifert, 2006).

The UK exported this issue. Prince Charles, an organic farmer, came out against GM crops in June 1998 in the *Daily Telegraph* by raising 10 pertinent questions. In early 1999, the Pusztai Affair sparked the 'great GM food debate' with experiments that showed animals at risk from eating GM crops. The *Daily Mail* supported the prince (see Durant and Lindsey,

188  Atoms, Bytes and Genes

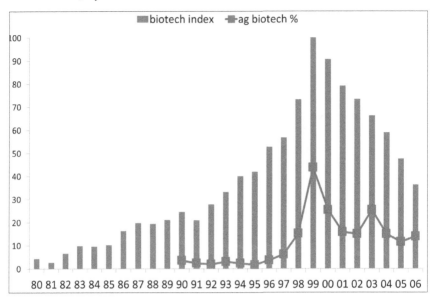

*Figure 8.2* Number of references in the British national press to biotechnology, 1980–2006 and to agricultural biotechnology as percentage of the total coverage. Index 2000 = 100 or 1,666 reference on a single news outlet. *Source*: Bauer et al, 2004; and Valentina Amorese for later updates.

2000). Figure 8.2 shows the issue cycle of genetic news in Britain between 1980 and 2002. At its peak, a single newspaper would carry 1,666 news items on 'genes etc.', of which 45% would be on GM crop and food or ag(ricultural) biotech. Hitherto, GM crops and foods were a small proportion of gene news, henceforth, they were a big part of the gene story. Elsewhere, the GM crop debate flared; in Austria over an illegal field test; in the US over the Monarch butterfly. A Cornell study suggested in June 1999 that caterpillars were at risk from the GMO Bt-corn, and in 2000, StarLink, the allergenic GM corn, was found at Taco Bell's.

In February 1997 *Nature* featured a story of 'viable offspring derived from . . . adult mammalian cells' (Wilmut et al., 1997). The ardent veterinary technicalities of adult nucleic transfer, needing over 200 trials to succeed, became the story of 'Dolly the sheep' leading into the global stem cell debate. Embryonic cell cloning on frogs had been done since 1958 (Nobel Prize, 2013), but the cloning of adult stem cells was a sensation. Cloning was a science fiction topic that resonated with public imagination: the cloning story went global instantaneously. Dolly was about animal husbandry for pharmaceutical purposes—her milk produced a particular protein—but became the yuck-symbol of human cloning. References to science fiction human cloning, misuse by dictators and amoral scientists, monsters and Frankenstein, the end of sex, and the lack of any regulation were instant concerns. Within weeks, presidents, the UN, the pope and national and international bioethics committees rushed to declare a ban on human cloning (see Einsiedel et al, 2002). Dolly aged faster than a normal sheep, developed arthritis and died of a retroviral lung disease in 2003, thus pointing to the health risks of a cloned existence. And, it triggered the global debate over stem cell cloning as a moral danger (Kolata, 1997; Nussbaum and Sunstein, 1998; Habermas, 2003).

GM soya and Dolly were triggers, stories that brought genetic technoscience to global attention. What until the mid-1990s was an occasional story of local relevance was now a global news value. Whatever the differences of plant engineering, nucleic transfer, genetic sequencing and mapping, genetic finger printing, testing and screening, IVF, stem cells and gene therapy, these topics would henceforth have a large public.

## Framing Big Topics

Salience of a field is one thing, but what are the topics being raised? Biotechnology coverage reports on *the science*, its breakthroughs and results continue to be a media topic. Here the major gatekeepers *Science*, *Nature* and *Nature Biotechnology* are a weekly source for journalists. Biology has become the leading science of the 21$^{st}$ century, replacing physics and chemistry in that role. On the science reportage frames like progress and economic prospect dominate, with occasional pointers to moral issues.

Biotechnology raises issues of economic *development*. Agricultural biotechnology brings home issues of farmers in Africa, India and Latin America, who assert themselves vis-à-vis multinational seed and agrichemical companies. Here the story is of corporate players who operate globally, developing seeds and drugs, rolling out field and clinical trials and marketing medicines, pesticides, seeds and crops. Counter-events, like the World Social Forum of Porto Alegre 2001 and later, foregrounded biotechnology. The anti-globalisation agenda of Seattle of 1999 pushed this theme and brought to public attention the role of the WTO, IMF, OCED, World Bank, WEF and regular G8 summits.

*Patents and property* cover all biotechnology. Highly technical, the patent system is an occasional news item. Central as it is as a real issue, it does not resonate in public imagination.[2] The conflicts of Monsanto with local farmers over seed rights, or the dispute over patent infringement and priority, as with the breast cancer genes BRCA1 and BRCA2, have passed the news threshold. Patenting gets the 'runaway' frame: the facts are in place and very little can be done about it.[3]

The grinding topic of biotechnology is governance, how to *regulate and secure public accountability*. Negotiations over regulation throw light on differences within Europe and across the Atlantic. Should regulation follow 'sound science evidence' or the 'precautionary principle' of acting before it is too late? Is regulation of GM food based on 'substantial' or on 'process' equivalence', on ingredients, or on how it is produced? Governance boils down to the questions: which risks are considered, and how shall they be defined? Public accountability is both an issue and a frame of the debate.

Much biotech is about food safety and environment. Food anxiety over adverse effects to human or animal health from the consumption of GM food stuffs gives opportunities for media coverage resonating consumer concerns about industrially processed foods. Potential for allergenic GM food, as in the transfer of a nut gene to a crop; antibiotic resistance, as in Bt-176-maize which used an antibiotic gene marker; abnormal organ growth, increased mortality of animals, and changes to the intestinal flora (Pusztai's claim of 1999), are some of the claims made and anxiously rebutted. Official reassurances had lost much credibility after BSE and other food scandals. Procedures for testing and certification were of public interest, so was the labelling of products. Much news is given to the prospects of new crops with added consumer value such as vitamin enhanced 'golden rice'. First generation crops were engineered to save on pesticides and benefited farmers; also an environmental plus. Food security, feeding the world population, is a key industry argument but with limited credibility: modern hunger crises are largely distribution problems. But growing pressures on food commodity prices have given it more credence. *Environmental impacts* on biodiversity in soil and above ground, microorganisms, insects, birds and weeds resonate with every gardener. Claims and counter-claims over agro-chemical use, biodiversity and yields populate the news; but until 2000 most of claims and counter claims were 'visioning' rather than research based (see Vain, 2007). The British field trials of the late 1990s showed that some GM varieties were beneficial, others not: it depends. Organic crops need protection; super-weeds and the low-tech alternatives to GM crops are news topics. Agricultural biotechnology sees the whole range of framing: much of it is seen as progress and economic prospects, but distinct from this is the sense of this technology as a 'Pandora's box', and in need of 'public accountability'.

There is a long-term trend towards the medicalisation of science news (Bauer, 1998). Here we find also news of genetic testing, gene therapy, old and neweugenics, embryonic and somatic stem cell research, genetic identity and ruminations on human nature, which merged with older sensitive topics such as abortion, reproductive technology and bioethics. *Genetic information* from profiling, testing, screening into large databases like envisaged by DECODE in Iceland raised questions of misuse (Greely, 1998). Genetic profiles capture public imagination. News of finally resolved crimes, of rapists convicted, the identified Tsar and Tsarina (1995) and Richard III (2013) make sure headlines (see Durant et al., 1996); so do concerns about the storage and usage of genetic information by employers, insurances or banks. Here the 'gene' merges with the IT frame of a surveillance state, and can become a matter of civil liberty. Genes for health, though, is mostly framed as progress and a bioethical issue, together with economic prospect and governance.

*Eugenics* with a focus on quality rather than quantity of the population dominated the inter-war years; this remained as a constant backdrop of commentary on genetic engineering and reproduction (see Barrett and Frank, 1999; Kevles, 1985). Predictive medicine and pre-symptomatic intervention opens the prospects of more dilemmas and aberrations; genetic diagnosis, gene therapy, enhancement, cloning and designer babies easily find the news pages. Eugenics is a prototypical moral framing.

Gene talk extends into reflection on *human nature*. The metaphors 'book of life' or 'code of life' resonate public imagination, both in words and visuals. The prospect of saving lives and reducing human suffering with genetic and stem cell research is blockbuster news. This ultimately touches on discussions on 'human nature', maximising fitness of the 'good life' people would want to live. Darwinian evolutionary psychology (see Cassidy, 2005; Linke, 2007; Rose and Rose, 2000) and Mendelian transhumanism (see Fuller, 2010) form a metaphysical sounding board for genetic engineering. This topic is framed as 'ethics' and 'nature/nurture'.

The great absent of public imagination is *biological warfare*. Only after 9/11/2001, the public gained awareness that biotechnology might have a proliferation problem analogical to atomic energy. Since the 1980s, any analogy between nuclear power and genetic engineering was carefully avoided (see Radkau, 1995). Then, in 2001, an US Army technician apparently drew attention to the neglected issue of dual use by releasing anthrax spores; ricin found in a North London flat in 2003 equally drew attention, 30 years after Asilomar. Terrorism and national security are the frame to bring this home. Genetic warfare is real, but not part of the public debate.[4]

*Framing Biotechnology Before and After the Watershed Years*

Frames put a topic in a particular light and suggest a particular course of actions. Building on Gamson and Modiglinai (1989), and how nuclear power was framed over the years of debate, our research typified eight frames of biotechnology (see Durant, Bauer and Gaskell, 1998; Nuffield, 2012, 81).

Much of genetic engineering is discussed under the headings of scientific and social *'progress'* and *'economic prospecting'* gathering unconditional support. Figure 8.3 shows the flow of framing in the long run. Scientific progress and economic prospect dominated the coverage right into the 1990s (see Plein, 1991). The challenges of morality are covered by several frames. *'Ethics'* refers to bioethical arguments that elaborate the utilitarian quest to reduce suffering. This is important in the early 1980s and in the stem cell debate after 2000. *'Pandora's Box'* admonishes heedless action and *'Runaway'* insinuates a fait accompli.

The latter characterised the rDNA debate in the 1970s: it is already too late to stop it. *'Nature/nurture'* of a naturalised culture and a cultivated nature was significant in the mid-1970s, and again into the new millennium in discussions on genetic enhancement. Restoring health is not good enough; in fact humans are not good enough (yet). *'Governance'* through public accountability rather than technocratic decisions became a mainstay of public opinion. This has been a matter of public demands, but also of appreciation and commentary on many formats of public engagement (see Chapter 7), an important way of discussing matters in the late 1970s, displaced by 'prospecting' in the 1980s, to re-emerge in the 'biotechnologisation of democracy' (Levidow, 2007), which refers to ways of innovating democratic processes to pave the way for biotechnology. Finally, biotechnology is part of *'Globalisation'* but not commonly addressed in the public debate.

Figure 8.3 visualises the changes in public discourse in these terms taking the watershed years as axial across Europe, US, Canada and Japan. Over time, the progress frame is less prominent, while ethical and governance arguments take its place. Clearly progress does not come automatically from innovation, but needs to be assured. European discourse reflects this by moving from seeking 'innovation', to seeking 'responsible innovation'. Before 1996/97, biotechnology and genetic engineering tended to be represented as the one big, new thing on the horizon. Under this sociological vision, research policies were reformulated, existing agencies reoriented, the stock market indexed the sector, *Nature* opened a new journal to map the field, and many countries legislated under the umbrella of a 'gene law'. The transition into bio-society was to be led by a new industrial sector, for which the OECD collected statistics to map its global progress. The terms covered many applications, GM crops and foods, genetic research in general, gene testing and therapy, xenotransplantation, animal models like the OncoMouse, rDNA insulin and many more. Figure 8.3 shows that most of this was mainly a matter of progress and a bit of ethics.

After 1996/1997 the world of biotechnology presents itself in a very different light. Two things are remarkable. First, what was a unified thing now splits. No matter whether funding agencies look after biotechnology or OECD collected its statistics, in public discourse, the 'thing' split into three different strands: red-biomedical biotechnology focused on providing innovations to human health; cloning or stem cell research, part of the same agenda, but more controversial because of the embryonic question, and green or agricultural biotechnology. These three discourses use very different framing, mobilise different groups of activist and cultivate different attitudes. These three 'biotechnologies' are positioned differently in social space. Figure 8.3 shows the different framing that emerged after the watershed years. Red stands for progress cum ethical issues. Cloning combines a mix of progress, ethics and governance, but also pessimistic elements of Pandora's box and runaway. Green has lost its

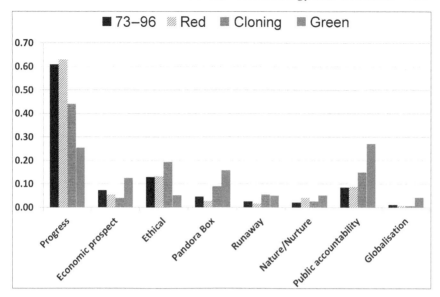

*Figure 8.3* Framing of biotechnology before and after the watershed years. The dark bar shows the framing for 1973 to 1996; the others for 1997 to 2002 for red, cloning and green biotech; n = 8,829 across Europe, US, Canada and Japan. *Source*: Bauer et al. (2004).

credibility as progress, but musters economic arguments. Its discourse is dominated by governance issues and radical 'hands-off' attitudes; the development is haunted by the image of a Pandora's box. It is the most diversely framed strand of genetic engineering.

By 2000, two large clusters of agricultural and biomedical biotechnology had emerged with complex interactions (see Bauer, 2005; Weingart, Salzmann and Wormann, 2008). This split was constitutive of a diversity in the debate across Europe, but in particular it allowed for the 'transatlantic reversal': a public stem cell debate in the US, where GM crops were a non-issue; in Europe GM crops were the big issue, while stem cell research could often rush ahead. Figure 8.4 shows how this split was cultivated in Europe by an influential elite press. Comparing pre- and post-1996 shows that coverage of biomedical genetics became more positive, while agricultural applications scored more negative. At the same time, attitudes shifted, and the gap between red and green increased, making this distinction a new common place. Castro and Gomes (2005) have demonstrated how this emerging duality played out on a more local level in Portugal. Biotechnology was no longer good or bad; it now depended on which one was talking about. The public ambivalence of modern biotechnology was overcome, by splitting different paths. With a choice, public discourse is no longer caught in an approach-avoidance conflict but in a clear hierarchy of options depending on which criteria apply. Thus, the global debate had created a new strategic context.

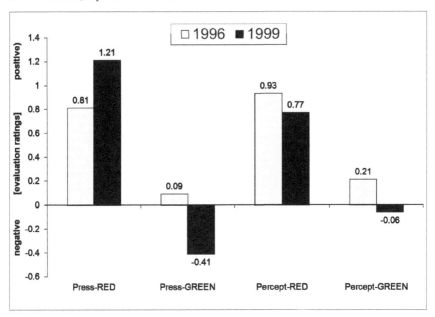

*Figure 8.4* Evaluation of different applications of biotechnology grouped into 'red' and 'green'. 'Press' stands for news items in the periods before and after 1996/97; 'percept' stands for public attitudes measured in 1996 and 1999 across fifteen Europen countries. *Source*: Bauer (2005).

## Public Perceptions: The Changing Strategic Context

Perceptions of biotechnology shifted during the 1990s, to surprise and chagrin of many protagonists. The continuous monitoring of attitudes on the part of the EC (with Eurobarometer) is an expression of anxieties over public opinion. By late 1996, 45% of Europeans took notice of biotechnology news, and 50% had talked about it with friends or family; this level of involvement had not changed by 2002. Despite the occasional flare-up, genetic engineering was never salient at the level of war, employment, health or education. Though people in Germany, Denmark, Sweden, Norway, Finland, and Switzerland took notice, in most other places, biotechnology remained the concern of an educated, vociferous minority (see Durant, Bauer, Gaskell, 1998; Gaskell et al, 2006).

Public opinion measures are themselves framed, and the framing of the questions often follows the public discourse very opportunistically. In the early 1990s, researchers expected that perceptions of genetic modifications would align on an evolutionary scale, the 'higher up'–from microorganism, plants, animals, to humans—the more sensitive (see Reiss and Straughan, 1996). Eurobarometer (EC 1991) was designed accordingly, but showed that new medicines like rDNA insulin and recombined micro-organisms to clear pollution were favoured while human tissue engineering and crop plants were not; GM foods and GM animal breeding were dubious (Marlier, 1992). The sensitivity to GM crops and the support for engeineering insulin violated the evolutionary scale expectation. In the Netherlands and UK, food and crop applications got early on a 'wait and see' (Heijs and Midden, 1995; Martin and Tait, 1992). Germans saw GM crops as tampering with nature, though attitudes had become more positive since the mid-1980s when Germany agonised over the Catenhusen enquiry (Hennen and Stoeckle, 1992; Radkau, 1995). By the end of 1996, genetic testing and new medicines had public support across Europe; in regards to GM crops or foods, most Europeans felt ambivalent. However, the Harvard OncoMouse or breeding animals for xenotransplantation were clearly out, seen as risky and morally dubious (see Durant et al., 1998). It became obvious to all but a few observers that negative attitudes and risk perceptions were not a matter of ignorance or deficient reasoning, but of moral qualms and doubts over the trustworthiness of the protagonists. The ambivalence on GM matters in Germany, Austria, Denmark, Sweden, Austria, France, the UK and Netherlands came to those with higher levels of biology knowledge (see Gaskell et al., 1997). The ethos and pathos of the biotechnology mobilisation was in tatters, not its logos. Moral qualms functioned as veto. There was a 'yack factor' on tampering with nature, potentially creating a brave new world of adulteration, pestilence and monstrosity—a veritable Pandora's box scenario. Many people defended 'Nature' as a numinous force or a complex system (see Chapter 4). These symbolic anchors absorbed uncertainties of things to come (Wagner et al., 2001 and 2002)

Regarding public perceptions a Europe-US gap came in evidence. In 1996, opinions matched on rDNA insulin type medicines on the positive and xenotransplants on the negative side. However, Europe worried about GM crops and food, the US more about genetic testing. The difference was small but significant because it correlated to larger issues of regulation and culture in the 'transatlantic puzzle' of such different receptions of innovation (see Gaskell et al., 1997 and 1999). The GM food debate and Dolly the sheep in the year that followed were able to resonate with public imagination because the public mind was already on that wavelength. In hindsight, protagonists of biotechnology wrestled over why they did not see that coming.[5]

This gap widened by 1999: in Europe, genetic testing, new medicines, bioremediation, and stem cell cloning was encouraged, despite the risks involved, while the GM crops, animal cloning and GM foods were in disfavour. What had emerged, three years into the debate, was a clear wedge between red biomedical and green agbiotech, which the elite press had anticipated (see Figure 8.3 above). Only Spain, Portugal, Finland and Ireland remained positive to agbiotech developments. Greece shifted radically from a great supporter to a great opponent of things 'genetically modified' (see Gaskell and Bauer, 2001).

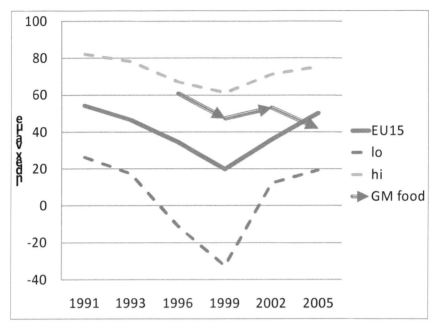

*Figure 8.5* The dynamics of European optimism over biotechnology from 1991 until 2005. Index is calculated as [%opt—%pess] divided by [%opt + %pess] excluding the DK responses. Lo and Hi tracks the upper and lower bounardy among EU15 countries. The GM food index is the sum of supporters and risk tolerant supporters in percent of about half the population that expressed an opinion. *Source*: Gaskell et al., 2006 based on Table 1.

In 1996, 70% of Europeans expected biotechnology to impact their lives positively; this rose to 75% in 2005. This seems paradoxical, considering the negative trend on GM crops from above 60% to below 40% support. 50,000 people across Europe opined every three years on whether 'biotechnology and genetic engineering would improve their lives over the next 20 years'. The balance of optimist minus pessimist declined from about 60 points in 1991 to 20 in 1999, at the peak of the debates over GM crops and Dolly the clone, only to recover by 2005 to the ex-ante level. Was this a blip in public opinion, recovering to previous levels, once hysteria cooled down? No. This reversal was made possible by the tectonic shift of representations of biotechnology.

Across Europe things differed; Figure 8.5 shows the bandwidth. Spain and Portugal remained more optimistic throughout. Denmark, Germany, Austria and Greece were more pessimistic. Also, Europeans were more optimistic about solar energy and IT through these years; and nuclear energy carries even less favour than biotechnology. Europeans are a far cry from anti-innovation technophobes, but selective on which innovation they support. Who are the sceptics on biotechnology in Europe? Older citizens, women, but also the politically interested, who take notice, keep informed and discuss things; the latter are more likely to be educated men (see Gaskell et al., 2006 and 2010).

How did the mass media coverage, exploding after 1996/97, affect the trajectory of public perceptions? The knowledge gap hypothesis suggests an increase in knowledge differences in the population, except when such an increase occurs in the context of controversy; this was the case in Europe (see Bonfadelli, 2005; Bauer and Bonfadelli, 2002). Mazur's expectation that increased reportage on a risky matter leads to a conservative shift in attitudes was not confirmed. Taking into account media exposure, there was no relation between increased coverage and a shift towards negative attitudes on biotechnology (Gutteling, 2005). The elite press cultivated the meta-framing into red and green biotechnology; the educated public took up this frame split across Europe (Bauer, 2005). What manifested itself in European public opinion, both press and perception, by 2005 created a new strategic situation: red biotech had a future in public discourse, green biotech had not.

In summary we might say: once biotechnology split its path, it became possible to imagine improvements to life. In answering this question in 2005, people were no longer ambivalent but most likely thought of new medicines, stem cell research to end suffering, genetic tests and genomics to cure cancer with a positive outlook. All the while GM crops and its traces in food, detested by many, were now travelling a different path with downward trajectory. The red-green split a) made red biotech recover favour in Europe, b) made possible the transatlantic crossover where agbiotch was favoured over stem cell cloning, and c) allowed the continuous expression of disfavour of GM crops while the 'good' biomedical biotechnology was favoured. The conspiratorial mind might suggest that whoever profited from this situation might be suspected of having brought it about. A weaker hypothesis might be that those who are likely to profit will do little to reverse it. Swiss biotechnology is a good demonstration of this logic.

*Excursion 8.4* Biotechnology Democracy in Switzerland

*An illustration of the strategic significance of the red-green split, are the five referenda on biotechnology in Switzerland. Mobilisation started in the late 1987 when the magazine* Beobachter *campaigned for a constitutional amendment (Art. 24, para. 10). A version of its proposal was adopted in May 1992 by 73.8% 'to protect humans and the environment against misuse of genetic and reproductive technology, to assert the dignity of life, and to empower the government to regulate the matter'. The vote was carried by the young, educated middle class with radical or green leanings; it was opposed by free-market liberals who anticipated too much regulation, and by value-conservatives for whom this regulation was not sharp enough (see Buchmann, 1995).*

What followed were 10 plus years of protracted discussion of the gene law 'Genlex'. In 1998 a further amendment to Article 24 came to a vote. The legislative process had stalled; the arrival of Monsanto's soya, Ciba-Geigy's (then Novartis, and later Syngenta) Bt-176 maize of 1995 and Dolly required a new ruling. The rule would ban GM animals, GMO releases, the patenting of life forms and GMO research would be allowed to find alternatives. The traditional pharma industry saw its base at risk. Through Interpharma, a lobby group mobilised the most costly campaign in the history of Swiss politics (see Cueni, 1999). Under the spectre of economic demise, scientists went to the streets to fight a 'ban on thought' led by Nobel Prize winners. Opponents mobilised among greens, farmers, bakers, butchers and the culture world. The amendment was cancelled with 66.6%. The left and greens were split by economic concerns (see Brown and Bieri, 1998).

March 2000 saw yet another vote 'against the manipulation of life', in fact a ban on IVF; this was rejected by 72% on a turnout of 42%, supported only by traditional religious orientations and some sectors of the left.

In 2004 another vote occurred on whether to allow regulated stem cell research. In November, 66.4% approved regulation on a turnout of 36%. Stem cell research was opposed by a coalition of traditional and religious voters on ethical and moral grounds and by greens mistrusting science and fearing adverse consequences.

Greens mobilised for the November 2005 vote on a moratorium on GM agriculture. This was narrowly approved with 50.6% on 42% turnout. Research showed that actual approvals were 13% higher, because some people voting 'no' to the moratorium registered 'in favour of GM research', when in fact they wanted to express disapproval (Hirter and Linder, 2005). The result was carried by the young to middle aged and an anti-free-market coalition of traditional farmers and left leaning voters. Syngenta prides itself of having opposed this referendum. For five years Switzerland was to grow no GM crops, except in research trials, and would debate the 'dignity of plants'. The government later extended the moratorium until 2015. The main counter-argument, that a moratorium is bad for research, did not carry; the plant scientists felt they were left hanging by an industry which, contrary to 1998, did not pull on its strength. Plant genetics does not have the resources to mobilise the voters, and biomedical biotechnology in Basel and elsewhere is well served with the current situation: the gate opened for red biotech, with a moratorium on green biotech. Switzerland became a key biotechnology hub.

## TECHNO-SCIENTIFIC RESPONSES TO THE CHALLENGE

Thus we come to the point: what were the reactions to public resistance and did this affect the trajectory of biotechnology? How did the biotechnology movement accommodate public opinion? The evidence for strategic adaptation can be found in the demise of 'life science vision', the fragmentation of regulatory regimes, delays in global diffusion of GM crops, in tactical concessions over genetic use restriction technology (GURT) and GM wheat, in shifts of research strategies from embryonic to somatic stem cells research, the found respect for donors in genetic research and finally in the loss of authority of bioethics.

## Strategic Learning: The End of the Life Science Vision

The American chemical company Monsanto (founded 1901) exemplifies the rise and fall of the *life sciences vision* of the biotechnology industry. The company's annual reports present this strategy of the 1990s as a commitment to 'finding solutions to the growing global needs for food and health by sharing common forms of science and technology among agriculture, nutrition and health'. This included to a) become the global leader in agricultural biotechnology, b) develop a world-class pharmaceutical company, c) build capabilities for science-based nutrition, and d) create genomics to support all areas of the life sciences (see Monsanto, 1999). The motto was 'food, health and hope'.

Throughout the 1990s, Monsanto aggressively acquired seed and genetic research companies like Agroseres (in 1997) in Brazil, Delta & Pine Land (in 1998), Calgene (in 1997), First Line Seeds (in 1998), DeKalb Genetics (in 1998) and Plant Breeding International (in 1998) and Cargill's (in 1998) seed operations in South America, Europe, Africa and Asia as the global platform to market its business model of GM-seed-cum-herbicide. This was successfully launched in 1995, first with Round-up Ready soya, and in the pipeline were corn, cotton, canola and wheat, the first generation GM crops. But the vision extended beyond agbiotech into pharmaceuticals, food and nutrition—second and third generation GM crops and nutriceutical with consumer and health benefits. 1998 the merger with American Home Products, a pharmaceutical company, failed; DuPont threatened with take-over; in 2000 Monsanto finally merged into Upjohn, Searle and Pharmacia to form a new Pharmacia.

However, in 1998 the life science vision had come unstuck in European public opinion as shown above. Emboldened by unprecedented success among US and Argentinean farmers and the official go-ahead in the EC, Monsanto responded with a 3 £million PR campaign in the UK to appease public doubts over GM crops and food. The campaign focussed on food security, failing to answer concerns about food safety and biodiversity. The PR effort backfired. The UK food debate moved globally, and Deutsche Bank discouraged investing in agbiotech venues. In June 1999, the president of Rockefeller Foundation publicly criticised Monsanto's pressing ahead to global markets when the risk-benefit analysis was incomplete and the ground rules unclear.[6] Rockefeller's initiatives in biotechnology were not the only ones disrupting. European companies and retailers who initially shared Monsanto's vision were equally dismayed by their competitor's pressing ahead of international negotiations at WTO, Codex Alimentarius and public opinion and spoiling the game.

By 1999, Europe kept a de facto moratorium on GM crop cultivation; never legally binding, but based on public commitments of retailers, continued discussion over EU regulations and various regions that declared GM-free, reminiscent of the nuclear-free zones of the 1980s. GMO field trials came to a halt (see EC, 2003), and the EU faced allegations of non-tariff trade barriers at the WTO, when in fact there was very little cultivation to be protected; at stake were consumers' rights to choose what to eat and environmental concerns. At the annual shareholder meeting of 1999, Monsanto's charismatic CEO Robert Shapiro was taken to task, unable to explain why the plot was lost in Europe and in Latin America, and he announced the merger with Pharmacia, who hastened to separate out its globally tainted agbiotech business. Shapiro, the visionary of the life science industry, left the company; in 2002, Hendrik Verfaillie, the engineer of the vision, followed suit. As Gerber took GM soya off its baby food range, and butterfly larvae suffered from Bt corn, many farmers went off GM crops again and the diffusion of new crops stalled in 1999 and 2000 in the US (see Figure 8.6 below). The *Economist* (23 December 1999), *New York Times* (25 January 2001) and the *New Yorker* (10 April 2000) published investigative pieces trying to unravel what had happened to a great vision. Commentators were sympathetic with the visionary who anticipated a world of neutriceuticals, where drugs came as food.

When in 2000 Monsanto merged into Pharmacia & Upjohn, this was no longer part of the life science strategy, but was instead preparation for many divorces. DuPont, Bayer, Novartis, ICI, Rhone-Poulenc and other global chemicals and pharma firms had equally pursued the life science idea (see Juma, 1989) and tinkered with 'pharming', i.e. value added crops with pharmaceutical traits and novel foods. Novartis, the product of Ciba and Sandoz of 1996, shed its agribusiness (seeds, plant protection), which together with Astra-Zeneca became Syngenta in 2000 (with turnover of 31% of Novartis).[7] Rhone-Poulenc, Hoechst and Schering became Aventis, who sold its agro parts to Bayer Crop Science in 2001. A campaign in 2000 by Aventis stressed human well-being, and the biotech industry lobby launched a concerted PR campaign to inoculate the US public against contagion by European opinions. By 2002, the life science vision had lost its gloss; commentators doubted its meaning as a concept beyond branding: 'some terms do not want to refer to anything, only want to create associations' (see Wenzel, 2002; my translation).

Strategic adaptation of the industry was in part a reaction to public resistance; the industry avoided integration and there remain two sectors. With the exception of DuPont and Bayer, the biotechnology world is now neatly split into agribusiness, green biotechnology, on the one hand and a drug business, red biotechnology, on the other. It could have all been different if Novartis, Aventis, DuPont and others had had their way, had Monsanto not pressed ahead blindly.

One might speculate on why Monsanto did what they did, and indeed many do. Shapiro, with a mixture of anger and disappointment, admitted naivety. Corporations must respond to challenges with their own resources, and Monsanto was proud of its culture. With a global mission rooted in Midwest America and focussed on the science, Monsanto was unable to make sense of what was going on, beyond being angry and blaming others (Specter, 2000). The lesson might be that: it is possible to fail if self-confidence is too strong. The biotechnology movement's costly lesson discovered the second hurdle, public opinion, and the importance of the secondary market. Knowing your primary market is not enough. Farmers depend on retailers who depend on consumers. The consumers worry about food safety, equity and the environment, and as citizens they influence regulations.

Echoes of Monsanto's debacle resonate today. St Louis became a biotech hub, but Monsanto dominates the local scene with a technology that has a reputation problem among investors; it seemed an unlikely environment for university spin-offs. The end of the life science vision changed expectations over functional foods and neutriceuticals; e.g. bread with an added drug effect. The GM controversy brought home that the business model must include the entire chain, from seed producers, growers, processors and retailers to consumers, to be convinced of benefits. The food industry is searching for natural alternatives while keeping the GM competence alive. Indeed, a study (Fernandez-Cornejo et al., 2006) concluded that consumer sentiment had impacted negatively on R&D, adoption rates and markets for GM crops and products globally. What is economically relevant is the resistance to genetic engineeering, not only the scientific soundness of its basis.[8]

## Diversity of Regulatory Regimes

Regulations are the target of corporate capture because they are a cost factor. Compliance costs include the efforts of providing the necessary information, and the time it takes to gain approval delays returns on a new product. Compliance costs are a major challenge for the biotech industry, and public opinion as it influences regulations is a major cost factor. The evidence on increasing compliance costs for GM crops is inconclusive, estimated in the area of $15 million per variety (see Kalaitzandonakes, Alston and Bradford, 2007). But the fear is that as in the case of nuclear power, safety regulations increase the entry hurdle of the new technology, and this makes public opinion relevant.

Public opinion contributed to the diversity of regulatory regimes on GM food. Clearly public opinion is only one factor among legal and political traditions which constrain regulatory outcomes (see Jasanoff, 1995 and 2005). Much has been written on the differences across the Atlantic, polemically contested in the international fora of WTO and Codex Alimentarius. Regulatory traditions confronted each other over the burdens of proof, proving danger or safety; the one bent on precaution, the other bent on deregulation and a narrow scientific risk analysis. The US looked for 'substantive equivalences' of new products (see Schauza, 2000) while the parliamentary logic blocked the public debate for agbiotech (see Sheingate, 2006), while the EU was accommodating diversity and building institutions with democratic legitimacy (Graber et al., 2001; Levidow et al., 1996; Cantley, 1995). The 2001 amendment to the EU directive on GM crops did not bring the deregulation which the industry had lobbied for, but reflected an interpretation of precaution (see Torgersen, 2001). So-called European trouble making and risk aversion turned into global leadership, to the dismay of many promoters of the technology. One constant line of defence for representatives both within and beyond the EU was the state of public opinion. EU citizens were clearly not in favour of GM food, and they still are not; not inclined to accept forced feeding, most people want labels on products with GM ingredients. Public opinion did not bring the precautionary principle to EU policy, but defended it against deregulation (see Hampel et al., 2006). Being able to reference public opinion was an important negotiation position countering the allegation of protectionism in national and international fora. And the search for legitimate positions brought about some old-new ideas regarding public participation, the consequences of which are being evaluated (see Levidow, 2007; Einsiedel and Kamara, 2006, also see chapter 7).

## Delaying GM crops and Creating Opportunities in Brazil

Within the framework of diffusion research, resistance manifests itself in two related ways: differential diffusion rates—some systems adopting slower than others—and in consequence of the geographical concentration of the innovation—some areas will have more of it than others at any moment in time. One could draw a map by colouring the density of the innovation in space.

Since 1996, ISAAA at Cornell University has consistently reported on the progress of GM crops globally within a diffusion framework. Year after year the reports highlighted the good news and tried to get around the bad news, but its database is illustrative of the effects of the GM food debate on global diffusion.

Herbicide tolerant RR soya and insect resistance Bt maize and cotton are the main varieties on a global roll-out since the mid 1990s. Figure 8.6 shows the diffusion of herbicide tolerant soya. In diffusion terms, Argentina reached the 50% mark in an incredible two years, the US reached that mark in 3 years and Brazil needed 10 years. The world took eight years to reach the 50% point and reaches for a plateau at 60–70% diffusion. Argentina had little resistance in the system, while the US saw staggered roll-out after 1999, when during the global GM food debate some farmers went off it again. The resistance was clearly manifest in Brazil. Here the diffusion researchers calculate the putative losses incurred as a result of lagging. Indeed, James (2007, 9) reports net losses incurred by Brazil to the extent of $4 billion until 2005 in comparison to Argentina. But is this the real story?

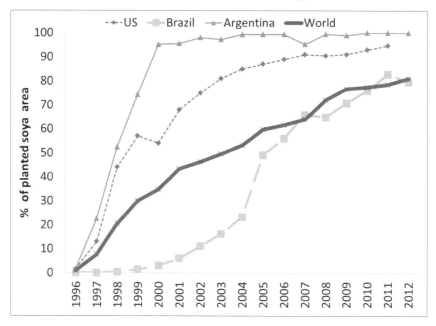

*Figure 8.6* Diffusion of GM soya among the three most important growers and globally. Index shows percentage of soya plantation that is GM in each context. *Source*: redrawn from data published by ISAAA (James, 2007; Bauer, 2006a).

---

*Excursion 8.5* The Brazilian Soya Miracle, 1996–2013

---

*Soya is a staple crop with a global market, with 80% produced by just three growers; in 1991 these were the US (52%), Brazil (19%) and Argentina (10%). By 2011, the US produced 35%, Brazil 28% and Argentina 19% of the world's soya. What happened? Brazil and Argentina expanded their share, while the US lost theirs; by 2013 Brazil has become the largest exporter of soya in the world, and is challenging the US as the leading producers (source: Folha de Sao Paulo, 12 Feb 2014). The Brazilian case is particularly illustrative, because this was achieved against and by delaying GM innovation. Rather than incurring losses, Brazil reaped the economic benefits by supplying those parts of the world that did not want GM food. Brazil was the strategic territory in the global battle over green biotechnology. This was clear to Monsanto and others, who increased their stakes in the country, to Greenpeace and their allies who campaigned locally, and to European food retailers who needed to source traditional produce to live up to their commitments.*

*A convoluted story unfolded in Brazil after 1998 when the high court imposed a moratorium over GM crops, protecting Greenpeace and the consumer organisation IDEC against the government's intent to approve transgenic crops. In the subsequent years, a confusing mix of public mobilisation, court actions, political ideology and economic calculus, representing all the complexity of Brazil, resulted in the above 'innovation gap'. GM crops were illegal until 2005; illegal planting* happened mainly along the southern border with Argentina where it was easy to smuggle seeds, so-called 'Maradonas'.

GM crop trials conducted by Monsanto came to a halt, though not Brazilian plant genetics; while soya production doubled from 30 million tonnes before 1998 to 80 million tonnes by 2013, to a large extent traditional soya to feed demand in Europe, Japan and China on a premium, which were reluctant to accept GM crops even as animal feed. The export of Brazilian soya sixtupled from 5 million tonnes in 1995 to 30 million tonnes in 2013, contributing $–15 billion plus annually to export value and saving the government's record in the early years of the Lula government. Where else could anyone buy non-GM soya in large quantities after 1998? European food retailers like Carrefour or Sainsbury's were busy tracing supplies in Brazil as early as 1998. Despite technophile and anti-left instincts, and with economic calculus, the Brazilian agribusiness celebrated a 'soya miracle', enabling the European moratorium. The soya planting areas more than doubled, productivity increased, and penetrated the rain forests of Mato Grosso and Para.

By 2003, it was clear to scenario makers that the future of green agbiotech hinged on Brazil; accordingly, the pressure on the government of Lula da Silva to lift the moratorium mounted. The bio-safety law of 2005 brought clarity: the issue of 'transgenicos' was buried in an emotive campaign to liberate stem cell research to reduce human suffering. This disarmed green and consumer resistance; they avoided an anti-science coalition with religious voices against embryonic stem cells (see Bauer, 2006).

The morals of this story are multiple: Brazil reaped the benefits by not jumping on the GM bandwagon. Global public sentiment over GM crops was a major factor in bringing about a constellation where this was possible. Contrary to international trends, it was the unification of green and red biotechnology under the banner of stem cell research that legalised GM crops, while in most other contexts food and pharma operate with quite different rule books.

## Operational Learning: Realigning the Portfolio

The path of technology is littered with dead ends and orphan inventions. What are of interest to the present argument are dead ends of biotechnology caused by public opinion. I will briefly describe some developments that promoters abandoned in a tactical move to appease public opinion. These moves are tactical because they constitute small changes in the portfolio and no structural change, accommodating public opinion nevertheless.

Genetic use restriction technology (GURT) became widely known under the polemical term 'terminator technology'. GURT technology was deemed to lead second generation value added crops; the first generation focussed on pest controls. GURT controls the genetic expression of traits by manipulating fertility or germination, or by conditioning gene expression on a particular chemical. The technology has three applications. Terminated seeds are sterile, but have attractive traits and the farmers buy new seeds every year. It creates a market and engineers customer loyalty. GURT increases yield by controlling the timing of germination and thus maximising harvesting seasons. Thirdly, GURT contains gene flow and thus avoids liability and legal risks of growers (see Einsiedel and Geransar, 2005). The power of GURT to control markets has put it into focus of campaigners. Sterile seeds undermine the seed production of farmers, deskill and create dependencies; sterile seeds are therefore immoral and dangerous. The Rockefeller Foundation urged Monsanto to restrain itself, which they did in October 1999. This retreat from GURT was presented with the new pledge to rebuild a reputation of 'dialogue, transparency, sharing, benefits and respect'.[9] However, this concession was no conversion on the road to Damascus. It was rather a temporary set-back of the biotech movement: terminator technology continues to be in the portfolio of other firms and patent holders. Campaigners suspect 'trojan horses' in second generation GM crops, a business model based on the potential to erase transgenic traces from crops (exorcists: see ETC Group, 2003a).

Another concession is witnessed in the fact that we have no GM flour in our bread. Primed by the experience of soya, US and Canadian farmers were reluctant to follow Monsanto into Round-up Ready wheat, which, in development since 1997, promised yield gains of 10–15%. Nearly half of US wheat goes to Europe and Japan, and farmers feared losing that market, as they did for soya. European millers were seeking alternative sources: bread made of GM flour was deemed unsellable. In May 2004 Monsanto made headlines by announcing a 'realignment of its R&D portfolio', deferring RR wheat, and focusing on corn, cotton and oilseeds (see Brown, 2004). Organic producers celebrated. Farmers reluctance, now aware of consumer and citizen sentiments around the world, made Monsanto to reshuffle its portfolio, a case of tactical learning: drop the path that might be a dead end.

## Changing the Research Focus: Somatic Rather Than Embryonic Stem Cells

The debate over Dolly the cloned sheep never was a debate on animal cloning for medicine, milk or meat. This it became only years later, when food safety of cloned cattle stock was queried in the US (see Rudenko, Mathesonm and Sundlof, 2007). The immediate significance of Dolly is the questions it raised on human cloning and embryo research in the 'stem cell wars' of 2000 to 2002 (see Weingart, Salzmann and Wormann, 2008). A culmination was reached in August 2001, when US President Bush announced a ban on federal funding for embryonic stem cell research (see Wigzell, 2007).

Political scientists explained why stem cell cloning was an issue in the US, and agbiotech was not. Lobbying the congressional process blocked the one but not the other. Stem cells aligned with abortion, which had already divided the country along party lines (see Sheingate, 2006). The moral issue was sourcing stem cells from early embryos, so called plastocysts which were later discarded.

In many contexts, the rules for embryo research were settled in the abortion debate, either on women's prerogative of a decision, or as in the UK with the case by case oversight of the Human Fertilisation and Embryology Authority (HFEA). As stem cells gained economic significance in the global frontier of science and a source of patents and industrial profit, the established consensus appeared in a new light. Most stem cell patents were publicly held, contrary to other gene patents, but this reflected the significance of public health funding for early stage research, and venture capital shied from the stem cell controversy (see Bergman and Graff, 2007). As stem cell research adopted the frame of therapeutic promise (see Rubin, 2008), women's choice was no longer the key frame of debating embryos. At issue was no longer the prerogative of women, but the prerogative of the industry to trade life forms; a concession to women does not extend naturally to industry.

A move to somatic or adult sourcing of stem cells would liberate this research from many moral dilemmas. In 2005, The International Society of Stem Cells Research defined the scientific challenge of reprogramming adults cells to produce stem cell lines that behave like embryonic ones. This would avoid the controversial use of donor eggs or creation of embryos for this purpose. Alternative sources such as blood, bone marrow, fat and other tissues provide stem cells that could be reprogrammed. The issue is the relative distribution of efforts between embryonic and adult stem cell research. As in choosing between nuclear or renewable energy: investment on the one seems lost on the other.

There is some evidence that the resistance to embryonic stem cell research has refocused research on somatic cells. Publications referring to 'human stem cell research' (excluding embryonic stem cells) grow faster than embryonic stem cell research, the latter stagnating briefly after the controversy of 2002 and again in 2006. Adult stem cell research makes much of the science news not least because it offers a solution to a moral dilemma and avoids reopening the abortion debate. Moral indictments can be an incentive for technological innovation, as can be shown in the case of Orthodox Jews, whose adherence to Sabbath rules finds ingenious ways of getting things done nevertheless, i.e. they invent procedures that have the same outcomes but do not involve actions prohibited by religious observation.

In the US, where the issue has been debated most intensely, stem cell research is nevertheless leading the world; producing 53% of inventions (see Bergman and Graff, 2007). The prospect of resisting research that falls within the frame of therapeutic hope is limited. By comparison, public opinion was more effective in curtailing GM technology with concerns over safety, choice and environmental risks.

## Tactical Shift: Respecting Tissue Donors

The Human Genome Diversity Project (HGDP) started in 1995 and took blood samples from tribal populations around the world to compare their genomes with a view to mapping genetic variety. This should complement the unified approach of sequencing and mapping the human genome. Rather than one 'book of life', there might be several revised editions.

Since 1993, the Indigenous Peoples' Council on Biocolonialism (IPCB) drawn attention to issues and interests of indigenous tribes who became the focus of this research. IPCB and others raised the question of consent. How should consent for tissue donation be obtained within tribal communities, through individual or collective consent? How can these communities secure benefits from this research? Fears emerged that research on genetic variation will revive notions of 'biological race' and fuel the age-old racism with a new science of a hierarchy among groups of people (see Smedley and Smedley, 2005). Furthermore, insights into the genetic make-up might give a lever for designing biological weapons against tribes, many of them involved in conflicts. The concerns were 'biopiracy', the looting of genetic resources in the developing world for the benefit of the developed world. Cavalli-Sforza, a leader of the project, described its efficient data collection as ethnocentric: one person can bleed 50 people and get on the airplane in one day. Henceforth, the HGDP was known as the 'vampire project', sucking blood from the many for the benefit of the few.

Between 2002–2006, the Haploid Mapping (HapMap) project, with a budget of >US$100 million, followed the first efforts to map genetic diversity. It incorporated some of these concerns. It reduced the samples to only four regions, Europe, Nigeria, Japan and China, and incorporated community consultation before samples were taken, to understand and consider potential concerns. And it excluded any linkage of genotype to phenotype from the project work. Association studies to show the functional utility of elements of the genome, which could be patented, are prohibited within the project, and all research data was to be public data at the end of the project (see M'charek, 2005). Similar projects targeting genetic variability proliferate in consortia such as the Genographic Project of the National Geographic Society and the Human Genome Structural Variation Working Group (HGSVWG, 2007). They identify variety and subscribe to open access. Others projects focus on association studies, like many of the national biobanks such as Iceland's DeCode, the '1000 genomes project' (2008), the Wellcome Trust Case Control Consortium (2007) and similar activities around the world.

Database projects face the dilemma of securing consent and enabling research; what shall and can people consent to (See Arnason, Nordal and Arnason, 2004)? Biobanks are worried about the motivation of donors. Blood donation has a long tradition, as in the UK National Health System, where it is done in solidarity with the community. This context is altered when biobanks are no longer projects of and for the community, but a private, profit-making business. New issues of privacy and equity arise of who uses the data and reaps the benefits (see Weldon and Levitt, 2004). In Iceland, the idea of merging three databases was changed in response to public resistance. When 10% of the population opted out, the government decided that deCode was to make do with only two databases and their licence be limited to 12 years and annually reimbursed.

Through the 1990s, the gene donor found respect in the rhetoric of research projects. Individual or community consent was, by 2008, de-rigueur; the equitable distribution of benefits is on the agenda and concerns over privacy and the misuse of genetic information cannot be easily dismissed. To what extent this is a reaction to public opinion is less clear. Debates on red biotechnology never had the reach of its green counter-part, except maybe over stem cell research in the US and in Germany. The promise of reduced suffering makes it for most people rather difficult to resist and mobilise against red biotechnology. The public opinion data reflects this state of affairs (see above): as opinions over biotechnology get negative, this was mainly due to green and less so over red biotechnology.

## The Loss of the Authority of Bioethics

My final observation concerns the changing status of bioethics. Bioethics is applied philosophy that traces its history to the Nuremburg Code of 1947 that prevents a repetition of the Nazi type abuse of medical research subjects. Ethical considerations of new technology moved centre stage in the context of genetic engineering. It is not entirely clear why this might have happened. Nuclear power saw debates over deterrence in the 1960s, but the main frame was technical risk and risk assessment. For some philosophers genetic engineering became the catalyst to think generally about technology in terms responsibility for future generations. Jonas defined the problem as dealing with dilemmas on the basis of principles not interests (Jonas, 1979, 50), but saw the principles of ethics based on a 'fixed human nature' that is undermined by progress in biological science. This defines the challenge of bioethics: how to determine principles when the foundations of these principles shift in line with what you are trying to assess? Gut feelings or 'yuck-factors' are rehabilitated as heuristics of judgement on such matters (see Gigerenzer, 2007).

Genetic engineering moves ethics centre stage not only among experts but also in public opinion. Eurobarometer surveys showed that by the mid-1990s moral concerns had gained veto character over applications of biotechnology (see above): moral soundness trumped risk perceptions. Many promoters had hoped to frame biotechnology as a risk issue and to keep morality out of the picture. Equally, in the European media ethics displaced the progress frame by 1999 (see above Figure 8.3; and Bauer, 2001).

The HGP had given bioethics the platform: 5% of its US$ 3 Billion budget for ELSI created the best opportunities for philosophers since Plato's academy. Universities rushed to set up bioethics units to secure a share of this bounty. The success of bioethics requires a sociological take (see Stevens, 2003). For bioethics 1997, the year of Dolly, is a turning point.

This year alone saw 50 bioethical declarations from governments and international organisations; national ethics committees were instituted if they did not exist already. A competition to ban human cloning, absorbing the universal yuck-factor, culminated in the UN's Universal Declaration on the Human Genome and Human Rights of 1998. This flurry reflected an increased presence and also opportunities for bioethics.

The issue of human cloning also shows the reputation problem of bioethics. Since the general outcry over human cloning in 1997, the goal posts have moved time and time again. An initial ban of human cloning became a ban of reproductive cloning, while therapeutic cloning should be allowed. Then reproductive cloning was to be unproblematic in embryonic stem cell research and some IVF procedures. Finally, human-animal hybrids are considered admissible for research purposes, at least in the UK. With every new technical possibility, the goal post moves on. How does this reflect on bioethics? One is once more reminded of Blumenberg's soliloquium: we need not, but we will do it. . . . (see chapter 1).

Bioethics moves with the time. But also bioethics is unable to block anything. To the contrary, it has an enabling role in technological change. The recent polemic over bioethics is not very flattering. 'Bioethicists have become nothing than sophisticated (and sophist) justifiers' declares Fukuyama (2002, 204). Bioethics has succumbed to regulatory capture, where the oversight agency (i.e. bioethics) gets caught up in what it supervises (i.e. biotechnology); its existence depends on what is overseen. In other words, bioethics will not bite the hand that feeds it.

Bioethics commissions attempt to generate public discussion. However, these are often not forums for effective public dialogue. They have become administrative tools to move forward controversial technologies, their specific criticisms notwithstanding. Bioethics is at risk of being co-opted (see Galloux et al., 2002; Stevens 2003). It has become a profession with interests in self-promotion, and as such has lost much of its authority. New versions of neuro-ethics or nano-ethics testify to their role in fostering the technologies on which they depend (deVries, 2007).

Bioethics became the latest ploy to absorb acceptance problems, as traditional instruments have become blunt. It could be argued that bioethics is the specific manner of the biotechnology movement to deal with a recalcitrant public: organise public meetings (procedural bioethics) and sharpen the arguments to carry universal support (utilitarian bioethics). Risk perception was the concepts developed to deal with a recalcitrant public in the context of nuclear power. An industry was created to reduce the lay-expert gap in perception with cognitive therapy. Also the risk perception approach has lost lustre and analytic power, but created a scientific community (see Chapter 7). In this sense, bioethics arguments take the role that 'risk perception' previously held, but suffers a similar fate of getting blunt. The attempt to delegate decisions to experts, of either risks or ethics, ultimately faces the problem of legitimacy. It is as if common sense reasserts itself. It might be that a 'critical' bioethics can rescue its case and assert local common sense against propaganda (see Kelly, 2006). It seems that old problem persists: to keep a watchful eye on the vanguard and their obsessions. Common sense reasserts itself against being disenfranchised by social engineering and technocracy, which is a legacy of modernity itself. The intensity with which the 'biotech revolution' is defended against its anti-science distracters 'in politics and protest' (e.g. Miller and Conko, 2004) testifies to the significance of the latter. But in the face of all Whiggish accounts, Biotechnology of 2015 is different from what promoters back in 1990 imagined it to be, and resistant public opinion played its notable and noble part.

# NOTES

1. OECD statistics of December 2012 reports a global total of 15,631 biotech companies, of which 6,636 are 'dedicated' biotech companies whose dominant activities are biotech (source: OECD key biotech indicators, December 2012).
2. An anecdote on this matter: in 1992, we conducted a series of focus groups in the greater London area to scope public perceptions of the human genome project; one of the groups consisted of health and legal professionals. We found very little awareness of the genome project, even among health professionals. The only person who knew everything about it was the one patent lawyer in the group. We later learnt that the lawyers had formed an informal discussion group on legal implications of genomic research.
3. The patent battle with Myriad Genetics over the rights to BRCA1 and BRCA2 was finally resolved in 2013, when the US Supreme Court unanimously decided that genes are products of nature and cannot be patented; synthetic genes would however be patentable (source: NZZ, 15 June 2013, no. 136, 20)
4. In the project 'Biotechnology and the Public' (Bauer, Gaskell & Durant, 1994; Bauer & Gaskell, 2002), we anticipated 'military uses' when we coded over 20,000 news items in 18 countries between 1973 and 2002; we found 66 main references (0.3%) to 'military uses'; clearly not something in the public mind before 9/11 2001.
5. Marc Cantley of the EU Commission had variously warned about and predicted difficulties over GM crops in Europe, which is why he instigated Eurobarometer to start its Biotech series in 1991. And Walter von Wartburg, the person responsible for communications at Novartis in the mid 1990s, agonised over why he and other protagonist did not read the signs of impending revolt when all looked like it was a smooth and easy path ahead for biotechnology (personal communications).
6. Monsanto later lobbied for years in Brussels and across Europe for the liberalisation of GM crops in Europe, with little success. These efforts formally came to an end in mid 2013 (source: NZZ, 3 June 2013, no. 125, 12)
7. Considering the performance of Novartis and Syngenta since the divorce, it showed that 10 years later Novartis had more than tripled its turnover, while Syngenta only doubled; the divorce clearly was more profitable for the one than for the other (source: reports in NZZ, 27 January and 10 February 2012).

8. By 2013, environmental activists go public with a change of mind. For example, Mark Lyans, at a conference in Oxford, came out as a converted supporter of GM crops: having studied the evidence he realized that it helps the productivity of farming (source: *Focus*, 5, February 2013, 95). For June 2014 a meeting at the Royal Society of London promises 'to start a new conversation on genetically modified crops' (funded by the Templeton Foundation).
9. The New Monsanto Pledge appeared on its website www.Monsanto.com as early as November 2000.

# 9  Some Further Observations on Resistance

'Resistance is futile.'

(*Star Trek: The Next Generation*)

In this book I have been tracing public opinion over nuclear power, over environmental consciousness, over information technology, and over genetic engineering and genomics. I have shown how public sentiment on these topics has changed, but also how concepts of 'public sentiment' have changed. It is a general feature of the social sciences to look at things with two eyes, one focussed on the facts, and the other on the framework which holds 'facts'. This makes any observation doubly unstable. In this sense, social psychology cannot be anything other than the history of the present. In that I concord with Gergen (1983) who suggested that only misplaced scientism can believe in discovering the universal laws of mentality, when what is uncovered, even with heavy experimental and statistical apparatus, are ultimately historically situated mindsets. By tracing the dynamics of public sentiment during the latter part of the 20$^{th}$ century, I hope to have shown how public opinion emerged as a factor to be reckoned with for science and technology. Crucial to this story is how the framing of mentality has changed.

A striking feature of this history is the differential in resistance across techno-scientific developments. Apart from contextual variations, it is fair to say that nuclear power received more critical sentiment and disruptions than genetic engineering ever did, and both developments faced still larger challenges than computers and IT. The absence of critical mass to challenge IT despite plenty of potential concerns is the surprising historical fact that calls for an explanation.

The social sciences, beyond social psychology, have grappled with 'resistance' all along this history of the present (see Bauer, Harré and Jensen, 2013). Major research programmes such as 'attitudes to science' and 'risk perception' were invented to contain public recalcitrance. The problem was defined as the gap between expert judgement and lay opinion, explained by a deficit of information (the more people know, the more they agree) or deficient information processing (if people knew probabilities, they would opine better) structured by socio-economic contexts. However, trying to explain the resistance to new technology by the 'causes' has limited reach. It poorly explains the differential resistance against nuclear, biotech and IT. More likely, disagreement with expertise is not the problem, but part of the solution; the problem is the techno-scientific project itself. While it is part of a toolbox to assimilate the public to the mobilisation effort, accommodation of public deliberation into the project promises a more sustainable solution.

In Chapter 2, Mobilising a Different Future and Chapter 5, Ten Propositions on Learning from Resistance, I developed a new look at dissent over techno-science: resistance is a necessary and primarily functional element of any mobilisation process. This framework theory specifies how the mobilising movement responds to challenges.

The key theoretical problems are these: when facing a challenge, how to turn external into internal attribution, how to avoid blame shifting and to take responsibility, and how to move from assimilation to accommodation of resistance? These problems are fruitfully elaborated in the 'pain analogy': resistance works like 'pain'. Resistance is functional for mobilisation efforts in several respects: it draws attention to neglected issues and stimulates reflexivity; it evaluates established directions and opens alternative paths, and enables structural learning. Resistance moves assimilation efforts into accommodation mode. Resistance as pain has transformative potential.

## TURNING TO THE NEXT TISSUE OR ABSTRACTION

How to end this book, brought me some agony. A maybe more 'natural' ending would be to extend the historical narrative to current affairs, and answer the question: so what? What can we learn from the controversies over nuclear power and biotechnology for nanotechnology, synthetic biology or the Human Brain Project?

Indeed, evidence abounds which illustrates that nanotechnology and synthetic biology have many features of social mobilisation processes. Nanotechnology has seen major resource mobilisation since the late 1990s. Movement organisations like the Foresight Institute have been drumming the 'nanofuture' since 1986. Dexler and Smiley were the prophets, and Crichton's novel *Prey* (2002) provided a dystopian vision of a world turned into 'grey goo'. In 1999, nanotech moblised the US government into funding an $8 billion research programme, and other governments followed suits. In 1992 the journal *Nanotechnology* had its first issue; in October 2006 *Nature Nanotechnology* added to this line of public recognition. Many universities have opened nanoscience centres. 2005 saw the creation of the LUX stock index, which tracks the movement's credibility on the capital market. Efforts to assimilate the public into nanotechnology take the form of trade shows, exhibitions and 'up-stream engagement' to pre-empt surprises down the line. Nano-opinions are being monitored (Gaskell et al., 2004; Lewenstein, 2005). Controversy flared briefly in 2002 when the activist group ITC raised the spectre of health and environment risks which Prince Charles doubled down on in 2004, and a counter-movement seemed to form as existing SMOs and issue entrepreneurs diversified their portfolio (ETC Group, 2003b). Learned societies and funding agencies commissioned risk assessment reports. A rhetorical sorting of science fiction, nano-nonsense and alarmism from the moderate voices followed the initial polemic. The nano bandwagon attracts free riders who capitalise on a new opportunity; Ecsite, the European Network of Science Centres and Museums, maintains the project Nanopinion, where they periodically sponsor supplements in newspapers which put nanotech to debate.[1] Material scientists, laser physicists and toxicologists continue to pursue their research much as before, but now under a new flag of convenience and with more public attention and funding. More recently, similar expeditions left base under the flag of 'synthetic biology'; inheriting structures of the human genome project in analytic techniques, researchers and labs and researchers wonder about its conflict potential (Torgersen and Hampel, 2012). The social sciences face high expectations to help avoid the so-called 'public opinion disasters' of GM food, stem cell research and nuclear power. Nanotechnology and synthetic biology should have a smoother future. Acceptance research is seeing a revival.

It is clearly relevant and tempting to add to these endeavours of dispensing well-paid advice to given agendas. But rather than jumping onto the next mobilisation bandwagon of 'emergent technologies' (see Nuffield, 2012), I opted to look sharper with the second eye, looking at the framework that holds the facts. What follows thus takes the route of further abstraction, looking to a new research programme for the social sciences.

## TOWARDS A GENERAL RESISTOLOGY

I seek to formulate the outlines of a general social resistology.[2] Academic probity requires I admit this is not an entirely new idea. Seidmann (1974), a psychiatrist, suggested such a trans-disciplinary field of enquiry some years ago. His concerns were twofold, a) comparing the resistance of the patient and of the guerrilla-partisan, and b) investigating the potential pathology which they share, i.e. the 'Herakles complex' of a nihilist power motive. Pathology of 'resistance' manifests itself in a denial of truth and moral responsibility, justification for which comes from adherence to norm-esoteric groupings, visionary aims and claims to metaphysical necessity. I share this quest for a 'resistology', but not Seidmanns's method of seeking to clarify normality through the pathological case.

Some years later, Nordal Akerman (1993) explored resistance starting from Clausewitz's analysis of warfare and the 'friction' of unforeseeable difficulties. Akerman invited commentators from different domains to reflect on 'what keeps you from realising your goals, what compromises all plans, sometimes making them unrecognisable. . . . constitutes the divide between dream and reality' (8). Resistance is the great leveller between human efforts across all cultures. He diagnoses a problem in the modern rationality project: to underestimate the problem of 'resistance'. Without friction, there is no movement (7ff). The contributors examined the social sciences, physics and metaphysics, the battlefield, the marketplace, play, industrial design and forecasting. The common thread is the ubiquitous resistance to rational design and a polemic against the persistent illusions of movement without friction.

In a similar but unconnected project, some years ago, I brought together historians and social scientists to explore the significance of resistance against new technology (Bauer 1995). The contributors reassessed the impacts of resistance against 19th century industrialism, and modern nuclear power and computing, and asked the then burning question: will there be resistance against biotechnology? History has since caught up with events, and the late 1990s brought global resistance to certain forms of genetic engineering. The questions raised at the time attempted to move the analysis from legitimacy to functionality: irrespective of being rational or irrational, informed or ignorant, what are the historical contributions of resistance?

The question of how resistance relates to rationality and irrationality in different fields of enquiry from physics, biomedicine, psychology and history to political science, criminology, law and military strategy, was again the topic of a recent collection of essays (Bauer, Harré, and Jensen, 2013). The evidence is clear across many different fields: the attribution of irrationality to resistance is unjustified, irrational in itself. Resistance is as much rational as it is irrational; prejudice is a risky way of thinking about 'resistance'. The contributors explored a key problem of any resistology: how to avoid the fallacy of an ideal language. Assuming that because there is the common word 'resistance' across different fields of enquiry, there must be a common reality to which it refers. There are very few words where sense and meaning coincide; many more are homonymic and polysemic, and in addition are used metaphorically. Thus the meaning of 'resistance' must be entangled in different language games using the term.

A 'resistology' will include the philosophical task of critiquing the language of 'resistance'. Such an effort might usefully reveal 'family resemblances' across different usages of resistance, and identify the to and fro of metaphorical entailment in the transfer between different fields. It might turn out that some particular use, e.g. in physics, is the root metaphor of thinking in the social sciences; as it seems the social sciences are indeed to a large extent a tribology of 'lubricating' social change understood in terms of forces (see Nowotny, 1983). This amounts to a comparative history of concepts. There are interesting example of such enquires: the notion of 'selection' moved to and fro between economics and evolutionary biology during the 19th century (see Young, 1971; Gale, 1972).

A phenomenology of 'resistance' will have to clarify this use of borrowed language. Semantically, terms *denote*, normally clarified by definition and convention, and *connote* an open set of associations that give emotional significance which differs across subgroups of language users. Depending on whether connotations are predominantly positive or negative, 'resistance' will be a self-description with a stigma.[3]

Rather than attempting to formulate a general theory of resistance, which would be premature or even a logical mistake, as the existence of a word does not imply that there must be a single corresponding slice of reality. It might be more useful to unfold a series of perspectives which reappear across different fields of enquiry. Three perspectives on resistance seem to be recurring: that of the actor, of resistance itself or of the observer.

## THREE PERSPECTIVES ON RESISTANCE

Resistance as an action can be analysed from three different perspectives. Each perspective offers a self-concept, a take on resistance and a view of alter-ego transactions. For the change agent who promotes change, resistance is the cause of failure and frustrations; the problem is external to his or her own actions, and the interaction with others is seen as a null-sum game of winning or losing. In the perspective of the resistor, the change agent abuses his or her power, resistance is a moral imperative, and conflict is necessary to bring these abuses to light. The bystander and analytic observer observes mindsets and self-observations. He or she sees conflicting actors who mind themselves and others in particular ways, and this is part of the problem. The analyst can map how actors behave and perceive each other with reference to the common conflict zone. The functionality of resistance is recognised, not as a self-serving bias, but in relation to the project and its insufficiencies. The bystander might be able to mediate the conflict and move the game towards a win-win situation. Resistance, frustrating on the one hand and morally justified on the other, is thus socially functional. Indeed, it is all of these things.

### The Change Agent's View: A Nuisance

For the change agent, resistance is a nuisance. The innovator thus tends to explain resistance in terms of deficiencies: deficits of cognition and information processing (e.g. bias, irrationality, risk perception, prospect theory), deficient personality dispositions (e.g. technophobia, nuclear phobia, cyberphobia, rigidity, conservatism) or collective dispositions of the social system (e.g. bureaucracy, anomie, routines, traditional culture). From the perspective of desirable change, traditions are obstacles to progress; and society is a target of intervention, manipulation and marketing strategies to change hearts and minds.

Resistance is analysed as a relation of five variables or questions: actors (who?), repertoire (how?), target of resistance (to what?), causal conditions (because-why?) and reasons (justification-why?). Change agents often see themselves in possession of the one best way, a 'Rationality' writ large; that is a source of confidence to overcome obstacles, but that can also become part of the problem.

Figure 9.1 shows the problem of resistance from the change agent perspective. The change agent promotes a project for the future, and might take confidence from a sense of revolution, being vanguard or a technocrat with a knowledge advantage. The change agent sees resistance as putting up a barrier, blocking progress in the projected direction. The analytic gaze is

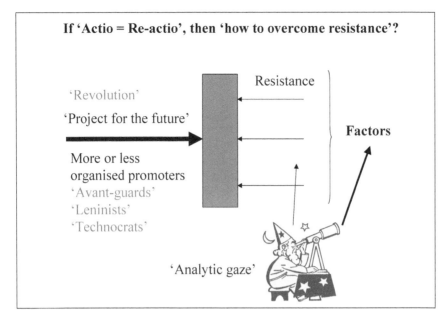

*Figure 9.1* Resistance as counter-force to change and its factors.

focussed on the factor of this barrier. The idea is that by analysing the factors of resistance it will be possible to control it: 'how to overcome resistance?' is the key question.

## Dynamic Fields of Forces and Cognitive and Motivational Deficits

In the social psychology of Kurt Lewin (1936 and 1952) one finds the explicit field theory of resistance that lives on in much experimental demonstration and analysis, though often implicitly. Lewin developed the 'hodological space' (Greek for ego-logical), the ego-centric perspective vis-à-vis others and the world. The *life space* is structured into regions, a topology of relevances exerting approach and avoidance forces. Each *region* is represented by a *valence* and a directed *vector*. Thus a field of forces emerges with the actor at its centre. The system tends towards equilibrium, thus *locomotion* takes place in the direction of balancing the forces. If negative valences exceed positive ones, the actor avoids, otherwise we approaches the target region. Conflicts of approach-avoidance arise. Equilibrium is reached with high or low *tensions* of forces. High tension leads to erratic behaviour, such as exits from the force field.

Regions between the actors and the goal are called *barriers*. Everything that inhibits reaching the desired state is a resistance or barrier; this includes other actors contrary actors with affective (do not like it), cognitive (do not believe it) and behavioural (will not do it) inclinations. These sources of frustrations are cognitive or *motivational deficiencies*. The barrier is the bundle of counter-forces. When the force of the barrier matches the force of goal attraction, there will be no movement across the field. Note, other actors enter the force field as potential frustration of the actor's ego-centric desire; the latter depends on the distance from the target; the closer the target, the larger the frustration.

Field theory suggests two strategies to overcome resistance (see Lewin, 1936; Coch and French, 1947; Lawrence, 1954). First, resistance is overcome by mobilising stronger forces for change, i.e. bulldozing the barriers. But because of the *dynamic law of action = reaction* in field of forces, this is a risky strategy. If resistance responds in kind, tensions will increase tensions and create uncertainty in the system. The second strategy seeks to decrease resistance in the first place, so that under the dynamic law of forces less energy is needed to move towards the target. Several tactics are described to undermine resistance, including confusion, distraction and persuasion (Knowles et al., 2001).

With psychological field theory, resistance seems controllable; other actors in the force field can be fixed. Take for example risk perception in the context of new technology. Assessment of experts, so-called 'objective risks', are compared with 'subjective' risks of non-experts, for example over nuclear waste disposal or GM crops, and deviations between expert and lay assessment are explained by *deficits of information and processing* on the part of the laity: lack of knowledge, inadequate knowledge, biased experience, biases of inference and information processing (see Kahnemann, Slovic, and Tversky, 1982). These deficits constitute counter-forces but manageable through training and de-biasing (Fischhoff, 1982), the provision of which becomes the business of debunking and popularising statistics. However, this technocratic perspective ignores that perception depends on emotional framing and tribal relations among people, and both emotional and tribal orientations and partiality are also the case for experts (see Chapter 7).

## Resistance Perspective: The Moral Imperative

Resistance is often reasonable in countering an abuse of power, an overstepping of the line and a limit of tolerance being reached. Resistance is thus justified by appeal to moral and ethical principles.

### The Rhetoric of Reaction

Hirschman (1991) has elucidated generic arguments that have been put forward in many different contexts against social reforms. The typical arguments against reforms of social practices are based on the law of unanticipated consequences of strategic action (see Merton, 1936); they are the arguments of perversity, of futility and of jeopardy.

The argument of *perversity* stipulates that purposive action intended to improve things only exacerbates the condition it wishes to remedy. The attempt to move things in a particular direction will move things, but ultimately in the opposite direction, due to volatile unpredictable environments. Reforms tend to bite back. The argument of *futility* stipulates that purposive attempts at social change will be unavailing and reach a null-effect because there is no grip on reality. A cybernetic and in that sense predictable environment will compensate for any wilful intervention. It works independent of human action, which has no grip on things. Finally, the argument of *jeopardy* suggests that the proposed social changes are too risky, as they endanger previous, precious accomplishment. A classic example is the liberties that might be under threat from further change ; a previous achievement is not to be sacrificed to a move too far that has predictable averse consequences.

A few years later, Hirschman (1995) reviews his evidence and, and in an act of self-subversion, concludes that there is often a kernel of truth to these 'reactionary' arguments. In effect, there is a progressive equivalence for each of them for all those who do not want to be 'reactionary'. The progressive uses perversity to show that inaction will not suffice, and decisive action is required to avoid disaster. One is reminded of the climate change debates. The progressives recognise futility when they argue only for those reforms that are within the tides of history, thus sorting tendencies from utopia. And finally, jeopardy is invoked when reform projects are to build on and to solidify earlier achievements.

## The Dignity of Resistance

The legitimisation of resistance often refers to civil disobedience and appeals to the tradition of non-violent protest (Sharp, 1973). *Philosophical reflections* express the resistance point of view (e.g. Saner, 1988 and 1994). To modern democracy, legitimate resistance poses a dilemma. The right to resist the law is a contradiction at the core of the law. The legal system is a process, and resistance signals shortcomings in that process such as slow reactions, legal gaps and the decay of democratic institutions under changing circumstances (Rhinow, 1985). Under separation of powers, the judges judge for themselves whether there is such a shortcoming or not, but they may need a reason to convene, and public resistance gives that occasion.

Resistance is grounded in the ethos of a revolt. Albert Camus' existential reassurance of the individual occurs among fellow human beings: 'Je revolte, donc nous sommes' (1951, 431) . Touraine varies this back to a Cartesian credo: 'je resist, donc je suis' (Touraine, 1992, 318). Resistance creates the 'subject', a distinct notion that is based on the triple distinction between 'soi' (= self), 'moi' (= me) and 'je' (= I). In becoming a modern subject humans face a neither-nor dilemma, rather than an either-or choice. We are called to resist the temptation of two evils: individual narcissism (the illusory 'soi' = self) on the one hand and collective self-denial (the internalised role 'moi' = me) on the other hand. Avoiding a romantic delirium of style and grand causes (the cultivated self) and a depersonalised modern bureaucracy (the role 'me', of duty), modern subjects (I) resist and hang in the balance as an existential condition. Both evils ultimately deny the 'je = I' as an autonomous subject (Touraine, 307ff). The 'I' gains strength and self-assurance only in equidistance between romantic narcissism and bureaucratic self-denial. The subjective individual, the 'I', must steer between the Scylla of particularism and the Charybdis of universalism. This tension cannot be resolved; it can only be endured with lucidity and resistance action. This is the condition of modernity to which a dual progress towards bureaucracy/efficiency and individualism has led us, overcoming both inefficient production and repressive collectivism. The dilemma defines the dignity of resistance in modern terms.

The social basis of resistance changes from one societal formation to another. In an industrial society the basis of resistance lies in old-style landed aristocracy (gentlemen) and those with small or no property holding at all ('proles', trade unions, communities) who at times form an alliance. In the coming knowledge society the major line of conflict is over the control of techno-science that confronts managerial experts and non-experts. Resistance against corporate patronage of scientific research, against public squalor and private splendour (Galbraith, 2005), may be a characteristic of post-industrial society. Sloterdijk (2005) rhapsodises on the new 'age of inhibition' where resistance has become normality again after 450 years of dis-inhibition of all restraints.

### The Right and Duty to Resist

The *right to resist* guards us against the abuse of power by 'tyranny', an authority out of bounds. Age-old myths and folktales justify such acts of challenging illegitimate authority. Antigone insists on the natural law of kinship against her law-making father and buries her outlawed brother. The moral of the story: natural law challenges statutory law. However, a legal system that includes the right not to comply is a paradox: the law cannot allow law breaking without undermining itself. Constitutional thinkers find ways out of this paradox. There is a need to tolerate some stringently defined deviance to compensate for imperfections in any system. Democratic societies need acts of resistance to test their systems of justice, to check on misconduct of their institutions, and to signpost lacunae in the operations of the law (e.g. Rhinow, 1985). Resistance might be illegal and therefore punished under the law, but it might be legitimate on moral grounds. This makes conscientious objection and civil disobedience risky business (e.g. Haskar, 1986; Bay and Walker, 1975). If resistance acts against positive law, punishment is the outcome. In court, however, the public display of moral conviction and the appeal to common principles has the power to convince others to join and for authorities to give way. There is a difference between the letter and the principle of the law. And according to Rawls (1972, 363ff) civil disobedience is justified as a last resort, after all legal means have proven to no avail, and with an appeal to justice. Resistance justified exclusively on religious or ideological grounds cuts these ties and falls outside the moral community of justice. There is thus tolerable and intolerable resistance against the law.

The right to resist includes by implication a *duty to resist*. The importance of such a construction lies in considerations on how to judge past actions: exclusively in relation to positive law or also in relation to a universal duty to resist injustice? In the light of a duty to resistance, it is thinkable that actors were wrong, despite having upheld the law. In conforming to existing law, one can violate natural law and thus be guilty on a different count. This controversial issue has preoccupied for example German law makers over how to call to account, if at all, former administrators of the Nazi regime—and those who continued in official positions after regime change in the former German Democratic Republic.

## Observer Perspective: Potential Functionality

The explicit analysis of root metaphors of existing models and mindsets will be a key problem of resistological research. We must ask whether our analyses of resistance are based on a physical analogy of 'friction' or a biological analogy to 'viral infection'; and what are the entailments of these analogies. For example Kruglanski et al. (2007) have demonstrated this by analysing the discourses of counter-terrorism. Practices in this field are rooted in metaphors of warfare, policing, infection or inter-group relations. The analysis shows that relying on any one of these mindsets is risky, as the analysis remains partial and becomes part of the problem rather than the solution. Any single metaphor misleads the intervention by overcommitting resources on a partial definition of the problem.

A different theoretical contribution to resistology is the elaboration of new metaphors. There might be a fundamental option between mechanistic and organic metaphors, which is an old dilemma for the social sciences. Where do the social sciences look for inspiration, in clockwork mechanisms or in ecological systems? The physics of free-fall and dynamic forces, based on the law of action and reaction, dominates much of resistology (Akerman, 1993). Much of this amounts to a social tribology considering resistance as a matter of energy conservation through effective lubrication.

To look at resistance within a framework of collective activity theory allows us to specify hypothetical functions (see Chapter 5). The movement approach allows us to draw analogies between individual and collective learning (Jost and Bauer, 2005; Bauer, 1991 and 1993). Through the analysis of social mobilisation I extend this analogy to the level of society and its projects. It is important to stress that the similarity between pain and resistance is not a homology. The pain metaphor of resistance does not imply that people resist because they are in pain. We can use the term 'pain' metaphorically for grievances, scandals and social deprivation. In a special case, like for example people experiencing RSI from overusing computers, people might mobilise against the computerisation of work and thus cause 'organisational resistance'. Resistance motivated by pain is a special case, but most of the time resistance is not motivated by other motives. But it will nevertheless be useful to understand this in terms of a pain process. The similarity between pain and resistance arises not from a common cause but from similar responses of movements: pain and resistance are challenges that lead to similar responses (Lorenz, 1974; Cranach, 1992).

*Excursion 9.1*  Folk Wisdom about Pain and Change

*Metaphorical use of 'pain'*

- *the pains of change*
- *no pain, no gain*
- *growing pains: learning to change and to adapt*
- *Growing pains in business: recognizing and assessing the need for organisational change*
- *Labour pains of change: China implemented*
- *Major social change brings growing pains*

*Common sayings on the pain experience*

- *It's an art to live with pain . . . mix the light into grey.*
- *Given the choice between the experience of pain and nothing, I would choose pain*
- *Pain is never permanent*
- *I was taught that pain is bad*
- *Any of us can speak frankly about pain until we are no longer enduring it*
- *Who, except the gods, can live through time forever without any pain?*
- *Life is pain; anyone who says differently is selling something*
- *Pleasure and freedom from pain, are the only things desirable as ends*

*The transformative potential of 'pain'*

- *We change when the pain to change is less than the pain to remain as we are*
- *Never let the pain from your past punish your present and paralyse your future.*
- *Everyone bares pain in their heart; it either makes you or breaks you*
- *Once we get the pain's message, and follow its advice, the pain goes away*
- *The great art of life is sensation, to feel that we exist, even in pain*
- *There's no comfort in the truth, pain is all you'll find*
- *Pain is the doorway to wisdom and to truth*
- *Those who become mentally ill often have a history of chronic pain*
- *Without pain, no suffering, and we would never learn from our mistakes*
- *With the growth of intelligence comes increased capacity for pain*
- *Evil being the root of mystery, pain is the root of knowledge*
- *Wisdom is nothing more than healed pain*

*Source:* http://www.litera.co.uk/deep_emotional_pain_sayings

## MOVING FROM METAPHOR TO FUNCTIONAL ANALOGY

While the 'pain metaphor' can be a rhetorical device, the 'pain analogy' is a heuristic for further research. The pain metaphor is common in talk of managing change: 'no pain, no gain' 'growing pains' (see Excursion 9.1). It stresses the inevitability and the need to endure pain. An appeal to stoicism seems to be the underlying wisdom. This is, however, not at all what the pain analogy of resistance is intended to achieve. The pain analogy does not dispense moral advice but conceptual elaboration and insight.

To elaborate the pain analogy, and derive from it testable hypotheses for the empirical study of 'resistance', is a theoretical and empirical challenge. Analogies are neither true nor false, but more or less explicit, and they are creative, stimulating and fruitful of new ideas and insights. But to compare acute pain and resistance one needs a framework which bridges the source (pain) and the target area (resistance) of this analogical transfer (see Figure 9.2). Analogy building is logically a partial mapping from one area to another. And what is mapped from one area to the other is selected by some abstract feature of this process which needs to be made explicit if the analogy is elaborated. This is an abstract feature of analogical reasoning (Holland et al., 1989, 287ff)L it is a framework of comparison or the 'tertium comparationis'. This framework determines what is relevant and what can be ignored in mapping between two areas. In this particular case I apply the conceptual notions of *self-monitoring* (Cranach and Ochsenbein, 1985) or *self-observation* (e.g. Luhmann, 1997, 866ff). Self-monitoring explores the effects of pain and resistance as far as they involve inner-directed attention and reflection,

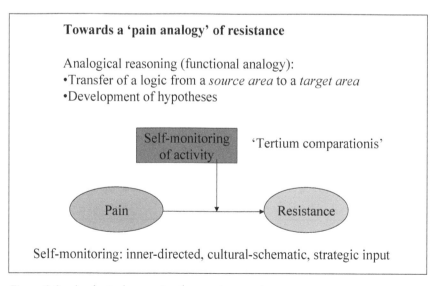

*Figure 9.2* Analogical reasoning from pain to resistance.

as these are culture specific, and they are strategically relevant. This elaborates the more fundamental point that this inner-direction is not epiphenomenal; it is not an irrelevant embellishment of action. The fact that this theory pre-existed had a double effect: it allowed me to 'discover' the potential analogy between pain and resistance in the first place by abductive insight (Eco, 1983); secondly, the theory of self-monitoring guides the elaboration of this particular analogy into testable hypotheses. Abductive logic allowed me to infer both the premise and the middle term in naming the empirical observations, and this invites further investigations:

- Premise (analogy): pain and resistance are self-monitoring
- Middle term (source area): pain affects movement
- Observed result (target area): resistance affects the mobilisation process

Self-monitoring explore the fact that complex and self-organising life forms develop introspection on the basis of a distinction self/other and system/environment. In this way the life forms are able to monitor their internal state of affairs. Living systems have two milieus to take reckon with, an internal and an external ecology. Human activity must therefore be seen as continuous adaptation to two contexts, inner and outer milieus. It is easy to see how we walk differently from A to B if we are in pain; being in pain might bring us to abandon the idea of reaching B; for this however, other things come into play, such as having painkillers at hand or having the mind set at enduring the pain. This suggests that acute pain that we experience it, determines the manner of walking, the direction and purpose of walking. Human activity builds on several self-monitoring processes that are integrated into each other: *conscious cognition, emotion and pain*. These processes pose constraints for any ongoing activity in addition to external environment. Introspection operates with specific codes, a more or less complex way of marking what is possible at any moment in time. At present, pain will be our focus.

*Table 9.1* Levels of Analysis and the Functional Analogy between Them

| | | **Cross-level functional analogy: self-monitoring** | |
|---|---|---|---|
| | Process | mediation<br>elaboration & modulation | area of effect |
| [source] | Pain ⟶ | conscious cognition<br>emotion | ⟶ activity |
| [target] | Resistance | informal conversations ↕ formal discourse | network<br>movement<br>global society |

*Source*: modifed after Bauer, 'functional analysis' 1995, p. 404.

The conceptual elaboration of the pain analogy for resistance in terms of self-monitoring involves two metaphorical transfers. In the first step, we move from pain to resistance and ask what we can deduce from that, secondly we move from individual behaviour to collective behaviour, and ask what we deduce from that. The intuition is that acute pain and resistance serve similar functions of self-observation albeit with different structures and at different levels of analysis. What pain does for individual activity through embodied experience, resistance does for collective activity through distributed awareness and networked communication. Table 9.1 draws the pain analogy across levels of analysis (on cross-level hypothesising, see Miller, 1986). On each level the signalling process is mediated by other processes. Pain affects us as an alarm call, coloured by feelings of anxiety, and some thinking to follow. Resistance affects the project and is communicated among challenged actors as mindsets that will guide their counter-measures. Resistance thus becomes a topic of formal and informal discourse, including mass mediation. In this book, I have explored these effects of resistance on the mobilisation of techno-science.

### The Theory of Self-monitoring of Activity: Introspection

Activities must be analysed on different levels: physiology, conscious experience and social and public communication. I talk on the phone while making my way home, and in my brain the neurons fire and the blood flows, at the same time my stomach is digesting lunch, and my kidneys cleaning the blood. Functional analogies compare structures by the similarity of their outcomes, and this can be within the same level of comparison (pain ~ emotion ~ cognition), or across levels (pain ~ resistance; pain experience ~ physiology of pain). Co-evolution is a characteristic of life; complex activities co-evolve with their internal and external milieus (Boulding, 1954; Cranach, 1992). Collective co-ordination and mobilisation efforts are 'autopoietic', self-organising and defined by the many problems that need to be resolved in order to continue in existence (e.g. Cranach, Ochsenbein and Valach, 1986). Failure to develop adequate structures to solve these problems jeopardises continuity, or 'Anschlussfaehigkeit' (Luhmann, 1984). The world of activity faces many problems that can be characterised as orientation-perception, *self-monitoring,* memory, feed-forward goal setting, feedback-back regulation, evaluation and consumption. These processes contribute to two overarching functions of self-organisation: giving direction and energising the effort. There is little movement without channelling energy, and effort without direction remains erratic and wasteful. For the analogy between pain and resistance, the structures and functions of self-monitoring explicate the tertium comparationis. Introspections focus on in the inner milieu and make available the needs as they are projected into awareness and communication, and this happens in coded and schematic form. Projects are self-descriptions of movement in form of a projected need. The project translates need into a blueprint for action.

Resistance is analysed in analogy to *acute pain* and its power to transform activity in order to be sustainable. Pain and resistance are self-preservation processes. Acute pain is a signal that something is going wrong. This signalling achieves the following further outcomes: (a) it focuses attention internally, (b) enhances reflexivity and the self-image, (c) evaluates on-going activity and (d) urges an alteration to the established course of action. Like pain, resistance has diagnostic and pragmatic value; however, like pain it is more reliably characterised by its pragmatic consequences than by its diagnostic value of causes (Wall, 1979).

Used diagnostically, resistance can locate a problem and suggest causes; but this is very much prone to error, not an error of judgement, but one to do with the very functioning of the self-monitoring. More reliable are the action implications of the pain experience: slow down and think, move out and do else! Ignoring and suppressing this signal is always an option, but a risky one. Being unable to feel pain is a lethal condition, because the pragmatic functions of pain are blocked. People with this rare condition die early from minor injuries which develop into major complications.

Finally, neither pain nor resistance is unmediated: the pragmatic significance, beyond the retraction reflex, arises from symbolic and cultural elaboration. Pain makes us go to the pharmacy or see a doctor, which is not a natural response, but one that is culturally formed. Pain makes us grit our teeth and shut up, or scream and shout loudly; whether one is a stiff-upper lip 'stoic' or not is again not natural, but a cultural condition.

## The Implication of the Pain Analogy of Resistance

The pain analogy in terms of self-monitoring has implications which in their totality might amount to a theory of resistance. The following observations on the *functional architecture of pain* can be made. My sources are a Royal Society meeting (Iggo, Iversen and Cervero, 1985), the classics Melzack and Wall (1988) and Wall's (1999), an unpublished text by Grafee (2001), the cultural history of pain (Morris, 1991) and Rollin (1989) on the historical struggle to recognise the pains of animals.

First, the relation between a noxious stimulus and pain experience is a tenuous one; there is no one-to-one mapping. There is continued pain 'in the leg' after having lost the leg (e.g. phantom pain). There is little correlation between stimulus intensity and pain: little stimulation and strong pain as in chronic pain (hypersensitivity) and heavy injury but no pain, as in battle wounds (endogenous opiates). States of trance and hypnosis can make pain go away, despite injury. More stable and predictable are the responses to the painful experience. The relation between behaviour and pain passes through attention; a strong affective-evaluative reaction is manifest in the motive to withdraw, seek hiding and rest. Researchers conclude that pain is a motivator rather than sensation (Wall, 1979). Pain is an atypical perception unlike vision, audition, smell or touch: it primarily serves the healing process, and is unreliable as sensory discrimination (for a contrary view see Algom, 1992). This suggests that the pragmatic-functional analysis should precede any causal analysis. The key is what comes afterwards, not what comes before. Pain requires attention, impedes a current course of action; it confuses, disorganises and distorts movements; in pain you move badly or not at all.

Secondly, pain is fundamentally *paradoxical*: in order for pain to have the positive functions of leading attention and healing, emotional evaluation and change of behaviour, it must be an unpleasant, namely 'painful' experience. The experience must be repulsive otherwise it would not raise alarm and motivate one to leave the danger area, slow down, seek care and promote healing. For pain as for mass communication 'only bad news is good news'. Pain can become a positive 'sensation seeking' motive, but this is the reconditioning of avoidance into an approach reaction by secondary association with pleasure. This is a particular cultural arrangement. Under normal circumstances, pain only works if it hurts. This paradox points to the inevitability of pain, which gives rise to a rich history of metaphysical, religious and existential reflections on pain; why me, why the innocent? Pain is a mystery that stimulates a quest for meaning (Lewis, 1977; Morris, 1991).

Thirdly, much of the dissociation between injury and pain experience is due to the *endogenous modulation*. The body's pain system includes mechanisms which modulate the injury into the pain experience and the expressive motor response. The *gate-control* mechanism in the dorsal horn of the spinal cord controls whether a noxious signal is projected into the brain areas and thus may lead to expressive activity (see Melzack and Wall, 1988). The nociceptive signal travels a *dual pathway*: the fast and noxio-specific A-delta fibres and unspecific and slow C-fibres without myelin. This dual system projects the stimulation upwards into the midbrain medulla and slower into the cortex, but also *descending* from the cortex back to the spinal gate where the noxious stimulus can be blocked. This architecture also generates *inhibitory neurotransmission*: nociception activates the body's own opiates, so-called enkephalins and endorphins, in the periaqueductal grey (PAG) and medulla regions of the brain, which serve as 'self-made painkillers'. This fourfold system of modulation—gate control, dual fast-slow pathway, descending pathways and endogenous 'pain killers'—moderates the intensity of the pain experience and thus mediates the reaction to any noxious stimulus. This system is complex enough to develop an Eigen-logic, *short- and long-term neuro-plasticity* leads to irreversible patterns and anatomical abnormalities which end up as chronic pain detached from any initial trigger. Indeed, chronic pain is explained by persistent alterations in the neural system of signalling pain. Chronic pain again is a secondary development from primary acute pain, where pain loses its functionality and becomes a 'pain'.

Fourthly, the *dissociation* between injury and pain due to endogenous modulation, gives rise to various phenomena of 'misrepresentation'. In extreme cases, people are totally insensitive to injury or excruciatingly hypersensitive to minute stimulation. The sensory threshold is conditioned by pain itself: being in pain makes you more sensitive to pain elsewhere; your whole body is mobilised and on alert. Memory of past pain increases the current pain, so does *nocipation*, the anticipation of impending pain. The more you dread your visit to the dentist, the more it will hurt. The affective colour of pain can lead to panic reaction, so called *catastrophising* about the potential implication of injury: a little pain in the stomach. . . and it must be cancer! The location of pain within the body image can be remote from the place of injury, so-called *referred* pain; in the extreme, the person feels the pain in the limb that is no longer there; the body image is still an intact frame of reference (on phantom limb pain see Howe, 1983).

Fifthly, the pain response is based on *symbolisation*, which combines the sensory-discrimination of 'pain' with an affective experience of 'painfulness'. What is taken- for-granted suddenly comes to attention. It leads to a reversal of attention from the 'world-out-there' to the 'self-in-the-world': heightened self-awareness is overwhelming. The painful tooth fills my attention span at the exclusion of everything else. Furthermore pain is over-loaded with *cultural meaning*. The pain system involves lower brain regions of emotional discharge and cortical regions of language and symbolic meaning processsing. The pain response is ultimately a culturally meaningful activity. Because higher cortical regions are involved in the pain experience and link back to the gate control mechanism through descending pathways, focussed cognitive activity, such as meditation and hypnosis, can control pain. The pain threshold is a cultural feature: people variably tolerate pain, and the pain response is strictly controlled through educational ideals such as the 'strong person', 'stiff upper lip' or a stoic ideal of controlling sensory experiences. The way we experience pain, endured as normal or as something to get rid of, is a cultural mindset: 'I have learnt that pain is bad' (see Excursion 9.1). *Pain expression* is a signal for human empathy; seeing others suffering elicits sympathy; seeing somebody in pain makes us inclined to help. Pain avoidance is the basis of utilitarian ethics: humans seek pleasure and avoid pain, and inflicting unnecessary pain is immoral. The capability of experiencing pain is a criterion for the status of a 'person' with rights and duties; it is controversial whether and when an animal or a growing foetus deserves that status (Rollin, 1989). Finally, what we do when in pain, whether we buy and swallow painkillers, call the emergency line, or meditate in a corner, calm down and care for ourselves, this all requires cultural *cognitive-motor schemata* of the 'diagnosis-treatment' type: 'if . . . in pain, then . . .do X'. Clearly these are mindsets with a rich cultural history. Pain gives rise to *'vertical thinking'*; it grounds people in existential awareness from which everything else derives; the metaphysical mind/body dualism disappears. Pain is the ultimate reassurance of reality. Pain is universal punishment as in the Augustinian doctrine of 'original sin'. Pain appears in rites of initiation giving status in the tribe. But pain also gives rise to *horizontal thinking*: it is simply unnecessary, particularly in its chronic state, an plague that needs to be overcome and controlled by whatever means available (Morris, 1991). The chronic pain community has been fighting and mobilising for the recognition of this problem and the resources to cope with chronic conditions.

Sixthly, pain has a strong connection to basic *learning*: namely aversive conditioning. Pain tells us strongly and immediately what not to do. Pain is the key to avoidance learning and to the attribution of 'superstitious causality' where there are only spurious correlations. In pain we come to believe in all possible causes. Having learnt to avoid a situation that is associated with pain we will not go back to that signal situation, and thus forfeit the opportunity to explore the situation with circumspection. In order for *higher-order insight learning* to occur, the avoidance response needs to be suspended. This is where cultural mindsets such as stoical endurance, jumping to diagnosis-action-schematas, or a sense for circumspection, come to play. Any refinement of meanings constitutes higher-order learning: how to think about pain and how not to, what to do about it?

Finally, this functional architecture gives the pain a characteristic time pattern, which is schematically illustrated in Figure 9.3. An initial hike of pain (the signal phase) stimulates the *endogenous pain control* and pain temporarily recedes (the modulation phase). This liberates attention to things like leaving the noxious situation, assessing and making sense of the situation and initiating care. The cultural mindset is activated: is this

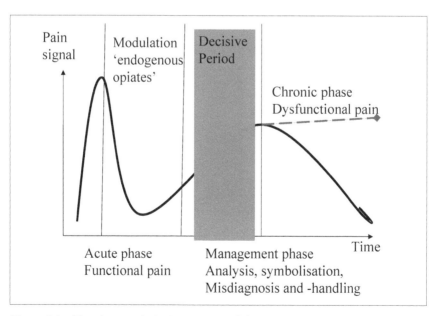

*Figure 9.3* The characteristic time pattern of the pain experience.

a 'punishment for my sins' or 'I have probably broken my arm because' or 'this was a mistake for which I am responsible'. During the elaboration phase pain reappears slowly (the decisive phase) and I will initiate remedial action; and this will determine whether my pain will ultimately go away or become chronic; pain will become chronic if I ignore my it , or kill it off with painkillers without consideration of what might cause it, or if I start moving in a contorted manner which temporarily brings relief but further muscle tension elsewhere (the outcome phase).

This functional architecture of is an evolutionary achievement of a complex organisms: to raise attention, to restructure the body image, to evaluate and to change action in line with cultural schemata and in response to noxious irritation. It signals the immediacy of danger, and urges intervention, to leave the situation and to care for self. The pain function is thus life-preserving, attention demanding and disruptive of normal functioning (Eccleston and Crombez, 1999); its expression invites empathy from others, and avoidance is a motive for learning, sociability and conformity.

Building a functional analogy between pain and resistance in these terms of self-monitoring has implications for the analysis of resistance. It shifts the analytic gaze as illustrated in Figure 9.4 from a focus on the drivers of resistance to a focus on the movers and shakers of the project. The star gazing magus moves from the right to the left of our

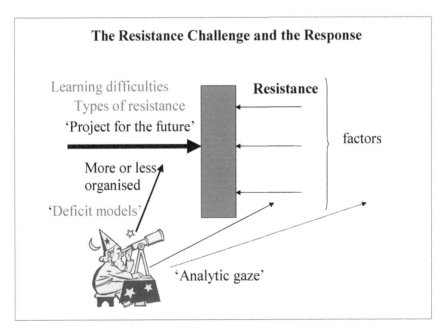

*Figure 9.4* The functional perspective of resistance.

constellation. The observing magus considers less the externalities, but puts additional focus on observing the reflexivity of the promoters. While factors, reasons and causes of resistance continue to be important for cause-effect reasoning, the analysis of the mindsets of *resistance* now takes a central space in the analysis.

The theories, concepts and narratives of resistance, the discourse of resistance and cultural repertoire of responses to the challenge, are now part of the problem, and thus, of the solution. It becomes clear, as the pain analogy suggests, that because the response to resistance can be a factor of escalation, the *control of the response* to resistance is as important as the control of the challenge itself. To control the response requires recognising and analysing the symbolic mindsets which direct and motivate these responses. In this vein, the analysis of resistance leads to reflective analysis and thinking. It analyses the factors of resistance, and at the same time brings to the fore the culturally conditioned 'gut reactions' of dealing with this challenge, including simplistic cause-effect models, which enables a higher-order learning process.

Learning must go beyond the avoidance of a tricky situation and how to improve the efficiency of established response patterns. This reflexive loop makes it more likely that we can redefine the situation and develop novel responses to the challenges posed by public resistance to techno-scientific mobilisation. Established cultural mindsets of resistance, in particular causal frames that externalise deficiencies, are more likely to be part of the problem than of the solution. Here lies an interesting entailment of the pain analogy: a general resistology must analyse 'resistance mindsets' as ways of dealing with the challenges of resistance. As such, it offers a self-referential theory of resistance that contains itself as a special case, and that enables us to recognise when models of resistance, such as the ones presented here, are part of the problem or part of the solution. By contrast we do not expect a friction or viral infection model of resistance to consider itself a 'friction' or 'infection' in the process. These models do not have a self-referential architecture.

The pain analogy thus invites a fundamental shift in theorising the resistance to social mobilisation. The reader might now be advised to go back to Chapter 5 and maybe also to Chapter 1 and continue to appendix 1 to get an even better sense of this and to close the theoretical loop on how techno-scientific mobilisation responds to the challenges of resistance. Gone are the times when resistance was futile and irrational. We can recognise this now as a simplistic mindset that becomes part of the problem.

## NOTES

1. For example see the sponsored supplement in the UK *Guardian*, 'Nanomaterials in our food', 27 April 2013.

2. This is still a modest beginning, but continues a stream of thinking of some years (see Bauer, 1991, 1993, and 1995; and Bauer, Harre & Jensen, 2013).
3. Connotation are very much an empirical matter. Some years ago I undertook a study of connotation of 'resistance' among management students in Germany, Switzerland and Britain. Free associations revealed mainly words which alluded to 'psychological deficit' such as anxiety, laziness and bureaucracy. Semantic differentials of 'resistance' however, revealed subtle differences across languages. Connotations vary on universal dimensions of evaluation, activity and potency (Osgood, 1969). In German 'resistance' connoted more activity and potency; in English more impotent passivity (Bauer, 1993, 141ff). Natural language reflects the cultural mindsets and makes a difference in management studies where the term generally serves self-interests to avoid blame and to attribute irrationality in the face of change.

# Appendix 1
## Notes on Social Movements and Social Influence

**NOTES ON SOCIAL MOVEMENTS**

The social movement literature is well summarised elsewhere (see Rucht and Neidhardt, 2002; Turrow, 1994; Kelly and Breinlinger, 1996). The literature generally refers to four perspectives—*grievance, mobilisation, framing and opportunity*—which jointly explain and highlight the characteristics of social movements and make use of both sociological and social psychological concepts.

The function of a social movement is to influence policy, translating societal preferences into policy, and making claims to public resources. Functionally equivalent to parties and pressure groups, social movements however achieve influence with different *structures, processes and contents*. They do politics by other means like the military do (see Zald and Useem, 1987, 250ff). Social movements arise when political parties and pressure groups fail to represent public opinion; they are an extra-parliamentary opposition. Social movement organisations (SMO) take up neglected grievances, mobilise resources and frame the issues for action when opportunities arise. Movements co-ordinate collective actions without formal membership and operate in cycles of ebb and flow, though not predictable from lunar tables.

### Grievances

Relative deprivation explains protest by the frustrations incurred by members of a social group and their confidence to do something about it. *Grievance* arises from a contrast between entitlement and reality, and in-group and out-group comparisons, and if this contrast is unjust or otherwise scandalous. Protest thus is reactive, and the perceived stimulus explains the reaction. However, lack of observed correlations between grievance and protest makes deprivation only necessary, but not a sufficient condition for protest. Societies create many frustrations, but protest is selective; hence other factors must come into play.

### Resource Mobilisation

The perspective of *resource mobilisation* offers additional considerations. SMOs are a stratagem derived from guerrilla experiences, but played out in civil society and within the law, to influence policy (McCarthy and Zald,

1987). 'Apathy' or 'false consciousness' do not explain the lack of protest, but which come with high costs of action. Protesters make rational decisions. Protest is costly and often risky; people get hurt and end up in police custody, and doctors and lawyers to fix it cost money. Protest happens when *symbolic and material resources* compensate the costs—if resources, then action. First aid needs to be provided to the injured, busses are needed for transport, and lawyers need to be at hand to defend those prosecuted or imprisoned, and somebody needs to co-ordinate all into time and location. Protest is not only reactive, but faces its own strategic decisions and dilemmas with risky consequences. Participating in *collective action is costly*; only a subset of those who might like to protest have the discretion of time and money, so costs need to be reduced and benefits increased. SMOs *channel a flow of resources* from the mass of giving believers and from elite sponsors who command large resources and who might give for different reasons. Resources are understood in a widest sense of money, ideas, skills and know-how, motivation, effort and access to the polity. With buses and food provided, more people will rally; but the rally also needs symbols that can be paraded. It costs to maintain an office, employ a professional cadre, and train staff who know how to mobilise volunteers and stage a protest. Society might raise the costs of movements; heavy policing for example increases the burden of protesting.

*Sponsors* of protest do not have to share the grievance. In politics, sponsoring the enemies of the enemy is a tactic whatever their preferences and beliefs. SMOs are legally NGOs who refer to grieving opinions and are able to capitalise on tactical alliances but need to balance grass-roots credibility and effectiveness in politics. The Taliban obtained weapons and training from the CIA to resist the USSR in Afghanistan; Hamas was sponsored by Israel to split and undermine the Fatah movement. In neither case is there much communality between sponsor and SMO beyond political expediency. Movements thus need a minimal organisation to deliver protest and influence in a segmented market of sponsors, beneficiaries, sympathisers and bystanders who might convert to the cause.

The resource perspective considers that discretionary expenditure is seasonal (no use to try to mobilise farmers during the harvest or planting season, or students during exams), more likely in a good economic climate, occurring among the somewhat affluent (new social movements are middle class affairs), and dependent on mood swings and psychological arousal. Protest is on offer from competing SMOs that form a social movement industry (SMI), including the entrepreneurialism of competent and determined leaders with a public profile. The resource perspective highlights competition among movements and the *mass media attention and issue cycle (Newig, 2004)*. Competition might be unfriendly; counter-movements aim to undermine the resources of those they oppose (Vander Zanden, 1959). All SMOs seek public attention to mobilise resources in numbers, reputation and political influence. But the media operate on a logic of their own: an issue goes in and out of the news not because it solved (that can happen), but because other issues take the space, media compete for a good story, and there is simple saturation: the public gets bored. *Competing SMOs* must brand themselves and seek public attention through strategic advertising: issue marketing and public relations come with product differentiation to manage the stages of the issue cycle. Considering the news routines of media organisations allows for free advertising; spectacular actions get media attention because they are newsworthy; they promote the cause without direct payment. SMO competition also creates a division of attention labour. Moderate protest often profits from the attention created by radicals with a sense for spectacle and drama.

But resource mobilisation theory runs the risk of becoming a content free formalism explaining everything from public protest to selling carpets on a market model of rational choice. Here the perspective of *strategic framing* comes in to admonish the power of ideas and the solidarity they can build. Particular symbols give meaning to issues and actions; and the symbols are the *content that matters*. By naming and framing the issues, social identities are created, and the words, metaphors and symbols resonate or do not resonate in the culture. People not only seek advantage but they also strive for belonging in solidarity, identity and recognition. But in the world of symbols, not everything works. The power of actions, words, ideas, sounds and images is contingent on local culture. The world is understood and actions are directed on the basis of social representations.

## Framing of Action

The key term of social movement research is *framing* (see Franzosi & Vicari, 2012; Polletta and Kai Ho, 2006; Benford and Snow, 1992; Gamson and Modigliani, 1989). The frame concept is social-psychological in origin and considers the fact that behaviour is constrained by mentality. Framing incorporates notions of Gestalt and pattern perception, action-production systems, mindsets, cognitive structure, script and schema which link perception to action: 'what is the case' links to 'what needs to be done' and vice versa. This is made explicit as implicit conditionals 'if X, then Y, unless C'.. Framing an action offers an attribution of causes and responsibilities that evoke responses, and point to what needs to be done.

*Frame analysis* shows how ideas, symbols and metaphors work in the (dis) service of social movements. Frames define the bias of social movements; bias is not optional, but a rhetorical necessity to maximise the power of persuasion. Because of this power, frames are contested both within movements, and between movements and counter-movements. Choice of framing is important to mobilise a narrow or wider support. For example to frame the issue of abortion as compassion may attract a wider constituency than focusing on autonomy and choice (see Ferree, 2003). Frames are not owned but they are part of a cultural stock of symbolic possibilities, often captured by metaphors and images that are well understood and that resonate in a particular historic context. In that sense frames are not constructed but sponsored and pushed.

The cultural analysis of social movements also highlights the fact that actions and symbols are a tool kit; they are a repertoire from a cultural stock. This *repertoire* of movement also includes other actions such as petitions, demonstrations, theatre, boycotts, sabotage, hunger strikes, bombing and attacks on people, non-violence, civil disobedience, sit-ins and barricades. These actions can be variously classified as violent or non-violent, more or less disruptive of normal life. Over historical time this repertoire changes and in different contexts it stacks up differently. Barricades were common in the 19th century, while organised marches are more prominent in the 20th century. Actions like hunger strikes or bombs seem adequate for some people, while disproportionate or out of bounds for others. Hunger strike and suicides might be part of a living history of protest in parts of the world, while absent and abhorred in others.

The cultural backdrop of actions and symbols is both resource and constraint, not everything goes everywhere, while overall the repertoire is large. Culture is the basis of rhetorical appeal and dramatisation of issues and guarantees the resonance and reception of certain images, sounds and words, while it excludes others. It is the repertoire that is learnt like a box of actions and symbols from which strategic actors might choose to raise attention. This repertoire grows and develops, including the mass media opportunities that arise with new technologies, genres and news routines. Social movements innovate, import and adapt symbols and actions, often by trial and error, vividly remembered when successful. In the repertoire are *meta-frames* such as 'progress', 'independence', 'risk' or 'justice', which bring many different issues together. Independence gives meaning to the nuclear power questions as well as to nation building; justice can challenge income distribution as well as policing; and progress has relevance for science as well as for social development. The symbols of past movements enter the toolkit for the next round of mobilisation. Movement and counter-movements learn and imitate each other and accumulate a common stock of symbols and actions to wield political influence outside normal channels.

## Opportunities

Social protest has roots in grievances; SMO organisations take it up; symbols and action repertoires define why and what shall be done, but this might still not suffice to explain the timing of protest. Here the structural analysis offers a fourth level of analysis: context makes social movement activities more or less likely and gives historical characteristics. The key term is *opportunity structure*. Protests are cyclical, they ebb and flow but less than predictable, explained by opportunities offered by elite disagreement and economic climate. Structure analysis also charts the transition from traditional lower class mobilisation focussed on distribution of wealth to new social movements with a focus on values, identity and culture, and more middle class based (Hechter, 2004). The segmentation of social milieus outside traditional classes leads to a fragmentation of political parties and declining party affiliation. New social cleavages reveal the deficits in the translation of public opinion into policy by established political representation and interest groups. Thus historical structural changes give rise to a society where social movements are here to stay (see Rucht and Neidhardt, 2002).

In this context it makes sense to analyse techno-scientific developments as social movements. Indeed, a literature is emerging which does just that. The history of techno-science is no longer a matter of tracing the change of ideas through genealogies of authors. The sociology of ideas turns into an analysis of scientific and intellectual movements (SIMs) (see Frickel and Gross, 2005). Such movements are contentious and *encounter resistance* which arises from grievance with established theory and paradigms, desire to gain access to resources of positions and elite attention and seek to grow by enrolling a wider constituency. SIMs are episodic in the sense that they come, institutionalise and go away either by fading or by being absorbed. Finally, SIMs vary in scope. They adopt neglected topics as in the case of genocide or gender studies; other SIMs pursue theories, concepts and methods, as often happens in the social sciences. Astronomy saw a bitter cleavage over observational and radio astronomy until the differences were resolved as a matter of wavelength within big bang theory. Other SIMs challenge the order of disciplines and innovate at the boundaries as did micro-biology between chemistry and biology or social psychology between sociology and psychology. SIMs also challenge the boundary between science and non-science, as in psychosomatic and alternative medicine, who bring healing techniques from the fringes into the standard repertoire of biomedicine. Other SIMs focus on that very boundary work, as for example in 'evolutionary psychology' (Cassidy, 2005). In order to do so researchers and intellectuals mobilise scholarly institutions to channel funding, positions, meetings, journals, publications, peer reviews and the public engagement of science (see Bauer and Jensen, 2011).

The ambition of SIMs is to analyse these changes not only by tracing ideas articulated by people on paper and patents, but also by tracing the mobilisation as it is motivated by frustrations, *encountering resistance*, action framing issues and making use of the opportunities as they arise. In this sense SIMs encounter resistance on the way, and the question remains open: how do SIMs deal with that challenges of resistance which they encounter?

## NOTES ON SOCIAL INFLUENCE

Social psychology has traditionally concerned itself with social influence. What the textbooks often report as a historical sequence of ideas might better be considered as different modalities of influence with contemporary significance (Sammut and Bauer, 2011). I will briefly review five modalities of social influence: crowds and leadership, contagion and imitation, conformity pressure, obedience to authority and conversion by minorities.

## Leadership: Attachment and Identification

Gustave Le Bon, a 19th-century French populariser of science, wrote a bestseller on crowd psychology in 1895. Crowd behaviour demonstrated human irrationality as de-individuation; the individual en masse is de-individuated. As soon as civilised individuals leave their splendid isolation they undergo a change of personality. They lose judgment and become hypnotised, intolerant and eruptive, the opposite of a well-tempered, civilised individual. If rationality is a characteristic of the civilised individual buffered from bodily emotions and from social influence, then people en masse are irrational and hysterical, 'effeminate' in the sexist language of the 19th century. The model of a 'buffered individual' as the seat of rationality is a key theme of Western thought (see Taylor, 2007). Le Bon expressed unease of the middle classes over the enfranchisement of working men (see Reicher, 2003; Ginneken, 1992).

However, crowd psychology did not rest with a diagnosis of human irrationality; it explored new opportunities. Masses needed leaders, like clay needed a potter to take form. A crowd has no telos, no purpose, but takes 'inspiration' from leadership; people en masse are open to suggestion, emotional identification with charisma (see Freud, 1922). These ideas were eagerly adopted by authoritarian leaders during the 20th century across ideological boundaries. This quest for charisma lives on in populism and personification of politics and techno-scientific innovations.

## Contagion and Imitation

Another 19th-century idea is Tarde's (1890) theory of social contagion. The similarities of human behaviour arise from either lineage or imitation. What kinship does not, imitation will explain. Invention is random, but imitation is lawful; it works like contagion, though without mutation. Tarde induced regularities from repeated facts. The rule that 'change of mentality precedes behaviour change' is illustrated by the fact that the 'outer' fashion of dress follows the 'inner' fashion of literature and reading. Similarly, he observed that 'social change follows the social hierarchy'; commoners imitate aristocracy. The prevalence of alcoholism over obesity becomes the law that 'drinking is more easily imitated than food habits'.

Tarde and Le Bon share the *doctrine of suggestion*: i.e. everyday life is somnambulant. People live in a state of 'primitive credulity', a non-pathological trance that makes us receptive to suggestion. In everyday social life, rationality is suspended and people have little or no autonomy (see Asch,

1952, 398ff for a critique of this doctrine). More recent echoes of this doctrine of suggestion seek to model the adoption of new ideas, services and products on the spread of germs (Caldwell, 2000); ideas go 'viral' because of their processing ease (i.e. stickiness or relevance, see Sperber, 1990) and their 'relative fitness' as memes (e.g. Lynch, 1996).

However, Tarde differed from Le Bon in appreciating historic events. Tarde distinguished crowds, co-present bodies in the street, from public opinion, the common attention of geographically dispersed spirits exposed to mass media, at the time this meant reading daily newspapers. He sees here a significant new phenomenon, *collective attention that sets the agenda* of politics.[1] The Dreyfus affair demonstrated to Tarde the power of this public conversation at a distance: instead of focussing on the war in Africa, French politics got focussed on this miscarriage of justice. This bias of policy is not explained by the problem, but by the power of public opinion. Where Le Bon only saw irrationality and a need for the charismatic leader, Tarde recognised a novelty in the process of democratic policy making (Tarde, 1901).

Theories of crowds and contagion are criticised because of the doctrine of irrationality and unconsciousness. These doctrinal assumptions are contingent cases rather than the universal. However, social intelligence reveals other modalities, where sociability is not a source of irrationality. The rationalisation of social influence arrives with experimental paradigms post WWII. These now famous laboratory dramas are less focussed on exact causal inferences than on demonstrating the power of social settings. Their value lies in admiration, shock and scandal of what people can do in situations of constraint.

## Normalisation and Compromise

The first paradigm demonstrates how conversations establish a common framework of judgement in the face of an ambiguous reality (Sherif, 1936). Group formation around a common frame of reference is the basis for coordinated action, and not the turning point of decline into barbarism. Experiments make the point. In a dark room a projected light oscillates (the auto-kinetic phenomenon); participants estimate the amplitude and estimates will vary. When people discuss the oscillation they reach a common understanding that persists for future judgements. The co-ordinated mindsets do not fall back to their individual judgments in isolation. The *frame of reference* is the outcome and the mind set for future action (see Reicher, 2003).[2]

Sherif's demonstration has affinities with Habermas's communicative action. The site of reason is not the isolated individual, but a safeguarded conversation where norms and artefacts emerge. Sherif showed that sociability is foundational, and Habermas focussed on the rules that safeguard discursive reason. The *rules of engagement* are inclusiveness, equality, no deception and self-delusion, and no pressures other than the better argument. Public speakers must be direct and to the point, and not speak with a hidden agenda. This explicitly counter-factual ethos offers a benchmark against which any real conversation leaves ample room for improvement (see Habermas, 2001).

Sherif's experiment does not consider a conflict of interest among the subjects; nobody seems to have a stake in the object of judgment; oscillating light seems rather a trivial issue. Other paradigms of social influence assume a conflict and an imbalance of influence between a powerful source A and a less powerful target B. The paradigms of majority and authority assume that the source of influence A aims to bring B in line; the source A seeks to assimilate the source B (conformity, authority), while A might end up accommodating B if the conditions are right (conversion).

## Assimilation by Peer Pressure and Conformity

Conformity is successfully dramatised by laboratory experiments (see Asch, 1952). Naïve persons join a group discussion with confederates of the experimenter in minority of one. The task is to match lines on visual display. Asch observes the naïve person's reactions to the confederates who, in a majority of 9 to 1, are briefed to assert an obviously incorrect claim. Most people, albeit after much emotional agony and to and fro, conform to the confederate majority view. However, conformity occurred only in public—people conformed amongst others but in private conversation they upheld the evidence of their own eyes. Conformity does not mean conviction. Asch dramatised the 'inner exile' of private dissent and public compliance.

However, granting a dissenting partner makes a difference; two hold out easier than one against the crowd. And reducing the majority to a 50/50 does away with compliance altogether, so does a split in the majority consensus. Reversing the game and confederates making false claims against a naïve majority dramatises the peer pressure: they face the ridicule, annoyance and disdain of the majority. Age strengthens against conformity pressure. Contrary to expectations, conformity neither depends on neither the sensory modality nor the absurdity of the claim made. Asch was shocked by the low probability of resistance (Asch, 1952, 450ff) and explained it with the implicit threat of sanction: dissent meant social exclusion, being an outcast. Our need for belonging overrides epistemic concerns.

The pull towards the group offers advantages that are forgone by dissent. Social animals have a natural tendency towards conformity. Rational choice would argue that conformity arises when the costs of non-compliance exceed the benefits of compliance. However, conformity involves the balancing of three considerations, one's relation to objective reality, to others qua normative expectations and to one's self-esteem.

This dilemma does not fall easily. By privileging objective reality, a person ignores social norms, incurs outcast status with consequences for a positive self-image. Social conformity thus reflects the predominance of social and ego concerns and over objective concerns. It is therefore not surprising that persons on the autistic spectrum, who experience difficulties in social relations, are advantaged in scientific-mathematical careers; they privilege objective over social intelligence (Wheelright and Baron-Cohen, 2001). A world dominated by only one form of intelligence might however be worrisome.

### Assimilation by Obedience and Compliance

Among monkeys, new practices spread by lower animals imitating the higher ranking female (see Whiten, Horner and de Waal, 2005). A similar facet of social intelligence is Stanley Milgram's famous set-up. He tested the power of reputation and authority to induce obedience to demands (see Milgram, 1974). Participants were instructed under false pretext to assist in a study on pain and learning. Participants had to apply rising levels of electroshock to a 'learner' behind glass whenever he made an error. The learner was an actor and briefed to err frequently and to simulate the increasing pains of the administered shocks. The experiment shows how far people go. The compliance rate defines the percentage of participants who go to the maximum of 400 volts. Debriefing interviews with participants revealed much distress about what they just went through, but also a tendency to defer responsibility to the scientific authority: I did what was my role. Milgram invariably appeared in a white coat, the regalia of scientific authority, and reminded any hesitating participant of the importance of the study for the progress of science. *Obedience to authority* is based entering an 'actant state' of doing things while deferring responsibility. Milgram does not explore the nature and origin of authority, in irrational identification with charisma (qua Le Bon) or in more reasonable arrangements. This interpretation remains controversial, because authority can be recognised with good reasons as a middle way between violence and persuasion (Arendt, 1961).

The Milgram experiments were replicated under many different conditions. If participants take instructions by phone, the compliance rate increases. Replication in different countries revealed variable compliance levels which can be interpreted as 'national characteristics', and changes in any one context as evidence for social change (see Blass, 2000).[3]

In dual-process models of persuasion, which distinguish fast, associative and peripheral from slow, central and elaborate information processing, conformity and reputation are perceptual cues that feed the peripheral route: identification with the group, or with a reputed authority, are makes for quick decisions under time pressure (Martin and Hewstone, 2003). And the 'wisdom of the crowd' suggests that market prices are herding signals which any individual would be unwise to resist (see Surowiecki, 2004).

## Accommodation: Minority Converts the Dominant Majority

But is it possible that a dissident minority influences the majority? How can the bias of conformity be stalled? In social psychology an examination of these questions started with the reversal of Asch's drama (Moscovici, Lage and Naffrechoux, 1969). The few confederates were briefed to make false claims about colour images varying on the green to blue spectrum in the face of the majority. Ambiguity arises from afterimages, a perceptual phenomenon. The 'naïve' majority found it difficult to accept the colour claims, but came to accept the insisting minority view. The search for factors of success in dissent has precursors in Gibbon's (1776) explanation of the fall of Rome by the rise of the 'Christian sect', in Weber's *The Protestantism Ethic and the Spirit of Capitalism*, in theories of the revolutionary vanguard, and in notions of active anomie see Moscovici, 1976 and 1985).

The power of the many can be shaken if there is public conflict, a desire for consensus, a positive perception of the minority arising from consistency of action: coherent, competent, and assertive, independence without rigidity in claims making (see Mugny and Perrez, 1991; Maass and Clark, 1984). Conflict polarises, clarifies positions, energises and establishes a new consensus away from initial the central tendency (Moscovici and Doise, 1994). Clearly, behavioural style cannot be taken for granted; it is a matter of mindset and organisation, hence the 'active minority' in contrast to a 'passive minority' without any sense of purpose.

Success of the minority hinges on perceptions: the few, identified and with an image of independence, will seed doubt, and become a source of information not easily dismissed, and thus make inroads into the majority views. Over time, public rejection, annoyance and disdains over dissent will give way to private doubts of the majority position. Overt rejection is undermined by *silent conversion* that shapes up in parallel. Minorities are subversive in that they change minds. Echoing Tarde's notion that 'private change will anticipate public change', people change their minds, and for good. However, the *'sleeper effect'* means that people, after having shifted their frame of mind, do not recall the source of information. The tragedy of minorities is *success without recognition*. The dissident view, at first annoying and absurd, is later self-evident and a new common sense.

A contingency of success lies in perceptions that the minority is in-group. Activism of people 'like us' is more convincing than activism 'by them'. Converted members of the majority are particularly convincing. However, 'psychologisation' is a tactic to attribute unconscious motives, for example the anti-nuclear protests of prominent physicists is explained by 'existential anxieties stemming from insecure childhood attachment' (see Weart, 1988, 246), or one diagnoses irrational 'technophobia' among techno-scientific dissenters (Bauer, 1995). Social categorisation and psychologisation are tactics to discredit the minority with eristic ad hominem arguments, which potentially undermine the success of minority activism (see Mugny, 1982).

Conformity under pressure works quickly; the majority view and the prestigious authority remain the focus of cognition and not the issue. The desire to avoid exclusion leads to overt compliance in combination with private dissent. Majority processes exert *normative influence*, secure a false consensus and avoid conflict; this outcome remains unstable and easily reversible with circumstances. By contrast, the conversion of the many by the few takes time, and is born out of conflict. It occurs in private before it is visible in public, and sticks through changing circumstances. Minorities exert *informational influence* and thrive on conflict that leads to a new consensus, but are easily forgotten.

Social psychology illuminates the heterogeneity of social influence, but its validity seems limited to the face-to-face situation of the laboratory. To understand techno-scientific developments in a modern public sphere, formal mass mediation needs to be considered (see Bauer, 2008). Informal conversations reflect and inform the formalised communication of the press, radio, film, television and the internet and its social media genres. Two levels of communication define modern public opinion, one being the ecological environment of the other, and the other being protected by law as 'freedom of expression' and 'freedom of the press'.

Unfortunately, there is yet no comprehensive model of the two levels of public opinion. The field is ghettoised by 'theorettes' that make and break academic careers. Social psychology focusses on message design and attitudes change, while media studies on the technologies of mass circulations. The research, originating in the war effort of psychologists such as Lewin, Hovland, and Lazarsfeld etc. (see Rogers, 1994), is a fragmented field of partial hypotheses on how to reach people's hearts and minds (see Bonfadelli, 2004).[4] There remains theoretical work to be done.

## NOTES

1. For Tarde the difference between crowds and public opinion marked the difference between collective psychology and social psychology. He can therefore be considered a founder of modern social psychology, and the common ancestor of public opinion research and of Actor-network theory (ANT) (Latour, 2002).
2. Henceforth, social psychology could state that group processes result in two kinds of outcomes: artefacts and social norms as frameworks for future actions; human interaction is framed by both artefacts and norms (Asch, 1952). It is a historical curiosity that subjective norms (in relation to judgments, attributions, attitudes) became its sole focus, while material artefacts

are ignored, which poses theoretical challenges for present day social psychology (see Bauer, 2008). In sociology this theoretical prejudice was addressed by the concept of 'inter-objectivity' (Latour, 1996).
3. Ironically, Milgram's drama can no longer be performed. Politically incorrect, it became a showcase of unethical human experimentation because participants were deceived. The experiment can cause harm to participants who suffer from harassment and guilt; and false pretext cannot be subject to consent. Replications of Milgram experiments will not pass the hurdles of university ethics committees focussed on informed consent. Only partical replications get through (see Burger, 2009).
4. The field is dominated by an interest in causal modelling of impact designs, following the logic of an artillery gunner: the mass media are the guns to be loaded and the audience is the target; choice of channels and message design, i.e. the ordnance, is supported by requisite target intelligence. How to maximise impact and minimise waste, given a strategy and cause-effect tables define the parameters of this 'artillery guidance system' of media effect models. Little research effort goes into the study of unintended collaterals, which is where the analysis of resistance finds it purpose.

# Appendix 2
## Chronologies of Atoms, Bytes and Genes

Table A2.1  Chronology of the Dual-path of the Atom

| Nuclear bomb proliferation | Nuclear combustion and energy |
|---|---|
|  | 1901–38 Before fission: imagination |
| **1939–45 Making the bomb (US)** |  |
| ▪ Aug 1945 Hiroshima, Nagasaki |  |
|  | **1939–52 Towards and beyond Japan** |
|  | ▪ 1945–54 nuclear hope |
| **1945–64 The Club ('hot Cold War')** |  |
| ▪ 1949 USSR fission bomb |  |
| ▪ 1951 US H-bomb (fusion) |  |
| ▪ 1953 British explosion | **1954–63 Euphoria: cheap energy** |
| ▪ 1954 Bikini fall-out incident | ▪ 1953/55 'Atom for Peace' |
| ▪ 1963 Test-Ban Treaty | ▪ IAEA, EURATOM, CERN |
| ▪ 1964 China & France join |  |
| **1963–89 Negotiating Constraints** |  |
| ▪ 1968 Non-proliferation Treaty |  |
| ▪ 1972–79 SALT agreements | **1964–94 nuclear industry challenged** |
| ▪ 1974 India joins | ▪ 1956–86 becoming a suspect new technology |
| ▪ 1980s START and ABM treaty | ▪ 1974: Nuclear confusion sets in |
|  | ▪ 1977–86 anti-nuclear Europe |
| **> 1989 The New World (Dis)Order** |  |
| ▪ 1998 Pakistan joins |  |
| ▪ 2006 North Korea joins (?) | **>1994 Keep nuclear power option** |
| ▪ 2007 UK seeks to renew Trident fleet (decision delayed to 2015) | ▪ clean energy renaissance (>2000) |
|  | ▪ after 2011: renaissance doubtful |
| ▪ 2010 new START treaty |  |

Sources: Goldschmidt, 1980; Mandelbaum, 1983; Weart, 1988; Cirincione, 2007

*Table A2.2* Key Phases of Anti-Nuclear Protest: Civil and Military

| | |
|---|---|
| 1957–1963 | nuclear disarmament, fall-out controversy; CND, PUGWASH etc. |
| Late 1960s: | in US against stationing of tactical nuclear missiles around US cities and siting of nuclear power plants in urban environments. |
| 1979–1983: | 'peace movement' against stationing of tactical nuclear missiles across Europe and US: peaks in spring 1982 after President Reagan elected in November 1981. |
| 1977–1986: | anti-nuclear power protest<br>Anti-nuclear protest in EU combines disarmament and anti-nuclear power protest; protest happens to a different clock in different countries. |

*Table A2.3* Major Nuclear Power Incidents and Accidents

| | |
|---|---|
| 1952: | 12 December, NRX Chalk River near Ottawa, Canada [INES 5] |
| 1957: | 7 October, Windscale-1 fire, Cumbria UK [INES 5]<br>South Ural mountains (Kyshtym evacuated) 1.5 x 10$^{15}$ Bq [INES 6] |
| 1961: | SL-1, USA, experimental military [three operators dead] [INES 4] |
| 1966: | Fermi-1, Detroit, USA, experimental breeder |
| 1969: | Lucens, Switzerland, experimental reactor [INES 5] |
| 1975: | Greifswald, East Germany [INES 3] |
| 1979: | 28 March, TMI, Harrisburg, PA (Evacuation) 2 x 10$^{14}$ Bq [INES 5] |
| 1980: | Saint Laurent-A2 France, 8 x 10$^{10}$ Bq [INES 4] |
| 1986: | 26 April, Chernobyl, Ukraine [31 dead] 11 x 10$^{18}$ Bq [INES 7] |
| 1987: | Goiana accident (Brazil; 249 people contaminated, 3 dead) [INES 5] |
| 1989: | Vandellos-1 Spain [INES 3] |
| 1999: | 30 September, Tokai-mura, Japan [2 operators dead] [INES 4] |
| 2011: | 11 March, Fukushima-Daiichi, 470,000 evacuees [INES 7] |

*Sources*:
http://www.johnstonsarchive.net/nuclear/radevents
http://www.iaea.org/newscenter/news/2011/fukushima120411.html

*Table A2.4* The Computer's Long Past and Short History

13th century: Ramon Lullus' Ars Magna and combinatorics of symbols

17th century: Leibniz explores formal languages

19th century: Babbage and Lady Lovelace invent software for the 'analytical engine'

Key actors of 20th century computing:
- IBM (hardware, mainframe)
- IBM & APPLE (Personal Computer)

*(continued)*

*Table A2.4* (continued)

- MICROSOFT (software)
- GOOGLE (internet)
- APPLE, FACEBOOK, Twitter (social media)

1936 Turing's seminal paper on 'computable numbers'
1939–1945 WWII brings machines such as ENIAC, COLOSSOS, MARK, Z3
1951 US census buys first commercial machine: UNIVAC
1952 IBM mainframe systems 701 and 601
1957 DEC founded (taken over by Compaq 1998)

1960 DEC PDP-1 first minicomputer
1961 IBM system 1401, big selling mainframe
1969 ARPANET started (US Defence; superseded by NSFNET in 1990)

1970s pocket calculators e.g. Hewlett-Packard
1971 Intel launched first single chip processor Intel 4004
1973 Ethernet launched by Xerox
1975 ALTAIR first Home computer
1975 MICROSOFT (IPO 1985, after 10 years)
1976 APPLE founded (IPO 1984, after 8 years)
1977 Commodore and APPLE II home computers
1978 DEC VAX computers launched

1981 IBM PC lunched: '$Billion Baby' becomes TIME 'man of the year'
1984 APPLE Mac(intosh) launched / PSION first Palmtop computer
1989 WWW (World wide web) invented by Tim Berners-Lee at CERN

1990 prototypes of internet: http, html; file sharing, e-mail
1994 Netscape, Yahoo internet search engines
1998 GOOGLE founded (IPO 2004, after 6 years)
1999 Y2K anxieties, widely exaggerated risk of computer failures
2001 Wikipedia created
2003 Blackberry, mobile smartphone
2004 FACEBOOK (IPO 2012, after 8 years)
2005 YouTube video sharing site is launched (2006, purchased by Google)
2007 APPLE iPhone
2008 Twitter launched, expanding social media
2010 APPLE launches iPad (integrated technology)
2013 NSA project PRISM of massive internet surveillance publicly known

*Table A2.5*  Dates in the History of Genetic Engineering and Biotechnology

1859 Charles Darwin publishes 'Origin of Species': theory of 'natural selection'
1866 Gregory Mendel experiments with peas and discovers 'laws of inheritance'
1882 Chromosomes are discovered

1902 Chromosomes found to carry hereditary information
1905 Identification of the sex chromosome
1926 X-ray is found to induce random genetic mutations

1943 The 'gene' is proposed to be the locus of inheritance
1944 Prove that DNA is hereditary

1951 X-ray diffractions of DNA by Rosalind Franklin
1953 Double Helix model of DNA; F Crick and J Watson in NATURE
1954 Report published on radiation effect of Atom Bombs
1956 Human chromosomes, 46 in 23 pairs: final number established
1959 Herman Muller advocates 'germinal choice' as a solution to 'genetic load' due to radiation; Down's syndrome identified as chromosomal anomaly; human sex determination established, Xs and Ys

1960 RNA is discovered to link DNA to proteins, defining the linear model; FDA approves hormonal oral contraception, the 'Pill' to control women's fertility.
1962 Crick, Watson and Wilkins receive Nobel Price for DNA; Ciba Conference London 'man in his future': new eugenics (H Muller, J Lederberg, F Crick, Medawar, Szent-Gyorgyi, J Huxley, JBS Haldane et al.); book published in 1963
1965 Jacob's paper on XYY syndrome and criminality
1969 Jensen's paper on 'IQ and heredity'

1971 Eysenck's book on 'IQ and race'
1972 Berg et al. splice genes from virus
1973 rDNA gene splicing technique patented by Boyer and Cohen
1975 Wilson's publishes 'Sociobiology'; Asilomar Conference takes place
1978 Louise Brown, the first IVF baby is born in UK
1978 Human insulin cloned

1982 FDA approves genetically modified (GM) insulin
1984 DNA finger printing developed by Alex Jeffreys in UK
1986 FDA approves GM vaccine for hepatitis B

*(continued)*

Table A2.5 (continued)

1988 First patent on GM animal: Harvard Onco Mouse
1989 IVF: first human embryo pre-implantation screening applied

1990 Start of HUGO: mapping and sequencing of the Human genome
1994 FDA approves GM food: the Flavr Savr tomato
1995 Breast cancer genes BRCA1 and BRCA2 identified; patent disputed until 2013
1996 Arrival of Round-up Ready GM soya in European ports
1997 Dolly the sheep is presented to mass media: cloned adult somatic cells
1998 Brazilian court blocks GM crops for commercial use

2000 First draft of human genome presented (in White House, Washington)
2002 US government (Bush) bans federal funding for embryonic stem cell research
2005 Brazil liberalises GM soya for commercial planting
2006 Professor Hwang (South Korea): his stem cell research found fraudulent
2007 First mapping of individual genome (Craig Venter's)
2009 US government (Obama) lifts ban on federal funding for stem cell research
2013 USA and Australian courts come to opposite conclusions as to BRCA patents

## BIOTECHNOLOGY AND THE PUBLIC: THE MASS MEDIA DATABASE

Various chapters of this book include data and graphics referring to longitudinal media coverage of nuclear power, the environment, computing and information technology, and biotechnology. These analyses are based the following datasets:

### Science and Technology in the British Press, 1946–1992

The database comprises newspaper coverage of science and technology between 1946 and 1992 in the British national press (see Bauer et al., 1995). The corpus is a probability cluster sample of articles, stratified by day, year and newspaper. The sample comprises contributions in the *Daily Telegraph*, *Daily Mirror*, *Daily Times*, *Daily Express* and *Guardian*. All relevant 'science and technology' articles appearing on a particular day are included; we selected 10 random weekdays for every second year between 1946 and 1992. Salience figures and contents are estimated from this sample.

*Population estimates*: a sample of 10 random weekdays for 23 years, every second year, yields 2,832 articles for the *Daily Telegraph* alone. The total number of articles in the *Telegraph* is estimated as 2,832 x 30 x 2 = 169,920 +/-10% error margin (153,000 < x < 187,000 articles). Three hundred days over 46 years (1946–1992), with an average of 28 pages per issue of which on average is 45% advertising, yield 300 x 46 x 28 x 0.55 = 212,520 pages of effective news space. We estimated that the Telegraph referred to science and technology on about every second page on average (ratio: 1.1—1.4 to 1).

The Media Monitor Archive, located at the Science Museum Library in London, is a resource open to researchers. It comprises more than *7,000 press articles* referring to science and technology over the period in quality and popular newspapers. The data is accessible as hard photocopy and in coded format in a SPSS data file. Source: Bauer et al., 1995; Project MACAS (mapping the cultural authority of science) is currently updating this database for the UK and elsewhere for the years 1990-2012, see *MACAS-project.com*.

## Biotechnology in the Press, 1973–2002

The international project 'Biotechnology and the Public' (Durant, Bauer and Gaskell, 1998; Gaskell and Bauer, 2001; Bauer and Gaskell, 2002; Gaskell and Bauer, 2006) conducted co-ordinated analyses of biotechnology coverage in the elite press between 1973 and 2002 in 16 countries across Europe and North America, along with Japan.

*Salience figures* count articles based in a single source outlet defined as the 'opinion leader', i.e. the newspaper which journalists might be reading (for example in the UK: the *Times* until 1987; the *Independent* after 1987). Until the 1990s, figures are based on searching physical news archives. In later years we accessed online resources and searched with keywords such as 'genes', 'biotechnology', 'cloning' 'stem cells' or 'DNA'. British newspapers show a high monthly correlation of topic salience, probably because of strong competition in the news market. This suggests that a single source is a good long-term indicator of public issue salience.

*Content* is coded for an annually stratified random sample of news items. The coding frame included actors, topics, frames and a rating of the evaluation of biotechnology in the news article (overall n = 20,514; UK: n = 1,615). The code 'evaluation' is based on two scales, negativity (0 = not applicable; 1 = slightly negative; 5 = discourse of great concern) and positivity (0 = not applicable; 1 = slightly positive; 5 = discourse of great promise). The evaluation index is the difference of positive and negative ratings. The code 'frames' is based on a scheme of eight typified argumentations: progress, economic prospect, ethics, Pandora's box, runaway, nature/nurture, public accountability/governance and globalisation. Only one frame is coded per news item. Source: Bauer and Howard, 2004; Bauer 2007.

# References

Adams, J (1985) *Risk and Freedom*, Cardiff, Transport Publishing Project.
Adams, J (1995) *Risk*, London, UCL Press.
Ahlemeyer, HW (1989) Was ist eine soziale Bewegung—Zur Distinktion und Einheit eines sozialen Phaenomens, *Zeitschrift fuer Soziologie*, 18, 3, 175–91.
Akerman, N (1993) (ed) *The Necessity of Friction*, Heidelberg, Physica-Verlag, 6–27.
Algom, D (1992) Psychophysical analysis of pain: A functional perspective, in: Geissler, HG, SW Link, JT Townsend (eds) *Cognition, Information Processing and Psychophysics: Basic Issues*, Hillsdale, LEA, 267–91.
Allum, N, P Surgis, D Tabourazi and I Brunton-Smith (2008) Knowledge and attitudes across cultures: A meta-analysis, *Public Understanding of Science*, 17, 1, 35–54.
Allerbeck, KR and WJ Hoag (1989) Utopia is around the corner: Computer diffusion in den USA als soziale Bewegung, *Zeitschrift fuer Soziologie*, 18, 1, 35–53.
Arnold L (1992) *Windscale 1957: Anatomy of a nuclear accident*, New York, St Martin's Press
Amann, M (2005) (ed) *Go.stop.act! Die Kunst des kreativen Strassenprotestes*, Frankfurt, Trotzdem Verlag.
Ambrose, SE (1990) *Eisenhower: Soldier and President*, New York, Simon & Schuster.
Anders G (2002 [1956]) *Die Antiquiertheit des Menschen. Ueber die Seele im Zeitalter der zweiten industriellen Revoluation*, Munchen, Volume 2, Beck's Reihe.
Arendt, H (1961) What is Authority?, in: *Between Past and Future: Eight Exercises in Political Thought*, New York, Viking Compass Book, 91–142.
Argyris, C and DA Schön (1978) *Organizational Learning: A Theory of Action Perspective. Reading*, MA, Addison-Wesley.
Arnason, G, S Nordal and V Arnason (eds) (2004) *Blood and Data: Ethical, Legal and Social Aspects of Human Genetic Databases*, Reykjavik, University of Iceland Press.
Arnstein, SR (1969) A ladder of citizen participation, *Journal of American Institute of Planners*, 35, 216–24.
Asch S (1987 [1952]) *Social Psychology*, Oxford, Oxford University Press.
Bandura, A (1969) *Principles of Behaviour Modification*, New York, Holt, Rinehart & Winston.
Barinaga, M (2000) Asilomar revisited: Lessons for today? *Science*, 287, 5458, 3 March, 1584–85.
Barrett, D and DJ Frank (1999) Population control for national development: From world discourse to national policies, in: J Boli and GM Thomas (eds) *Constructing World Culture: International Non-governmental Organizations since 1875*, Stanford, SUP, 198–221.

Bar-Tal, D (2000) *Shared belief in society. Social Psychological Analysis*, London, SAGE.
Barthes, R. (1988). 'The Old Rhetoric—an aide-memoire', in: *The Semiotic Challenge*. Trans. Richard Howard, Challenge, New York: Hill & Wang, pp. 11–93.
Bateson, G (1972) *Towards an Ecology of Mind*, San Francisco, Chandler.
Bauer MW (1991) Resistance to change—a monitor of new technology? *Systems Practice*, 4, 3, 181–96.
Bauer MW (1993) Resistance to change. A functional analysis of responses to technical change in a Swiss bank, unpublished PhD thesis, London School of Economics and Political Science, January.
Bauer MW (1995a) Industrial and post-industrial public understanding of science, paper presented to the Chinese Association for Science and Technology, Public Understanding of Science Conference, 15–19 October 1995, Beijing.
Bauer, MW (1995b) Technophobia: A misleading conception of resistance to new technology; in: Bauer, M. (ed) *Resistance to New Technology: Nuclear Power, Information Technology, Biotechnology*, Cambridge, CUP, 97–122.
Bauer, MW (1995c) Towards a functional analysis of resistance; in: Bauer M (ed) *Resistance to New Technology: Nuclear Power, Information Technology, Biotechnology*, Cambridge, CUP, 393–418.
Bauer, MW (1998) La longue duree of popular science, 1830–present, in Deveze-Berthet, D (ed) *La promotion de la culture scientifique et technique: ses acteur et leurs logic*, Actes du colloque des 12 et 13 decembre 1996, 75–92
Bauer, MW (1998) The medicalisation of science news: From the rocket-scalpel to the gene-meteorite complex, *Social Science Information*, 37, 731–51.
Bauer, MW (2000) Science in the media as cultural indicator: Contextualising surveys with media analysis, in: Dierkes, M and C von Grote (eds) *Between Understanding and Trust: The Public, Science and Technology*, Amsterdam, Harwood Academic Publishers, 157–78.
Bauer, MW (2001) Biotechnology: ethical framing in the elite press, *Notizie di Politeia* (Milano), 17, 63, 51–66.
Bauer, MW (2002) Areans, platforms and the biotechnology movement, *Science Communication*, 24, 2, 144–61.
Bauer, MW (2004) The vicissitudes of 'public understanding of science': From 'literacy' to 'science in society', in: *Science meets Society*, Lisbon, Gulbenkian Foundation, 37–63.
Bauer, MW (2005) Distinguishing red from green biotechnology: cultivation effect of the elite press, *International Journal of Public Opinion Research*, 17, 1, 63–89.
Bauer, MW (2006a) Paradoxes of resistance in Brazil, in: Gaskell, G and MW Bauer (eds) *Genomics and Society: Legal, Ethical and Social Dimensions*, London, Earthscan, 228–49.
Bauer, MW (2006b) Towards Post-industrial Public Engagement with Science: Revisiting a 10-year old hypothesis; paper presented at EASST conference Lausanne, Switzerland, 23–26 August 2006.
Bauer, MW (2007) The public career of the 'gene', *New Genetics and Society*, 26, 1, 29–46.
Bauer, MW (2008a) Social influence by artifacts, *Diogenes*, 217, 55, 1, 68–83.
Bauer, MW (2008b) Paradigm change for science communication: Commercial science needs a critical public, in: Cheng, D, M Claessens, Toss Gascoigne, J Metcalfe, B Schiele and S Shi (eds) *Communicating Science in Social Contexts: New Models, New Practices*, New York, Springer, 7–26.
Bauer, MW (2012) Public attention to science 1820–2010—a 'longue duree' picture, in: Rödder, S, M Franzen and P Weingart (eds) The sciences' media connection—public communication and its repercussions. *Sociology of the Sciences Yearbook* 28, Dordrecht, Springer, Chapter 3, 35–58.

Bauer, MW (2013a) Knowledge society favours science communication, but puts science journalism into the clinch, in: Baranger, P and B Schiele (eds), *Science Communication* Today: International Perspectives, Issues and Strategies, Paris, CNRS Editions, 145–165.

Bauer, MW (2013b) Social influence by artefacts: norms and objects as conflict zones, in: Sammut, G, P Daanen and FM Moghaddam (eds) *Understanding the Self and Others: Explorations in Inter-subjectivity and Inter-objectivity*, London, Routledge, 189–205.

Bauer MW, R Shukla and P Kakkar (2009) *The integrated data on public understanding of science* [EB_PUS_1989_2005]—Codebook and unweighted frequency distributions, LSE (London), NCAER (Delhi) and GESIS (Koln).

Bauer, MW, N Allum, and S Miller (2007) What have we learnt from 25 years of PUS research—liberating and widening the agenda, *Public Understanding of Science*, 15,1, 1–17.

Bauer, MW and M Bucchi (eds) (2007) *Journalism, Science and Society. Science Communication between News and Public Relations*, New York, Routledge.

Bauer, MW and H Bonfadelli (2002) Controversy, media coverage and public knowledge, in: Bauer, MW and G Gaskell (eds) *Biotechnology: The Making of a Global Controversy*, Cambridge, CUP, 149–75.

Bauer, MW, J Durant, G Evans (1993) European public perceptions of science, *International Journal of Public Opinion Research*, 6, 153–86.

Bauer, MW and G Gaskell (1999) Towards a paradigm for research on social representations, *Journal for the Theory of Social Behaviour*, 29, 2, 163–86.

Bauer, MW and G Gaskell (2002) The biotechnology movement, in: Bauer, MW and G Gaskell (eds) *Biotechnology: The Making of a Global Controversy*, Cambridge, CUP, 379–404.

Bauer MW and G Gaskell (2008) Social representations—a progressive research programme for social psychology, *Journal for the Theory of Social Behaviour*, 39, 4, 335–53.

Bauer MW, G Gaskell & J Durant (1994) *Biotechnology and the European Public—An international study of policy making, media coverage and public perceptions, 1980–1996*. Unpublished project outline (the Hydra Paper. 10th June), London, LSE & Science Museum.

Bauer, MW, G Gaskell and N Allum (2000) Quality, quantity and knowledge interests: avoiding confusions, in: MW Bauer and G Gaskell (eds) *Qualitative Researching with Text, Image and Sound*, London, SAGE, 3–18.

Bauer, MW and J Gregory (2007) From journalism to corporate communication in post-war Britain, in: Bauer, MW & M Bucchi (eds) *Journalism, Science and Society: Science Communication between News and Public Relations*, London, Routledge, 33–52.

Bauer, MW, R Harré and C Jensen (2013) *Resistance and the Practice of Rationality, New Castle*, Cambridge, Scholars Publishers.

Bauer, MW and S Howard (2004) Biotechnology and the Public [Media module]—Europe, North America and Japan, 1973–2002. Integrated Codebook, LSE-report, May 2004.

Bauer, Martin W. and Jensen, Pablo (2011) The mobilization of scientists for public engagement *Public Understanding of Science*, 20, 1, 3–11.

Bauer, MW, M Kohring, A Allansdottir and J Gutteling (2001) The dramatisation of biotechnology in the mass media, in: Gaskell, G and MW Bauer (eds) *Biotechnology 1996–2000: The Years of Controversy*, London, Science Museum, 35–52.

Bauer, MW, Petkova K, P Boyadjieva and G Gornev (2006) Long-term trends in the representations of science across the iron curtain: Britain and Bulgaria, 1946–95, *Social Studies of Science*, 36, 1, 97–129.

Bauer, MW, A Ragnarsdottir, A Rudolfsdottir and J Durant (1995) *Science and Technology in the British Press, 1946–1990: A Systematic Content Analysis of the Press*, vol 1–4, London, Science Museum & Wellcome Trust.
Bauer, MW, R Shukla and N Allum (2012) (eds) *The Culture of Science—How the Public Relates to Science across the Globe*. New York, Routledge—Vol 15 Routledge Science, Technology & Society Series
Bay, C and CC Walker (1975) *Civil disobedience: Theory and Practice*, Montreal, Black Rose Books.
Baylis, J (1997) *Anglo-American Relations since 1939: The enduring Alliance*, Manchester, MUP.
BBC News (2009) *The statistics of CCTV*, 20 July, accessed @ news.bbc.co.uk/2/hi/uk/8159141.stm
Beck, U (1992 [1989]) *Risk Society: Towards a New Modernity*, London, SAGE.
Bell, A (1994) Climate of opinion: Public and media discourse on the global environment, *Discourse & Society*, 5, 1, 33–64.
Bell, D (1967) Notes on the post-industrial society, *The Public Interest*, 24–35.
Bell, D (1973) *The Coming of the Post-industrial Society: A Venture in Social Forecasting*, New York, Basic Books.
Bendix, J and CM Liebler (1991) Environmental degradation in Brazilian Amazonia: perspectives of US news media, *The Professional Geographer*, 43, 4 (November), 474–85.
Benford, TD and DA Snow (2000) Framing processes and social movements: an overview and assessment, *Annual Review of Sociology*, 26, 611–39.
Beniger, JB (1986) *The Control Revolution*, Cambridge, MA, HUP.
Berg, P, D Baltimore, HW Boyer, SN Cohen, RW Davis, DS Hogness, D Nathans, R Roblin, JD Watson, S Weissmann and ND Zinder (1974) Potential biohazards of recombinant DNA molecules, *Science*, 1985, 303.
Bergmann, K and GD Graff (2007) The global stem cell patent landscape: Implications for efficient technology transfer and commercial development, *Nature Biotechnology*, 25, 4, 419–24.
Beuzekom, B van and A Arundel (2006) *OECD Biotechnology Statistics—2006*, Paris, OECD.
Bigelow, JH (1982) A catastrophe model of organisational change, *Behavioral Science*, 26, 26–42.
Biotechnology and the European Public Concerted Action (BEPCA) (1997) Europe ambivalent on biotechnology, *Nature*, 387, 26 June, 845–47.
Blass, T (2000) (ed) *Obedience to Authority: Current Perspectives on the Milgram Paradigm*, Mahwah, NJ, Erlbaum.
Bloch, E (1986 [1959]) *The Principle of Hope*, Oxford, Blackwell.
Blumenberg, H (1981 [1959]) Lebenswelt und Technisierung unter dem Aspect der Phaenomenologie, in: *Wirklichkeiten in denen wir leben*, Stuttgart, Philipp Reclam jun, 7–54.
Blumenberg H (1990 [1979]) *Work on Myth*, Cambridge MA, MIT Press
Blumenberg H (1997) Alles uber Futurologie—ein Soliloquium, in: *Ein moegliches Selbstverstaendis*, Stuttgart, Philipp Reclam jun., 29
Boden, MA (1988) *Computer Models of the Mind*, Cambridge, CUP.
Boecker, F and H Gierl (1988) Die Diffusion neuer Produkte—eine kritische Bestandesaufnahme, *Zeitschrift fuer Betriebwirtschaftliche Forschung*, 40, 1, 32–48.
Boehme, H (1996) Die technische Form Gottes. Ueber theologische Implikationen von Cyberspace, NZZ, 13/14 April, 86, 53.
Boesch, C (1996) The emergence of cultures among wild chimpanzees, *Proceedings of the British Academy*, 88, 251–68.
Boesch, EE (2005) The sound of a violin, in: Cole, M, Y Engestrom, and O Vasquez (eds) *Mind, Culture and Activity: Seminal Papers from the Laboratory of Comparative Human Cognition*, Cambridge, CUP, 164–84.

Boholm, A (1996) Risk perception and social anthropology: critique of cultural theory. *Ethnos* 61, 1-2, 64-84.
Bonfadelli, H (2004) *Medienwirkungsforschung I—Grundlagen*, Stuttgart, UTB, 3rd edition.
Bonfadelli, H (2005) Mass media and biotechnology: Knowledge gaps within and between European countries, *International Journal of Public Opinion Research*, 17, 1, 42-62.
Borup, M, N Brown, K Konrad and H van Lente (2006) The sociology of expectations in science and technology, *Technology Analysis & Strategic Management*, 18, 2/3, 285-98.
Botelho, AJJ (1995) The politics of resistance to new technology: Semiconductor diffusion in France and Japan until 1965, in: Bauer, MW (ed) *Resistance to New Technology*, Cambridge, CUP, 227-54.
Boulding, K (1956) *The Image*. Ann Arbor, U of Michigan P.
Bowcock, AM (2007) Guilt by association, *Nature*, 447, 7 June, 645-46.
Braun, K and S Schulz (2010) A certain amount of engineering involved: Constructing the public in participatory governance arrangements, *Public Understanding of Science*, 19, 4, 403-19.
Brey, P (2005) Artifacts as social agents, in: Harbers, H (ed) *Inside the Politics of Technology*, Amsterdam, AUP, 61-84.
Broderick, M (1988) *Nuclear Movies: A Filmography*, Northcote, Post-Modem Publishing
Brookes, SK, AG Jordan, RH Kimber and JJ Richardson (1976) The growth of the environment as a political issue in Britain, *British Journal of Political Science*, 6, 245-55.
Brosnan, M (1998) *Technophobia: the Psychological Impact of Information Technology*, London, Routledge.
Broszat, M (1981) Resistanz und Widerstand, in: Broszat, E Froehlich, A Grossmann (eds) *Bayern in der NS-Zeit IV*, Munchen, Oldenbourgh Verlag, 691-709.
Brown, P (2004) Monsanto abandons worldwide GM wheat project, *Guardian*, Tuesday, 11 May 2004.
Brown, R and F Bieri (1998) Lessons from the Swiss Biotech referendum, Briefing paper no 8, Brussels, efb. [www.efb-central.org/images/uploeads; accessed 25 August 2008].
Brumfield, G (2009) Supplanting the old media, *Nature*, 19 March, 274-77.
Bruner, J (1991) The narrative construction of social reality, *Critical Enquiry*, 18, Autumn, 1-21.
Bucchi, M (2010) *Beyond Technocracy: Science, Politics and Citizens*, New York, Springer.
Bucchi, M and RG Mazzolini (2007) Big science, little news: science overage in the Italian daily press, 1946-1997; in: Bauer, MW and M Bucchi (eds) *Journalism, Science and Society: Science Communication between News and Public Relations*, London, Routledge, 53-70.
Buchmann, M (1995) The impact of resistance to biotechnology in Switzerland: a sociological view on the recent referendum, in: Bauer, M (eds) *Resistance to New Technology*, Cambridge, CUP, 207-226.
Bud, R (1991) Biotechnology in the 20th century, *Social Studies of Science*, 21, 415-57.
Burger JM (2009) Replicating Milgram: Would People still obey today, *American Psychologist*, 64, 1-11.
Burgess, A (2003) *Cellular Phones, Public Fears, and a Culture of Precaution*, Cambridge, CUP.
Cabral, R (1990) The nuclear technology debate in Latin America, STIC (Science, Technology, Ideology, Culture) a comparative research project, no 1, Gothenburg, Gothenburg University.

Camus A ([1951] 1981) L'homme revolte, in: *Essais Paris*, Bibliotheque de la Plêiade.
Cantley, M (1995) The regulation of modern biotechnology: A historical and European perspective, in Rehm, HJ, G Reed, D Brauer (eds) *Biotechnology: A Multivolume Treatise*, Weinheim, Wiley-VCH, vol xii, Chapter 18, 505–681.
Card, SK, TP Moran and A Newell (1983) *The Psychology of Human-computer Interaction*, Hillsdale NJ, LEA.
Carlisle, RP (1997) Probabilistic risk assessment in nuclear reactors: engineering success, public relations failure, *Technology and Culture*, 38, 920–41.
Carloppio, J (1988) A history of social psychological reactions to new technology, *Journal of Occupational Psychology*, 61, 67–77.
Carson, R (1962) *Silent Spring*, Boston, Houghton Muffin.
Carty, V and J Onyett (2006) Protest, cyberactivism and new social movements: the reemergence of the peace movement post 9/11, *Social Movement Studies*, 5, 1, 229–49.
Cassidy, A (2005) Popular evolutionary psychology in the UK: An unsual case of science in the media? *Public Understanding of Science*, 15, 175–205.
Castells, M (1996) *The Rise of the Network Society*, Oxford, Blackwell.
Castro P and I Gomes (2005) Genetically modified organisms in the Portuguese press: thematisation and anchoring, *Journal for the Theory of Social Behaviour*, 35, 1, 1–17.
Chadarevian, S de (2003) Portrait of a discovery: Watson, Crick and the Double Helix, *ISIS*, 94, 90–115.
Chase A (2003) *Harvard and the Unabomber: The education of an American Terrorist*, New York, WW Norton & Company.
Cassidy A (2006) Evolutationary psychology as public science and boundary work, *Public Understanding of Science*, 15, 2, 175–205.
Chernousenko, WM (1991) *Chernobyl—Insights from the Inside*, Berlin, Springer.
Churcher, J and E Lieven (1983) Images of nuclear war and the public in British civil defense, *Journal of Social Issues*, 39, 117.
Ceruzzi PE (2003) *A History of Modern Computing*, Cambridge MA, MIT Press, 2$^{nd}$ edition
Cirincione J (2007) *Bomb Scare: The History and Future of Nuclear Weapons*, New York, CUP.
Clausewitz, C von ([1832] 1983) *Vom Kriege*, Frankfurt, Ullstein Verlag.
Coch L. and J.R.P. French (1948) Overcoming resistance to change, *Human Relations*, 1, 512–532.
Cohn SM (1997) *Too Cheap to Meter: an Economic and Philosophical Analysis of the Nuclear Dream*, Albany, State University of New York Press
Collingwood, RG ([1945] 1960) *The Idea of Nature*, Oxford, OUP.
Collins, HH (1990) *Artificial Experts: Social Knowledge and Intelligent Designs*, Cambridge, MA, MIT Press.
Collinson, D and S Ackroyd (2005) Resistance, misbehaviour and dissent, in: Ackroyd S et al. (eds) *The Oxford Handbook of Work and Organisation*, Oxford, OUP, 305–26.
Computer Industry Almanac (2010) *Worldwide PC market*, Press Release; accessed at www.c.i.a.com/worldwideuseexec.htm
Conan Doyle, A (1894) *The memoirs of Sherlock Holmes*, George Newnes
Consumer International (2014) *Digital must be at the heart of consumer rights*, 4$^{th}$ April, access @ http://www.consumersinternational.org/
Converse, JM (1987) *Survey Research in the United States: Root and Emergence 1890–1960*, Berkeley, University of California Press.
Cook-Deegan, R (1994) *The Gene Wars: Science, Politics, and the Human Genome Project*, New York, Norton & Company.

Cortada, JW (1993) *The Computer in the US: From Laboratory to Market, 1930 to 1960*. Armonk, NY, ME Sharp.
Cowe, R and S Williams (2000) Who are the green consumers? Part 1, Cooperative Bank Green Consumerism Report no 1.
Crafts N (2001) *The Solow Productivity Paradox in Historical Perspective*, LSE working paper, November.
Cranach, M von (1976) (ed) *Method of Inference from Animal to Human Behavior*, The Hague, Mouton.
Cranach M von (1987) *Makroskopische Ansichten. Essays ueber die Entwicklung der Welt, ueber den Menschen und die Gesellschaft*. Research reports from the Department of Psychology, University of Berne
Cranach, M von (1992) The multi-level organisation of knowledge and action: An integration of complexity, in: Cranach, M von; W Doise, and G Mugny (eds) *Social Representations and the Social Bases of Knowledge*, Lewiston, Hogrefe and Huber, 10–22.
Cranach, M von and G Ochsenbein (1985) Selbstueberwachungssysteme und ihre Funktion in der menschlichen Informationsverarbeitung, *Schweizerische Zeitschrift fuer Psychologie*, 44, 221–35.
Cranach, M von, G Ochsenbein and V Valach (1986) The group as self-active system, *European Journal of Social Psychology*, 16, 193–229.
Crick F and JD Watson (1953) Molecular Structure of Nucleic Acids: a structure for Deoxyribose Nucleic Acid, *Nature* (25 April), 171, 737–738.
Cueni TB (1999) *The Swiss biotech referendum: a case study in science communication*, 11th International workshop on nuclear public information in practice, Berne, European Nuclear Energy Society, pp75–78.
Daamen, DDL, IA van der Lans, CJH Midden (1990) Cognitive structures in the perception of modern technologies, *Science, Technology and Human Values*, 15, 2, 202–25.
Dagwell, R and R Weber (1983) System designers' user models: a comparative study and methodological critique, *Communication of the ACM*, 26, 11, 987–97.
Dalton, RJ (2005) The greening of the globe? Cross-national levels of environmental group membership, *Environmental Politics* 14, 4, 441–59.
Day JC, A Janus and J Davis (2005) *Computer and internet use in the United States: 2003*. Current Population Reports P23–208, Issued October, Washington DC, US Census Bureau.
DeGreene BK (1988) Long wave cycles of sociotechnical change and innovation: a macropsychological perspective, *Journal of Occupational Psychology*, 61, 7–23.
Defra (2007) Survey of Public Attitudes and Behaviours Toward the Environment, London, National Statistics Press Release, 14 August 2007.
Delli Carpini, MX, FL Cook, LR Jacobs (2004) Public deliberation, discursive participation, and citizen engagement: a review of the empirical literature, *Annual Review of Political Science*, 7, 315–44.
Dertouzos, ML and J Moses (eds) (1979) *The Computer Age: A Twenty-year View*, Cambridge, MA, MIT Press.
Derrida, J ([1967] 2003) *Die Stimme und das Phaenomen*, Frankfurt, Suhrkamp Edition.
Devall, B (1985) *Deep Ecology: George Sessions*, Salt Lake City, Gibbs M. Smith Inc.
deVries, R (2007) Who will guard the guardians of neuroscience? *EMPO Reports*, 8, special issue, 65–69.
Dierkes, M, U Hoffmann and L Marz (1996) *Visions of Technology: Social and Institutional Factors Shaping the Development of New Technologies*, Frankfurt, Campus.

Dinello, D (2005) *Technophobia: Science Fiction Visions of Post-human Technology*, Austin, University of Texas Press.
Dora, C (2006) (ed) *Health, Harzards and Public Debate: Lessons for Risk Communication from the BSE/CJD Saga*, Copenhagen, WHO.
Dorn H (1991) *The Geography of Science*, Baltimore, John Hopkins University Press.
Douglas JD (1971) *Technological Threat*, Upper Saddle River NJ, Prentice Hall.
Douglas M (1992) *Risk and Blame: Essays in Cultural Theory*, London, Routledge.
Douglas, M and AB Wildavsky (1982) *Risk and Culture: An Essay on the Selection of Technical and Environmental Dangers*. Berkeley, UCP.
Downs, A (1972) Up and down with ecology—the 'issue-attention cycle', *Public Interest*, 28, 38–50.
Dreyfus, HL (2001) *On the Internet*, London, Routledge.
Dreyfus, H and S Dreyfus ([1986] 1991) 'Why computers may never be like people', in: Forester, T (ed) *Computers in the Human Context*, Cambridge, MA, MIT Press, 125–45.
Duclos, D (1994) Etymologies du risque, in Beckenback, N, W Van Treeck (eds), *Soziale Welt, Umbrüche Gesellschaftlicher Arbeit*, Sonderband 9, VIII/ 664, 1994.
Duijn, JJ van (1983) *The Long Wave in Economic Life*, London, George Allen & Unwin.
Durrenmatt F (1962) *Die Physiker*, Zurich, Verlags AG 'Die Arche' (English in 1964).
Durant J, M Bauer, G Gaskell (1998) (eds) *Biotechnology in the Public Sphere: A European Source Book*, London, Science Museum [part I, III and IV].
Durant, J, Bauer M, C Midden, G Gaskell, and M Liakopoulos (2000) Two cultures of public understanding of science, in: Dierkes, M and C von Grote (eds) *Between Understanding and Trust: The Public, Science and Technology*, Reading, Harwood Academics Publisher, 131–56.
Durant J, Hansen A, Bauer M (1996) The public understanding of new genetics, in: Marteau, T and MPM Richards (eds) *The Troubles Helix: Social and Psychological Implications of the New Human Genetics*, Cambridge, CUP, 235–47.
Durant, J and N Lindsey (2000) The great GM food debate—a survey of media coverage in the first half of 1999, London, Parliamentary Office of Science & Technology, POST-Report no 138.
Dutton, WH and EJ Helsper (2007) *The Internet in Britain 2007*, Oxford, Oxford Internet Institute OII.
Duveen, G (2001) Genesis and structure: Piaget and Moscovici, in: Buschini, F and N Kalampalikis (eds) *Penser la vie, le social, la nature*. Melanges en l'honneur de Serge Moscovici, Paris, MSH, 163–74.
Eagly, AH and S Chaiken (1993) *The Psychology of Attitudes*, Fort Worth, Harcourt Brace College Publishers.
Ebel, A (1994) Wer verschmutzt wessen Luft in Europa? Schadstofftransport in der Atmosphaere, *Naturwissenschaften*, 81, 49–64.
EC, European Commission (1982) The Europeans and their environment, Brussels.
EC, European Commission (1986) The Europeans and their environment in 1986, Brussels.
EC, European Commission (1991) Public Transportation and Biotechnology (35.1), Brussels, April (eds KH Raif & A Melich).
EC, European Commission (1992a) Europeans and the environment in 1992, Brussels, Spring.
EC, European Commission (1992b) Europeans, Science and Technology (38.1), Brussels, Eurobarometer,

EC, European Commission (2001) Europeans, Science and Technology (55.2), Brussels, Eurobarometer,
EC, European Commission (2003) Review of GMOs under research and development and in the pipeline in Europe, Brussels, EC Joint Research Centre.
EC, European Commission (2005)—Europeans, Science and Technology (Special 224, 63.1), Brussels, Eurobarometer, April. EC,
European Commission (2008) European's attitudes to climate change, (Special 300), , Brusssel, Eurobarometer, March–May 2008.
EC, European Commission (2010) Europeans, Science and Technology (Special 340, 73.1), Brussels, Eurobarometer.
Eccleston, C and G Crombez (1999) Pain demands attention: a cognitive-affective model of the interruptive function of pain, *Psychological Bulletin*, 125, 356–66.
Eco, U (1983) Horn, Hooves and Insteps: Some hypotheses on three types of abduction, in: Eco, U and TA Sebeok (eds) *The Sign of Three: Dupin, Holmes and Peirve*, Bloomington, IUP, 198–220.
Eder, K (1996) The institutionalisation of environmentalism: ecological discourse and the second transformation of the public sphere, in: Lash, S, B Szerszyinski and B Wynne (eds) *Risk, Environment and Modernity*, London, SAGE, 203–23.
Edgerton, D (2006) *The Shock of the Old: Technology and Global History since 1900*, London, Profile Books.
Edwards, PN (2012) Entangled histories: Climate science and nuclear weapons research, *Bulletin of the Atomic Scientists*, 68, 4, 28–40.
Ehrlich, P (1968) *The Population Bomb*, New York, Ballantine Books.
Einsiedel E (1993) Mental maps of science: Knowledge and attitudes among Canadian adults, *International Journal of Public Opinion Research*, 6, 35–44.
Einsiedel E (2001) Citizen Voices: Public Participation on Biotechnology, *Politeia*, 17, 63, 94–104.
Einsiedel, E, A Allansdottir, N Allum, MW Bauer, A Berthomier, A Chatjouli, S deCheveigne, R Downey, JM Gutteling, M Kohring, M Leonarz, F Manzoli, A Olafsson, A Przestalski, T Rusanen, F Seifert, A Stathopoulou and W Wagner (2002) Brave new sheep—a clone named Dolly, in: Bauer, MW and G Gaskell (eds) *Biotechnology: The Making of a Global Controversy*, Cambridge, CUP, 313–47.
Einsiedel, E and Cochlan E (1993) The Canadian press and the environment: reconstructing social reality, in: Hansen, A (ed) *The Mass Media and Environmental Issues*, Leicester, LUP, 124–49.
Einsiedel, E and RM Geransar (2005) GURT in context. A preliminary analysis of the impacts of genetic use restriction technologies on developing countries, Calgary, Project GENOME Pairies.
Einsiedel, E and MW Kamara (2006) The coming age of public participation, in: Gaskell, G and MW Bauer (eds) *Genomics and Society: Legal, Ethical and Social Dimensions*, London, Earthscan, 95–112.
Entman, RM (1993) Framing: Toward clarification of a fractured paradigm, *Journal of Communication*, 43, 51–58.
Epstein, S (1996) *Impure Science: AIDS, Activism, and the Politics of Knowledge*, University of California Press, 1996
Ernst & Young (2002) Beyond Borders—the Global Biotechnology Report 2002, E&Y UK, June.
ETC Group (2003a) Terminator technology—five year later, 79, May/June.
ETC Group (2003b) The Big Down: Atomtech—Technologies converging at the Nanoscale, Winnipeg, ETC Group.
Etzkowitz, H and L Leydesdorff (2000), The Dynamics of Innovation: From National Systems and 'Mode 2' to a Triple Helix of University-Industry-Government Relations. Research Policy, 109–23.
Eysenck, HJ and L Kamin (1981) *Intelligence: The Battle for the Mind*, London, Pan Books.

Farr, RM (1996) *The Roots of Modern Social Psychology*, Oxford, Blackwell.
Fernandez-Cornejo, J, M Caswell, L Mitchell, E Golan and F Kuchler (2006) The first decade of genetically engineered crops in the US, Washington DC, USDA Economic Research Service / April 2006 / EIB-11.
Ferree, MM (2003) Resonance and radicalism: Feminist framing in the abortion debates of the US and Germany, *American Journal of Sociology*, 109, 2, 304–44.
Festinger, L, HW Riecken and S Schachter ([1956] 2008) *When Prophecy Fails*, London, Pinter & Martin.
Festinger L (1983) *The human legacy*, New York, Columbia University Press
Finlay M (1987) *Powermatics—a discursive critique of new communication technology*, London, Routledge.
Fischhoff, B (1982) Debiasing, in: Kahnemann, D, P Slovic and A Tversky (eds) *Judgement under Uncertainty: Heuristics and Biases*. Cambridge, CUP, 422–44.
Fischoff, B (1998) Risk perception and communication unplugged: twenty years of process, *Risk Analysis*, 15, 2, 137–45.
Fischoff, B, P Slovic, S Lichtenstein, S Read and B Combs (1978) How safe is safe enough? A psychometric study of attitudes towards technological risks and benefits, *Policy Sciences*, 9, 127–52.
Flam, H (ed) (1994) *States and Anti-nuclear Movements*, Edinburgh, EUP.
Fogg, BJ (1998) Persuasive computers: perspectives and research directions, CHI 98, 18–23 April, Los Angeles.
Forsyth, T (2007) Are environmental social movements socially exclusive? An historical study from Thailand, *World Development*, 35, 12, 2110–30.
Foster, KR (1986) The VDT debate, reprinted in: Forester, T (1991) (ed) *Computers in the Human Context*, Cambridge, MA, MIT Press, 188–97.
Frankfurt, HG (2005) *On Bullshit*, Princeton, PUP.
Franklin, J (2007) The end of science journalism, in: Bauer, MW and M Bucchi (eds) *Journalism, Science and Society: Science Communication between News and Public Relations*, New York, Routledge, 143–56.
Franzen, M, S Roedder and P Weingart (2007) Fraud: causes and culprits as perceived by science and the media, *EMBO Reports*, 8, 1, 3–7
Franzosi, R (2004) *From Words to Numbers: Narrative, Data and Social Science*, Cambridge, CUP.
Franzosi R and S Vicari (2012) What's in a Text? Answers from Frame Analysis and Rhetoric for Measuring Meaning, unpublished manuscript , Emory University.
Frayn, M (1998) *Copenhagen—The Play*. Copenhagen, London, Methuen Drama.
Freire, P ([1974] 2005) Extension or communication, in: *Education for Critical Consciousness*, London & New York, Continuum, 87–146.
Freud S (1922) Group Psychology and the Analysis of the Ego, New York, Boni and Liveright (translated by James Strachey).
Frese, M (1987) Human computer interaction in the office, in: Cooper, CL and IT Robinson (eds) *International Review of Industrial and Organisational Psychology*, 117–65.
Frickel, S and N Gross (2005) A general theory of scientific/intellectual movements, *American Sociological Review*, 70, 2, 204–32.
Fukuyama F (2002) *Our Posthuman Future—Consequences of the Biotechnology Revolution*, London, Profile Books.
Fuller, S (2010) *Science* (The Art of Living Series, Ed. Mark Vernon), London, Acumen.
Gale, BG (1972) Darwin and the concept of struggle for existence: a study in the extra-scientific origin of scientific ideas, *ISIS*, 63, 321–44
Galbraith, JK (2005) *The Economics of Innocent Fraud*, London, Penguin.

Galloux, JC, H Gaumont, and E Stevers (1998) Europe, in: Durant, J, MW Bauer and G Gaskell (eds) *Biotechnology in the Public Sphere: A European Sourcebook*, London, Science Museum, 177–87.
Galloux, JC, AT Mortensen, S deCheveigne, A Allansdottir, A Chatjouli and G Sakellaris (2002) The institutions of bioethics, in: Bauer, MW & G Gaskell (eds) *Biotechnology: The Making of a Global Controversy*, Cambridge, CUP, 129–48.
Galvan, CG (1983) A difusao da industria nuclear, *Revista de Economia Politica*, 3, 4, 107–25.
Gamson, WA and A Midigliani (1989) Media discourse and public opinion on nuclear power: A constructivist approach, *American Journal of Sociology*, 95, 1–37.
Gans, HJ (1993) Reopening the black box: Towards a limited effect theory, *Journal of Communication*, 43, 4, 29–35.
Gaskell, G et al. (1997) Europe ambivalent on biotechnology, *Nature*, 387, 345–47 (June; Biotechnology and the European Public Concerted Action Group).
Gaskell G, A Allansdottir, N Allum et al. (2011) The 2010 Eurobarometer on the life sciences, Nature Biotechnology, 29, 2, 113–14.
Gaskell G and MW Bauer (2001) (eds) *Biotechnology, 1996–2000: The Years of Controversy*, London, Science Museum.
Gaskell G and MW Bauer (2006) *Genomics & Society. Legal, Ethical and Social Dimensions, London*, Earthscan.
Gaskell, G, MW Bauer, J Durant, and NC Allum (1999) Worlds apart? The reception of genetically modified foods in Europe and the US, *Science*, 285, 16 July, 384–87.
Gaskell G, S Stares, A Allansdottir, N Allum, C Corchero, C Fischler, J Hampel, J Jackson, N Kronberber, N Meijgaard, G Revuelta, Ca Schreiner, H Togersen and W Wagner (2008) Europeans and Biotechnology in 2005: patterns and trends. Final report on Eurobarometer 64.3, London Brussels, LSE and DG Research.
Gaskell, G, T TenEyck, J Jackson and G Veltri (2004) Imagining Nanotechnology: Cultural support for technological innovations in Europe and the United States, *Public Understanding of Science*, 14, 81–90.
Gauchat, G (2012) The politicization of science in the public sphere: a study of public trust in the US, 1974 to 2010, *American Journal of Sociology*, 77, 2, 167–87.
Gehlen, A (1980) *Man in the Age of Technology*, New York, Columbia University Press.
Gehlen, A ([1953] 1965) Die Technik in der Sichtweise der Anthropologie, in: Anthropologische Forschung—Zur Selbstbegegnung und Selbstentdeckung des Menschen, Hamburg, Rowolt.
Gelperin, D and Bill Hetzel (1988) The growth of software testing, *Communications of the ACM*, 31, 6, 687–95.
Gerbner, G and L Gross (1976) Living with Television: The Violence Profile, *Journal of Communication*, 26, 2, 173–99.
Gergen, KJ (1983) Social psychology as history, *Journal of Personality and Social Psychology*, 26, 309–30.
Gervais, MC (1997) Social representations of nature: The case of the 'Braer' oil spill in Shetland, unpublished PhD thesis, London School of Economics.
Geyer, M (1992) Resistance as ongoing project: visions of order, obligations to strangers, struggles for civil society, *The Journal of Modern History*, 64, suppl. December, 217–41.
Giami, A (1991) de Kinsey au Sida: l'evolution de la construction du comportement sexual dans les enquetes quantitatives, *Sciences Sociales et Sante*, 9, 4, 23–55.

Giami, A (1996) The influence of an epidemiological representation of sexuality: the ADSF survey questionnaire, in: Bozon, M and H Leridon (eds) *Sexuality and Social Sciences: An Analysis of the French Sexual Behaviour Survey*, London, Dartmouth, 57–82.

Gibbon, D ([1776] 2004) *The Christians and the Fall of Rome*, London, Penguin.

Gibson, W (1984) *Neuromancer*, London, Victor Golancz.

Gieryn, TF (1983) Boundary-work and the demarcation of science from non-science: strains and interests in professional ideologies of scientists, *American Sociological Review*, 48, 781–95.

Gigerenzer, G (2007) *Gut Feelings*, New York, Viking.

Giles, J (2007) Breeding cheats—misconduct special, *Nature*, 445, 7125, 18 January, 242–43.

Ginneken, J van (1992) *Crowds, Psychology and Politics, 1871–1899*, Cambridge, CUP.

Gladwell, M (2000) *Tipping Point: How Little Things Can Make a Big Difference*, London, Abacus.

Glaser, A and F N von Hippel (2006) Thwarting nuclear terrorism, *Scientific American*, February, 38–45

Godin, B (2005) *Measurement and Statistics on Science and Technology: 1920 to the Present*, London, Routledge.

Godin, B (2010) 'Meddle not with those that are given to change', Project on the intellectual history of innovation, Working paper no 6, Montreal.

Godin, B (2012) The culture of science and the politics of numbers, in: Bauer, MW, R Shukla and N Allum (eds) *The Culture of Science: How the Public Relates to Science across the Globe*. New York, Routledge, 18–35.

Godin, B and Y Gingras (2000) What is scientific and technological culture and how to measure it? A multi-dimensional model. *Public Understanding of Science* 9, 43–58.

Goepfert, W (2008) The strength of PR and the weakness of science journalism, in: Bauer, MW and M Bucchi (eds) *Journalism, Science and Society: Science Communication between News and Public Relations*, New York, Routledge, 215–26.

Goldschmidt, B (1980) *Le complex atomique. Histoire politique de l'energie nucleaire*, Paris, Fayard.

Golin, T (1999) A Guerra Guaranitica. Como os exercitos de Portugal e Espanha destruiam os sete povos dos jesuitas e indios guaranis no Rio Grande do Sul (1750–1761), Porto Alegre, editora UFRGS.

Gooch, GD (1996) Environmental concerns and the Swedish press, *European Journal of Communication*, 11, 2, 107–27.

Graber, P, J Hampel, N Lindsey and H Torgersen (2001) Biopolitical diversity: The challenge of multi-level policy making, in: Gaskell, G & MW Bauer (2001) (eds) *Biotechnology, 1996–2000: The years of controversy*, London, Science Museum, 15–34.

Grafee C (2001) Pain Asymbolia [unpublished manuscript; 159pp; available as pdf-file]

Graumann, CF and L Kruse (1990) The environment: Social construction and psychological problems, in: Himmelweit, H and G Gaskell (eds) *Societal psychology*, London, SAGE, 212–29.

Grayson, L (1995) *Scientific Deception: An Overview and Guide to the Literature of Misconduct and Fraud in Scientific Research*, London, The British Library

Greely, HT (1998) Legal, ethical and social issues in human genome research, *Annual Review of Anthropology*, 27, 473–502

Gregory, J and MW Bauer (2003) PUS Inc: l'avenir de la communication de la science, in: Schiele, B and R Jantzen (eds) *Les territories de la culture scientifique*, Montreal and Lyon: Press Universitaire de Lyon, pp41–65.

Gregory, J and S Miller (1998) *Science in Public: Communication, Culture and Credibility*, New York, Plenum Trade.

Grimston, MC and P Beck (2002) *Double or Quits? The Global Future of Civil Nuclear Energy*, London, Royal Institute of International Affairs & Earthscan Books.

Gross, A.G. (1990) *The Rhetoric of Science*, Cambridge, MA, HUP.

Gutteling, J (2005) Mazur's hypothesis on technology controversy and media, *International Journal of Public Opinion Research*, 17, 1, 23–41.

Haber, E (1996) Industry and the university, *Nature Biotechnology*, 14, 4, April, 441.

Habermas, J ([1962] 1989) *The Structural Transformation of the Public Sphere: An Enquiry into a Category of Bourgeois Society*, Cambridge, Polity Press.

Habermas, J (1990) Vorwort zur Neuauflage 1990, in; 'Strukturwandel der Oeffentlichkeit, Frankfurt, Suhrkamp, 11–50.

Habermas, J (2001) *Kommunikatives Handeln und detranszendentalisierte Vernunft*, Stuttgart, Munchen, Reclam Jun.

Habermas J (2003 [2001] *The Future of Human Nature*, Cambridge, Polity Press.

Hacker, W (1986) Complete vs incomplete working tasks—a concept and its verification, in: Debus, G and H W Schroff (eds) *The psychology of work and organisation*, North Holland, Elsevier, 23–36.

Hadley Centre (1992) Transcient climate change experiment: A numerical experiment in which atmospheric concentration of carbon dioxide were increased by 1% per annum (compound) over 75 years, Bracknell, Berks, Meteorological Office, August

Hafner, K and J Markoff (1991) *Cyberpunk: Outlaws and Hackers on the Computer Frontier*, Reading, Corgi Books.

Hagerstrand, T (1967) *Innovation Diffusion as a Spatial Process*, Chicago, CUP.

Hampel, J, P Graber, H Torgersen, D Boy, A Allansdottir, E Jelsoe and G Sakellaris (2006) Public mobilisation and policy consequences, in: Gaskell, G and MW Bauer (eds) *Genomics and Society: Legal, Ethical and Social Dimensions*, London, Earthscan, 75–94.

Hannemyr, G (2003) The internet as hyperbole: a critical examination of adoption rates, *The Information Society*, 19, 111–21.

Hansen, A (1993) (ed) *The Mass Media and Environmental Issues*, Leicester, LUP.

Hansen, A (1994) Journalistic practices and science reporting in the British Press, *Public Understanding of Science*, 3, 111–34.

Harré, R and M von Cranach (1982) *The Analysis of Action*, Cambridge and Paris, CUP and MSH.

Harvie, DI (2005) *Deadly Sunshine: The History and Fatal Legacy of Radium*, Strand, Tempus.

Haskar, V (1986) *Civil Disobedience, Threats and Offers. Gandhi and Rawls*, Delhi, Oxford University Press.

Hecht, G (1998) *The Radiation of France: Nuclear Power and National Identity after WW II*, Cambridge, MA, MIT Press.

Hechter, M (2004) From class to culture, *American Journal of Sociology*, 110, 2, 400–45.

Hedgecoe, A (2004) Critical bioethics: Beyond the social science critique of applied ethics, *Bioethics*, 18, 2, 120–43.

Heijden, H-A vander, R Koopmans and M Giugni (1992) The West-European environmental movement, in: Finger, M and L Kriesberg (eds) *Research in Social Movements, Conflict and Change: The Green Movement Worldwide*, London, Jai Press, 1–40.

Heijs, WJM and CJH Midden (1995) *Biotechnology: Attitudes and Influencing Factors, Third Survey*, Eindhoven, TU.

Heims, SJ (1975) Encounter of behavioural sciences with new machine-organism analogies in the 1940s, *Journal of the History of the Behavioural Sciences*, 11, 368–73.
Heims, SJ (1991) *The Cybernetics Group*, London, MIT Press.
Hennen, L (1994) *Ist die (deutsche) Oeffentlichkeit 'technikfeindlich'? Ergebnisse der Meinungs- und Medienforschung.* TAB-Arbeitsbericht Nr. 24, Bonn, TAB.
Hennen, L (1997) *Monitoring Technikakzeptanz und Kontroversen ueber Technik. Ambivalenz und Widersprueche: die Einstellung der deutschen Oeffentlichkeit zur Technik*, TAB-Arbeitsbericht Nr 54, Bonn, TAB.
Hennen, L (2002) *Monitoring Technikakzeptanz und Kontroversen ueber Technik*, TAB Arbeitsbericht nor 83, Bonn, TAB.
Hennen, L and T Stoeckle (1992) *Gentechnologie und Genomanalyse aus der Sicht der Bevolkerung*, Bonn, TAB-Dikussionspapier no 3, December 1992.
Human Genome Structural Variation Working Group (HGSVWG) (2007) Completing the map of human genetic variation, *Nature*, 447, 10 May, 161–65.
Hilgartner, S and Ch L Bock (1988) The rise and fall of social problems: A public arenas model, *American Journal of Sociology*, 94, 53–78
Hilty, LM, A Köhler, F Van Schéele, R Zah and T Ruddy (2005) *Rebound Effects of the Progress in Information Technology*, Berlin, Springer-Verlag.
Hinde, RA and Stevenson-Hinde J (1973) (eds) *Constraints on Learning*. London: Academic Press.
Hirschman, AO (1989) Two-hundred years of reactionary rhetoric: the case of the perverse effect, The Tanner Lecture on Human Values, Michigan University, 8 April 1988.
Hirschman, AO (1991) *The Rhetoric of Reaction: Perversity, Futility and Jeopardy*, Cambridge, MA, HUP.
Hirschman, AO (1995) *A Propensity to Self-subversion*, Cambridge, MA, HUP.
Hirter, H and W Linder (2005) VOX analyse der eigenossischen Volkabstimmung vom 27 November 2005, Bern, gfs.
Hofstadter, DR (1979) *Gödel, Escher, Bach: An Eternal Golden Braid*. Harmondsworth, Penguin.
Holland, JH, KJ Holyoak, RE Nisbett and PR Thagard (1989) *Induction: Processes of Inference, Learning and Discovery*, Cambridge, MA, MIT Press.
Home Office (2006) Strategic Plan for the National Identity Scheme—Safeguarding your identity, London, HMSO, December 2006. [http://www.Identity cards.gov.uk/downloads/Strategic_Action_Plan.pdf, accessed 11 September 2007].
Hood, C and H Rothstein (2001) Risk regulation under pressure: problem solving or blame shifting? *Administration and Society*, 33, 1, 21–53.
Hood, C, H Rothstein, J Spackman, J Rees and R Baldwin (1999) Explaining risk regulation regimes: Exploring the minimal feasible response hypothesis, *Health, Risk and Society*, 1, 151–66.
House of Commons Select Committee on Science and Technology (2010) The disclosure of climate data from the Climate Research Unit at the University of East Anglia, 8th report of session 2009–10; London, HC 387–1, Stationary Office.
House of Lord Select Committee on Science and Technology (2000) Science and Society, 3rd Report, London, HMSO
Houts, PS, PD Cleary and The-Wei Hu (1988) *The Three Mile Island Crisis: Psychological, Social and Economic Impacts on the Surrounding Population*, University Park, Pennsylvania University Press.
Howard, GS (1986) *Computer Anxiety and Management use of Microprocessors*, Ann Arbor, UMI Research Press.
Howe, JF (1983) Phantom limb pain—a re-afferentation syndrome, *Pain*, 15, 101–107.
Hunger, F (2007) Der kurze Aufstand gegen die Vorherrschaft des Binaersystems, NZZ (International), 33, 9 February 2007, 30.

IAEA (1994) *The Nuclear Power Option*, Vienna, IAEA.
IAEA (2006) *International Nuclear Power Statistics*, Vienna, IAEA
Iggo, A, LL Iversen, and F Cervero (1985) Nociception and pain, proceedings of a discussion meeting held on 24–25 May 1984, London, Royal Society.
International HapMap Consortium (IHMC) (2007) A second generation human haplotype map of over 3.1 million SNPs, *Nature*, 449, 18 October, 851–61.
Inglehart R (1990) *Cultural Shift in Advanced Industrial Society*. Princeton, NJ: Princeton University Press.
Inglis, DR (1969) Civil uses of nuclear explosives, in: Barnaby, CF (ed) *Preventing the Spread of Nuclear Weapons*, Pugwash Monograph I, London, Souvenir Press, 77–89.
Initiative D21 (2007) *(N)onliner Altas 2007—Gemeinsam fuer die Digitale Gesellschaft*, D21 & TNS Infra-test [access @ initiativd21.de]
Intergovernmental Panel on Climate Change (IPCC) (1990) Scientific assessment of climate change, Report of Working Group 1, Cambridge, CUP, July.
IPPNW (1991) *Radioactive Heaven and Earth. The Health and Environmental Effects of Nuclear Weapons Testing In, On, and Above the Earth*, London, Zed Books.
Irwin, A and B Wynne (1996) (eds) *Misunderstanding Science? The Public Reconstruction of Science and Technology*, Cambridge, CUP.
Isaacs, J and T Downing (2008) *Cold War*, London, Abacus.
Isaacson, P and E Juliusson (1981) IBM's billion dollar baby: The personal computer (market research report), *Future Computing Inc*; [http://www.futurecomputing.com/ All/pubs/8108_billion.aspx; accessed 27 August 2007]
ITU (2003) World telecommunication indicators 2002, Geneva, ITU, 9 October 2003 [http://www.itu.int/ITU-D/ict/; accessed: 15 May 2007].
Iyengar, S and D Kinder (1987) *News that Matters: Television and American Opinion*, Chicago, CUP.
Jackson, T (2003) Sustainability and the struggle for existence: the critical role of metaphor in society's metabolism, *Environmental Values*, 12, 289–316.
Jacques, J and D Raichvarg (1991) *Savant et ignorants: une histoire de la vulgarisation des sciences*, Paris, Seuil.
James, C (2007) Global Status of Commericalised Biotech/GM Crops: 2007, Ithaca, ISAAA Brief 37.
Jamison, A (1996) The shaping of the global environmental agenda: The role of non-governmental organisations, in: Lash, S, B Szerszyinski and B Wynne (eds) *Risk, Environment and Modernity*, London, SAGE, 224–44.
Jasanoff, S (1990) *The Fifth Branch: Science Advisors as Policy Makers*, Cambridge, MA, HUP.
Jasanoff S (1995) Product, process or programme: three cultures and the regulation of biotechnology, in: Bauer, MW (ed) *Resistance to New Technology*, Cambridge, CUP, 311–34.
Jasanoff S (2003) Technologies of humility: Citizen participation in governing science, *Minerva*, 41, 223–44.
Jasanoff S (2005) *Designs on Nature: Science and Democracy in Europe and the United States*, Princeton, PUP.
Jasanoff S (2007) Science & politics: Technologies of humility, *Nature*, 450, 1 November, 10–33.
Jasper, J M (1988) The political life cycle of technological controversies, *Social Forces*, 67, 357–77.
Jaufmann, D and E Kistler (1986) Technikfreundlich?—technikfeindlich?—empirische Ergebnisse im nationalen und internationalen Vergleich, *Aus Politik und Zeitgeschichte*, 48, 35–53.
Jaufmann, D, E Kistler, and G Jaensch (1989) *Jugend und Technik*, Frankfurt, Campus.

Jermier, JM, D Knights and WR Nord (eds) (1994) *Resistance and Power in Organisations*, London, Routledge.
Jewell, R, B Hedges, P Lynn, G Farrant and A Heath (1993) The 1992 British election: the failure of the polls, *Public Opinion Quarterly*, 57, 238–63.
Johnson-Laird, P (1988) *The Computer and the Mind: An Introduction to Cognitive Science*, London, Fontana Press.
Johnston, R. (2006) *Number of Nuclear Explosions by Conducting Country and Year* [http://www.johnstonsarchive.net/nuclear/nuctestsum.html; accessed 4 December 2006].
Jonas, H (1979) *Das Prinzip Verantwortung*, Frankfurt, Suhrkamp.
Jones, H and JH Soltren (2005) Facebook: threats to privacy [groups.csail.mit.edu/mac/classes/6.805/student-papers/fall05-papers/facebook.pd f; accessed February 2013]
Jordan, T and P Taylor (2004) *Hacktivism and Cyberwars: Rebels with a Cause*, London, Routledge.
Jost, G and MW Bauer (2005) Organisational Learning by Resistance. Working Paper, Department of Social Psychology, London School of Economics.
Jovchelovitch, S (2007) *Knowledge in Context: Representations, Community and Culture*, London, Routledge.
Jowell R, B Hedges, P Lynn, G Farrant and A Heath (1993) The polls—a review. The 1992 British election and the failure of the polls, Public Opinion Quarterly, 57, 238–263.
Juma, C (1989) *The Gene Hunters. Biotechnology and the Scramble for Seeds*, London, Zed Books.
Jungk, R (1969) *The Nuclear State*, London, John Calder Ldt.
Kahan, DM, D Braman, P Slovic, J Gastil and G Cohen (2007) The second national risk and culture study: Making sense of—and making progress in—the American culture war of fact. The Culture Cognition Project at Yale Law School, report released 27 September 2007.
Kahnemann, D, P Slovic and A Tversky (eds) (1982) *Judgement under uncertainty: Heuristics and Biases*, Cambridge, CUP.
Kalaitzandonake N, JA Alston and KJ Bradford (2007) Compliance costs for regulatory approval of new biotech crops, *Nature Biotechnology*, 25, 509–511.
Kaldor M (1980) Disarmament: the armament process in reverse, in: Thompson, EP and D Smith (eds) *Protest and Survive*, Harmondsworth, Penguin, 203–20.
Karasek, RA (1979) Job demands, job decision latitude, and mental strain: Implications for job design, *Administrative Science Quarterly*, 24, 285–308.
Katz, D (1949) An analysis of the 1948 polling predictions, *Journal of Applied Psychology*, 33, 15–28.
Katz, JE and AR Tassone (1990) Public opinion trends: Privacy and information technology, *Public Opinion Quarterly*, 54, 125–43.
Kelly, C and S Breinlinger (1996) *The Social Psychology of Collective Action: Identity, Injustice and Gender*. London, Taylor & Francis
Kelly, SE (2006) Toward an epistemological luddism of bioethics, *Science Studies*, 19, 1, 69–82.
Kepplinger, HM (1992) Artificial horizons: how the press presented and how the population received technology in Germany from 1965–1985, in: Rothman, S (ed) *The Mass Media in Liberal Democratic Societies*, New York, Paragon House, 147–76.
Kepplinger, HM (1995) Individual and institutional impacts upon press coverage of sciences: The case of nuclear power and genetic engineering in Germany, in: Bauer, MW (ed) *Resistance to New technology: Nuclear Power, Information Technology and Biotechnology*, Cambridge, CUP, 357–78.
Kevles, D (1995) *In the Name of Eugenics*, Cambridge, MA, HUP.

Key, VO (1961) *Public Opinion and American Democracy*, New York, Knopf.
Kielbowicz, RB and C Scherer (1986) The role of the press in the dynamics of social movements, in: Lang, G and K Lang (eds) *Research in Social Movements, Conflicts and Change*, Greenwhich, CT, JAI Press, 71–96
Kipphardt, H (1964) *In der Sache J. Robert Oppenheimer*, Frankfurt, Suhrkamp.
Kirby, DA (2011) *Lab Coats in Hollywood: Science, Scientists and Cinema*, Cambridge, MA, MIT Press.
Kish, L (1965) *Survey sampling*, New York, John Wiley & Sons.
Kitcher, P (2003) *In Mendel's Mirror: Philosophical Reflections on Biology*, Oxford, OUP.
Klein, HK and DL Kleinman (2002) The social construction of technology: Structural considerations, *Science, Technology and Human Values*, 27, 1, 28–52.
Kling, R and I Iacono (1988) The mobilisation of support for computerization: the role of computerization movements, *Social Problems*, 35, 226–43 (reprint in: Leigh Star, S (ed) (1995) *Ecologies of Knowledge*, Albany, SUNY Press, 119–53)
Knight, D (2006) *Public Understanding of Science: A History of Communicating Scientific Ideas*, London, Routledge.
Knowles, ES, Butler, S, & Linn, JA (2001). Increasing compliance by reducing resistance. In JP Forgas & KD Williams (Eds.), Social influence: Direct and indirect processes. Philadelphia PA, Psychology Press, 41–60.
Kohring, M and J Matthes (2002) The face(t)s of biotechnology in the nineties: how the German press framed modern biotechnology, *Public Understanding of Science*, 11, 143–54.
Kolata, G (1997) Clone. The road to Dolly and the path ahead, London, Penguin.
Kraut, R, M Patterson, V Lundmark, S Kiesler, T Mukopadhyay, W Scherlis (1998) Internet paradox: A social technology that reduces social involvement and psychological well-being, *American Psychologist*, 53, 9, 1017–31.
Kriesi, HP (1998) The transformation of cleavage politics: the 1997 Stain Rokkan lecture, European Journal of Political Research, 33, 165–185.
Krige, J (2005) Critical reflections on the science-technology relationship, *Transactions of the Newcomen Society*, 76, 259–69.
Krimsky, S (2003) *Science in the Private Interest. How the Lure of Profits Corrupted Biomedical Research?* Lanham, Rowman & Littlefield.
Krippendorf, K ([1980] 2005) *Introduction to Content Analysis*, Beverley Hills, SAGE.
Kroll, G (2001) The silent springs of Rachel Carson: Mass media and the origins of modern environmentalism, *Public Understanding of Science*, 10, 4, 403–20.
Kruglanski, AW, M Crenshaw, JM Post and J Victoroff (2007) What should this crisis be called? Metaphors of Counter-terrorism and their implications, Psychological Science in the Public Interest, 8, 3, 97–133.
Kruglanski A (2004) The psychology of closed mindedness, New York, The Psychology Press;
Küppers, G and H Nowotny (1979) (eds) The impact of controversy on decision-making structures, University of Bielefeld, Science Studies Report 15.
Kuhn, TS (1962) *The Structure of Scientific Revolutions*, Chicago, CUP.
Langer, M (1995) Why the atom is our friend: Disney, General Dynamics and the NSS Nautilus, *Art History*, 18, 63–96.
Lassen, J, A Allansdottir, M Liakopoulos, AT Mortenson and A Olofsson (2002) Testing times—the reception of Roundup Ready soya in Europem, in: Bauer, MW and G Gaskell (eds) *Biotechnology: The Making of a Global Controversy*, Cambridge, CUP, 279–312.
Latour, B (1988) The prince for machines as well as for machinations, in: Elliott, B (ed) *Technology as Social Process*, Edinburgh, EUP

Latour, B (1996) On inter-objectivity, *Mind, Culture and Activity*, 2, 4, 228–45.
Latour, B (2002) Gabriel Tarde and the end of the social, in: Joyce, P (ed) *The Social in Question: New Bearings in History and the Social Sciences*, London, Routledge, 117–32
Latour, B (2005) *Von der Realpolitik zur Dingpolitik*, Berlin, Merve Verlag.
Lawrence, PL (1954) How to overcome resistance to change, *Harvard Business Review*, 32,3, 49–57.
Lawrence, S (2007) State of the biotech sector: 2006, *Nature Biotechnology*, 25, 7, July, 706.
Lawrence, S and R. Lahteenmaki (2008) Public biotech 2007: The numbers, *Nature Biotechnology*, 26, 7, July, 753–62.
Le Bon, G ([1895] 1966) *The Crowd: A Study of the Popular Mind*, New York Viking/Compass.
Le Bon, G (1895) *Psychologie des Foules*, Paris, Alcan.
Leahy, PJ and A Mazur (1980) The rise and fall of public opposition in specific social movements, *Social Studies of Science*, 10, 259–84.
Lear, LJ (1992) Bombshell in Beltsville: The USDA and the challenge of 'silent spring', *Agricultural History*, 66, 2, 151–70.
Lee, S (1970) Social attitudes and the computer revolution, *Public Opinion Quarterly*, 34, 53–59
Leiserowitz AA, EW Maibach, C Roser-Renouf, N Smith and E Dawson (2010) Climate gate, public opinion and the loss of trust, working paper School of Forestry, Yale University
Lenhart, A (2000) Who's not online, PEW Internet in American Life Project [http://www.pewinternet.org/PPF/r/21/report_display.asp; 20 September 2007].
Lenhart, A et al. (2003) The ever shifting internet population, PEW Internet in American Life Project [http://www.pewinternet.org/pdfs/ PIP_Shifting_Net_Pop_Report.pdf; 20 September 2007].
Levidow, L (2007) European public participation as risk governance: Enhancing democratic accountability for agbiotech policy, *East Asian Science, Technology and Society: An International Journal*, 1, 19–51.
Levidow, L, S Carr, R von Schomberg and D Wield (1996) Regulating agricultural biotechnology in Europe: Harmonisation difficulties, opportunities, dilemmas, *Science and Public Policy*, 23, 3, 135–57.
Levy S (1984) *Hackers: Heroes of the Computer Revolution*, Anchor Press/Doubleday.
Lewenstein, B (2005) Introduction: Nanotechnology and the Public, *Science Communication*, 27, 2, 169–74.
Lewin, K ([1936] 1966) *Principles of Topological Psychology*, New York, McGraw-Hill.
Lewin, K (1952) *Field Theory in the Social Sciences*, London, Tavistock.
Lewis, CS ([1940] 1977) *The Problem of Pain*, Glasgow, Found Publications.
Lewontin, RC (1991) *The Doctrine of DNA: Biology as Ideology*, London, Penguin.
Linke, S (2007) Darwins Erben in den Medien. Eine wissenschafts- und mediensoziologische Fallstudie zur Renaissance der Soziobiologie, Bielefeld, transcript Verlag.
Litmanen, T (2004) The temporariness of societal risk evaluation. Understanding the Finnish nuclear decisions, Jyvaskyla [manuscript].
Liu, JT and K Smith (1990) Risk communication and attitude change: Taiwan's National debate over nuclear power, *Journal of Risk and Uncertainty*, 3, 331–49.
Livingstone, S (2002) *Young People and New Media: Childhood and the Changing Media Environment*, London, SAGE.

Livingstone, S (2006) Reflections on the games families play, Argyle Lecture at the BPS conference, *The Psychologist*, 19, 10, 604–07.
Lodge, D (1975) *Changing Places*, London, Penguin.
Lorenz, K (1974) Analogy as a source of knowledge, *Science*, 185, 229–34.
Lovins, AB (1979) Is nuclear power necessary? London, Friends of the Earth, Energy Report 3.
Lovins, AB (2013) The economics of a US civilian nuclear phase out, *Bulletin of the Atomic Scientists*, 69, 2, 44–65.
Luebbe, H ([1962] 1978) Zur politischen Theorie der Technokratie, in: *Praxis der Philosophie*, Stuttgart, Reclam, 35–60.
Luhmann, N (1984) Soziale Systeme. *Grundriss eine allgemeinen Theorie*. Frankfurt, Suhrkamp.
Luhmann, N (1992) *Wissenschaft der Gesellschaft*, Frankfurt, Suhrkamp.
Luhmann, N (1997) Protestbewegungen, in: *Die Gesellschaft der Gesellschaft*, Vol2, Frankfurt, Suhrkamp, 847ff.
Luhmann, N (1998) Familiarity, confidence, trust: Problems and alternatives, in: Gambetta, D (ed) *Trust: Making and Breaking Cooperative Relations*, Oxford, Basil Blackwell.
Luhmann, N (1990a) *Oekologische Kommunikaton*, Opladen, Westdeutscher Verlag.
Luhmann, N (1990b) The improbability of communication, in: *Essay on Self-reference*, New York, Columbia University Press.
Luhmann, N (2000) *Die Politik der Gesellschaft*, Frankfurt, Suhrkamp.
Lynch, A (1996) *Thought Contagion: How Beliefs Spreads through Society*, New York, Basic Books.
Lyon D (1994) The Electronic eye: the Rise of Surveillance Society, Cambridge, Polity Press.
M'charek, A (2005) *Human Diversity Genome Project: An Ethnography of Scientific Practice*, Cambridge, CUP.
Maass, A and RD Clark III (1984) Hidden impact of minorities: Fifteen years of minority influence research, *Psychological Bulletin*, 95, 428–50.
MacCulloch, D (2003) *Reformation: Europe's House Divided 1490–1700*, London, Allan Lane.
Machlup, F (1962) *The Production and Distribution of Knowledge in the US*, Princeton, PUP.
Macleod, R (1995) Resistance to nuclear technology: Optimists, opportunists and opposition in Australian nuclear history, in: Bauer, MW (ed) *Resistance to New Technology*, Cambridge, CUP, 165–88.
Maddison, A (1991) *Dynamic Forces in Capitalist Development*, Oxford, OUP.
Mahajan, V and RA Peterson (1985) *Models for Innovation Diffusion*, London, SAGE.
Mandelbaum, M (1983) *The Nuclear Future*, Ithaca, Cornell UP.
Mangano, J (2004) Three Mile Island: Health study meltdown, *Bulletin of the Atomic Scientists*, Sept/Oct, 60(5), 31–35.
Mannheim, K (1986) *Conservatism: A Contribution to the Sociology of Knowledge*, London, Routledge & Kegan.
Markus, ML and N Bjorn-Andersen (1987) Power over users: Its exercise by system professionals, *Communication of the ACM*, 30, 6, 468–504.
Marlier, E (1992) Eurobarometer 35.1: Opinions of Europeans on biotechnology in 1991, in: Durant, J (ed) *Biotechnology in Public: A Review of Recent Research*, London, Science Museum, 52–108.
Marris, E (2008) Almost in bloom. *Nature Biotechnology*, 26, 4, 471–72.
Martin, WK (2011) Envisioning technology through discourse: A case study of biometrics in the National Identity Scheme in the UK, PhD thesis, London School of Economics and Political Science.

Martin, R and M Hewstone (2003) Social-influence processes of control and change: Conformity, obedience to authority, innovation, in: Hogg, MA and J Cooper (eds) *The SAGE Handbook of Social Psychology*, London, SAGE, 347–66.

Martin, S and J Tait (1992) Attitudes of selected public groups in the UK to biotechnology, in: Durant, J (ed) *Biotechnology in Public: A Review of Recent Research*, London, Science Museum.

Martin, R (1995) New technology at fleet street, 1975–80, in: Bauer, MW (ed) *Resistance to New Technology*, Cambridge, CUP, 189–206.

Martineau B (2001) *First Fruit: the Creation of the Flavr Savr Tomato and the Birth of Biotech Food*, New York, McGraw-Hill

Marx Ferree, M (2003) Resonance and radicalism: Feminist framing in the abortion debates of US and Germany, *American Journal of Sociology*, 109, 2, 304–44.

Maslin, M (2004) *Global Warming: A Very Short Introduction*, Oxford, OUP.

Matusow, H (1968) *The Beast of Business*, London, Wolfe.

May, J (1990) *The Greenpeace Book on the Nuclear Age: The Hidden History of Human Costs*, London, Victor Gollancz Ldt.

Mayer, RN (1989) *The Consumer Movement: Guardians of the Market Place*, Boston, Twayne Publishers.

Mazur, A (1975) Opposition to technological innovation, *Minerva*, 13, 58–81.

Mazur, A (1984) Media influences on public attitudes towards nuclear power, in: Freudenberg, WR and EA Rosa (eds) *Public Reactions to Nuclear Power: Are There Critical Masses?*, Washington, AAAS, 97–114.

Mazur, A (1990) Nuclear power, chemical hazard, and the quantity of reporting, *Minerva*, 28, 294–323.

Mazur, A and J Lee (1993) Sounding the global alarm: Environmental issues in the US national news, *Social Studies of Science*, 23, 681–720.

McAllister, I and DT Studlar (1993) Trends in public opinion on the environment in Austrialia, *IJPOR*, 5, 4, 353–61.

McCarthy, J (1966) Information, in: Scientific American Book (ed) *Information*, San Francisco, 1–16.

McCarthy, JD and MN Zald (1987) Resource mobilization and social movements: A partial theory, in: Zald, MN and JD McCarthy (eds) *Social Movement in an Organizational Society: Collected Essays*, New Brunswick, Transaction Books, 15–48

McCombs, M (2004) *Setting the Agenda: The Mass Media and Public Opinion*, Cambridge, Polity Press.

Meadows, DH, DL Meadows, J Randers and WW Behrens III (1972) *The Limits to Growth*, New York, Universe Books.

Medlock, J, R Downey and E Einsiedel (2007) Governing controversial technologies: Consensus conferences as a communication tool, in: Brossard, D, J Shanahan and TC Nesbitt (eds) *The Public, the Media and Agricultural Biotechnology*, Wallingford, CABI, 308–26.

Mejlgaard, N, C Bloch, L Degn, MW Nielsen, and T Ravn (2012) Locating science in society across Europe: Clusters and consequences, *Science and Public Policy*, 398, 741–50.

Mellor, F (2010) Negotiating uncertainty: Asteroids, risk and the media, *Public Understanding of Science*, 19,1, 16–33.

Melzack, R and P Wall (1988) *The Challenge of Pain*. Revised edition, Harmonthworth, Penguin.

Mensch, G (1975) *Das technologische Patt*, Frankfurt, Umschau Verlag.

Merton, RM (1936) The unanticipated consequences of purpose social action, *American Sociological Review*, 1, 894–904.

Merton, RM and P Kendall (1946) The focussed interview, *American Journal of Sociology*, 22, 129–52.
Meyer, M (2008) *Principia Rhetorica*, Paris, Fayard.
Michael, M and N Brown (2005) Scientific Citizenship: self-representations of xenotransplantation's publics, *Science & Culture*, 14, 1, 39–57.
Milgram, S (1974) *Obedience to Authority*, New York, Harper & Row.
Miller, HI and G Conko (2004) *The Frankenfood Myth: How Protest and Politics Threaten the Biotech Revolution*, Preager Publications.
Miller GA, E Galanter and KH Pribram (1960) *Plans and the Structure of Behaviour*, New York, Holt, Rinehard & Winston
Miller, JG (1986) Can systems theory generate testable hypotheses? From Talcott Parsons to Living Systems Theory, *Systems Research*, 3,2, 73–84.
Miller, S, P. Caro, V. Koulaidis, V. de Semir, W. Staveloz and R. Vargas (2002). Benchmarking the Promotion of RTD culture and Public Understanding of Science, (Brussels, Commission of the European Communities).
Mirowski, P and EM Sent (2005) The commercialization of science, and the response of STS (draft version of chapter to appear in the New Handbook of STS, MIT Press).
MIT (2003) *The Future of Nuclear Power: An Interdisciplinary MIT Study*, Cambridge, MA, MIT Press.
Mokyr, J (1990) *The Levers of Riches: Technological Creativity and Economic Progress*, Oxford, OUP.
Mokyr, J (2000) Innovation and its enemies: the economic and political roots of technological inertia, in: Olson, M and S Kahonen (eds) *A Not-so-Dismal Science: A Broader View of Economics and Society*, Oxford, OUP, 61–91.
Monbiot, G (2006) The denial industry, *Guardian*, Tuesday, 19 September 2006.
Monsanto (1999) *Annual Report 1988*, St Louis, Monsanto.
MORI (1999) Stress at work already affects 60% of the working population, 3 November 1999: accessed @ Ipsos MORI research archive ipsos-mori.com
Morgan, M and J Shanahan (1997) Two decades of cultivation research: An appraisal and meta-analysis, in: *Communication Yearbook 20*, London, SAGE, 1–45.
Morris, DB (1991) *The Culture of Pain, Berkeley*, UCP.
Moscovici, S (1976) *Social Influence and Social Change*, New York, Academic Press.
Moscovici, S (1977) *Essai sur l'histoire humaine de la nature*, Paris, Flammarion.
Moscovici, S (1985) Social influence and conformity, in: Lindszey, G and E Arsonson (eds) *Handbook of Social Psychology*, 3rd edition, vol. 2, 347–412.
Moscovici, S ([1961] 2008) *Psychoanalysis: Its Image and its Public*, Cambridge, Polity.
Moscovici, S and W Doise (1994) *Conflict and Consensus: A General Theory of Collective Decisions*, London, SAGE.
Moscovici, S, E Lage, M Naffrechoux (1969) Influence of a consistent minority on the responses of a majority in a color perception task, *Sociometry*, 32, 365–79.
Mugny, G (1982) *The Power of Minorities*, London, Academic Press.
Mugny, G and JA Perez (1991) *The Social Psychology of Minority Influence*, Cambridge, CUP.
Mulkay, M (1997) *The Embryo Research Debate: Science and the Politics of Reproduction*, Cambridge, CUP
Muller, JZ (1997) *Conservatism: An Anthology of Social and Political Thought from David Hume to the Present*, Princeton, PUP.
Myers, DJ (1994) Communication technology and social movements: Contributions of computer networks to activism, *Social Science Computer Review*, 12, 250–60.

Nader, R (1965) *Unsafe at Any Speed, The Designed-In Dangers of The American Automobile*, New York, Grossman Publishers.
Naess, A (1973) The shallow and the deep, long-range ecology movements: A summary, *Inquiry*, 16 (Oslo), 95–100.
Nash, HT (1980) The bureaucratization of homicide, in: Thompson, EP and D Smith (eds) *Protest and Survive*, Harmondsworth, Penguin, 62–74.
Nature (2006) Learning from Chernobyl—special report, *Nature* 440, 7087, 982–94.
Nealey, S, BD Melber and WL Rankin (1983) *Public Opinion and Nuclear Power*, Lexington, Lexington Books.
Neidhardt, F (1993) The public as a communication system, *Public Understanding of Science*, 2, 339–50.
Neisser U (1967) *Cognitive Psychology*, East Norwalk, CT, Appleton-Century-Crofts.
Nelkin, D (1987) *Selling Science: How the Press Covers Science and Technology*, New York, WH Freeman.
Nelkin, D (1995) Forms of intrusion: Comparing resistance to information technology and biotechnology in the USA, in Bauer, MW (ed) *Resistance to New Technology*, Cambridge, CUP, 379–92.
Nelkin, D and M S Lindee (1995) *The DNA Mystique: The Gene as a Cultural Icon*, New York, Freeman.
Nepstad, SE (2004) Persistent resistance: Commitment and community in the Plowshares Movement, *Social Problems*, 51, 1, 43–60.
Newell, A and HA Simon (1972) *Human Problem Solving*, Englewood, Prentice-Hall.
Newig, J (2004) Public attention, political action: The example of environmental regulation, *Rationality and Society*, 16, 2, 149–90.
Nisbet, MC and T Myers (2007) The polls trends: 20 years of public opinion about global warming, *POQ*, 71, 3, 444–70
Noelle-Neumann, E (1974) The spiral of silence: A theory of public opinion, *Journal of Communication*, 1, 43–51.
Noelle-Neumann (1984) *The Spiral of Silence: Public Opinion, Our Social Skin*, Chicago, Chicago University Press.
Noelle-Neumann E (1991) The theory of public opinion: The concept of the spiral of silence, in: Anderson, J (ed) *Communication Yearbook 14*, Newbury Park CA, SAGE, 256–87.
Norman, DA (1984) Categorisation of action slips, *Psychological Review*, 88, 1, 1–15.
Norman, DA (1998) *The Invisible Computer*, Cambridge, MA, MIT Press.
Norman, DA (2004) *Emotional Design. Why We Love or Hate Everyday Things*, New York, Basic Books.
Norris, P (1997) Are we all green now? Public opinion on environmentalism in Britain, *Government and Opposition*, 32, 320–39.
Norris, RS and HM Kristensen (2010) Global nuclear weapons inventories, 1945–2010, *Bulletin of the Atomic Scientists*, July/August, 69 (5) 75–81.
Novas, C (2006) The political economy of hope: Patient organisations, science and biovalue, *Biosocieties*, 1, 3, 289–305.
Novotny, H (1983) Rediscoving friction: All that is solid does not melt in air; in Akerman, N (ed) *The Necessity of Friction*, Heidelberg, Physica-Verlag.
National Radiological Protection Board (NRPB) (2004) Mobile Phones and Health, National Radiological Protection Board, 15, 5, Ditcot [www.nrpb.org].
NSF (2006) Science Indicators, Washington, DC, National Science Foundation
Nuffield Council (2012) Emerging technologies: Technology, choice and public good, Swindon, Nuffield Council on Bioethics.
Nussbaum, MC and CR Sunstein (1998) (eds) *Clones and Clones: Facts and Fantasies about Human Cloning*, New York, WW Norton.
OCHA (2000) *Chernobyl: A Continuing Catastrophe*, New York, United Nations.

O'Connor, C and H Joffe (2013) How has neuroscience affected lay understandings of personhood? A review of the evidence, *Public Understanding of Science*, 22, 3, 254–68.
OECD (1992) Technology and the economy: The key relationships, Paris: OECD.
OECD (2005) A framework for biotechnology statistics, Paris, 2005 [http://www.oecd.org/dataoecd/5/48/34935605.pdf; accessed 13 August 2007].
OECD (2012) Key Biotechnology Indicators, Paris, December.
Office of National Statistics (ONS) (2006) Internet access households and individuals, London, Office of National Statistics.
Osgood, CE (1969) On the whys and wherefores of E, P, and A, *Journal of Personality and Social Psychology*, 12, 3, 194–99.
Osgood, K (2006) *Total Cold War: Eisenhower's Secret Progaganda Battle at Home and Abroad*, Kansas, University Press of Kansas.
OST (1985) Automation of America's offices, 1985–2000, Washington, US Congress, OST.
OST (1987) The electronic supervisor. New technology, new tensions. Washington, US Congress, OST..
OTA (1991) Biotechnology in a global economy, OTA-BA-494, Washington, DC, US Congress & US Government Printing Office, October 1991.
Otway, H and B Wynne (1989) Risk communication: Paradism and paradox, *Risk Analysis*, 9,2, 141–45.
Oudshoorn, N and T Pinch (2003) Introduction: How users and non-users matter, in: Oudshoorn, N and T Pinch (eds) *How Users Matter: The Co-construction of Users and Technology*, Cambridge, MA, MIT Press, 3–25.
Paichelier, G (1988) *The Psychology of Social Influence*, Cambridge, CUP.
Pakulski, J (1993) Mass social movements and social class, *International Sociology*, 8, 131–58.
Palmer, B and M Sharp (1993) The battle for biotechnology: Scientific and technological paradigms and the management of biotechnology in Britain in the 1980s, *Research Policy*, 22, 463–78.
Pálsson, G (2007) How Deep is the Skin? The Geneticization of Race and Medicine, **BioSocieties**, 2, 2, June 2007, 257–72
Papert, S (1980) *Mindstorm: Children, Computers and Powerful Ideas*, New York, Basic Books.
Patterson, W (1976) *Nuclear Power*, Harmondsworth, Penguin.
Perkovich, G (1999) *India's Nuclear Bomb: The Impact on Global Proliferation*, Berkeley, UCP.
Perrow, C (1984) *Normal Accidents: Living with High-risk Technologies*, New York, Basic Books.
Peters, HP (2012). Scientific sources and the mass media: Forms and consequences of medialization, in: Rödder, S, M Franzen and P Weingart (eds), *The Sciences' Media Connection: Public Communication and its Repercussions*. Sociology of the Sciences Yearbook 28. Dordrecht (NL), Springer, 217–40.
Peters, HP (1995) The interaction of journalists and scientific experts: Co-operation and conflict between two professional cultures. Media, Culture & Society, 17(1), 31–48.
Peters, T and RH Waterman (1982) *In Search of Excellence*, New York, Harper Business.
Phillips, DP, Kanter, EJ, Bednarczyk, B, Tastad, PL (1991) Importance of the lay press in the transmission of medical knowledge to the scientific community, *The New England of Medicine*, 325, 1180–83.
Piaget J (1972) *The Principles of Genetic Epistemology*, London, Routledge & Kegan Paul.

Pinch, T (1998) The social construction of technology: A review, in: Fox, R (ed) *Technological Change: Methods and Themes in the History of Technology*, Harwood Academic Publishers, 17–35.

Pisano, GP (2006) *Science Business: The Promise, the Reality and the Future of Biotech*, Cambridge, MA, Harvard Business School Press.

Plein, LC (1991) Popularising biotechnology: The influence of issue definition, *Science, Technology and Human Values*, 16, 4, 474–90.

Polletta, F and M Kai Ho (2006) Frames and their consequences, in: Goodin, RE and CH Tilly (eds) *Contextual Political Analysis*, Oxford, OUP, 187–209.

Powell, K (2007) Functional foods from biotech: An unappetizing prospect? *Nature Biotechnology*, 25, 5, 525–31.

Powell, WW, KW Koput and L Smith-Doerr (1996) Interorganisational collaboration and the locus of innovation: Networks of learning in biotechnology, *Administrative Science Quarterly*, 41, 116–45.

Priest, S (2001) *Grains of Truth: The Media, the Public, and Biotechnology*, Lanham, Rowman & Littlefield.

Pylyshyn, ZW (1986) Computation and Cognition: Towards a Foundation for Cognitive Science, Cambridge, MA, MIT Press.

Rabinovici, D (2008) *Der ewige Widerstand. Ueber einen strittigen Begriff*, Graz, Syria Verlag.

Radkau, J (1983) *Aufstieg und Krise der deutschen Atomwirtschaft, 1945–1975. Verdraengte Alternativen in der Kerntechnik und der Ursprung der nuclearen Kontroverse*, Hamburg, Rowolt.

Radkau, J (1995) Learning from Chernobyl for the fight against genetics? In: Bauer MW (ed) *Resistance to New Technology*, Cambridge, CUP, 335–56.

Radkau J (2011) *Die Aera der Oekologie. Eine Weltgeschichte*, Munchen, CH Beck.

Ramberg, B (1996) Learning from Chernobyl, *Foreign Affairs*, 304–28.

Ramana, MV (2013) Policy responses to Fukushima: exit, voice and loyality, *Bulletin of the Atomic Scientists*, 69, 2, 66–76.

Raschke, J (1988) *Soziale Bewegungen. Ein historisch-systematischer Grundriss*, Frankfurt, Campus Verlag.

Rawls, J (1972) *A Theory of Justice*, Oxford, Clarendon Press.

Reicher, S (2003) The psychology of crowd dynamics, in: Hogg, MA and S Tindale (ed) *Blackwell Handbook of Social Psychology: Group Processes*, Oxford, Blackwell, 182–208.

Reiss, MJ and R Straughan (1996) *Improving Nature? The Science and Ethos of Genetic Engineering*, Cambridge, CUP.

Renn, O (1990) Public responses to the Chernobyl accident. *Journal of Environmental Psychology*, 10, 151–67.

Rhinow, AR (1985) Widerstandsrecht im Rechtsstaat? *Staat und Politik*, 30, Bern, Haupt.

Rhodes, R (1987) *The Making of the Atomic Bomb*, NY, Simon & Schuster.

Ribeiro, D (1972) *Os Brasileiros—teoria do Brasil*, Sao Paulo, Editora Paz & Terra.

Richards, MPM (1996) Lay and professional knowledge of genetics and inheritance, *Public Understanding of Science*, 5, 217–30.

Rifkin, J (1998) *The Biotech Century: Harnessing and Remaking the World*, New York, J Tarcher/Putnam.

Rogers, EM (1983) *Diffusion of Innovations*, 3rd edition, New York, Free Press.

Rogers, EM (1994) *A History of Communication Studies*. New York, Free Press.

Rogers, EM (1996) *Diffusion of Innovations*, 4th edition, New York, Free Press.

Rogers, EM (2004) Diffusion of the internet, in: Lee, PSN, L Leung, CYK So (eds) *Impact and Issues in New Media: Toward Intelligent Societies*, Cresskill, NJ, Hampton Press, 21–34.

Rollin B (1989) *The Unheaded Cry—Animal consciousness, animal pain and science*, Oxford, Oxford University Press.
Rosa, EA and RE Dunlap (1994) Nuclear power: Three decades of public opinion, *Public Opinion Quarterly*, 58, 295–325.
Rose, H and S Rose (eds) (2000) *Alas, Poor Darwin: Arguments against Evolutionary Psychology*, London, J Cape.
Rose, S, RC Lewontin, LJ Kamin (1984) *Not in Our Genes: Biology, Ideology and Human Nature*, London, Penguin.
Rosen, LD & P Maguire (1990) Myths and realities of computerphobai: A meta-analysis, *Anxiety Research*, 2, 175–91.
Rothman, S (1990) Journalists, Broadcasters, Scientific Experts and Public Opinion, *Minerva Vo.* 28, 117–33.
Rothschild, E (1980) The American arms boom, in: Thompson, EP and D Smith (eds) *Protest and Survive*, Harmondsworth, Penguin, 170–85.
Rothstein, H, M Huber and G Gaskell (2006) A Theory of Risk Colonisation, *Economy and Society*, 35, 1, 91–112.
Rowe, G and LJ Frewer (2004) Evaluating public-participation exercises: A research agenda, *Science Technology Human Values*, 29, 512–56.
Royal Society (1985) the public understanding of science, London, RS
Rubin BP (2008) Therapeutic promise in the discourse of human embryonic stem cell research, *Science as Culture*, 17, 1, 13–27.
Rubinstein, MI (1954) *Soviet Science and Technique in the Service of Building Communism in the USSR*, Moscow, Foreign Language Publishing House.
Rucht, D (1995) The impact of anti-nuclear power movements in international comparison, in: Bauer MW (ed) *Resistance to New Technology*, Cambridge, CUP, 277–92.
Rucht, D and F Neidhardt (2002) Towards a movement society? On the possibilities of institutionalizing social movements, *Social Movement Research*, 1, 1, 7–30.
Rudenko, L, JC Mathesonm and SF Sundlof (2007) Animal cloning and the FDA—the risk assessment paradigm under public scrutiny, *Nature Biotechnology*, 25, 1, 39–43.
Rüdig, W (1990) *Anti-nuclear Movements: A World Survey of Opposition to Nuclear Energy*, Harlow, Longman.
Rusinek, BA (1993) Kernenergie, schöner Götterfunken. Zur Kontextgeschichte der Atomeuphorie, *Kultur & Technik*, 4, 15–21.
Sammut G and MW Bauer (2011) Social influence: modes and modalities, in: Hook, D, B Franks and MW Bauer (eds) *The Social Psychology of Communication*, London, Palgrave, 87–106.
Sandbach, F (1978) The rise and fall of the 'limits of growth' debate, *Social Studies of Science*, 8, 495–520.
Saner, H (1988) *Identitaet und Widerstand. Fragen in einer verfallenden Demokratie*. Basel, Lenos.
Saner, H (1994) Im Vorschein der Apokalypse. Grossrisiken und die Herausforderung an die Demokratie, *NZZ, Beilage 'Technologie und Gesellschaft'*, 26 January, 36.
Schaefer, MS (2008) Diskurskoalitionen in den Massenmedien. Ein Beitrag zur theoretischen und methodischen Verbinding ovn Diskursanalyse und Offentlichkeitssoziologie, *Kolner Zeitschrift fuer Sociologie und Sozialpsychologie*, 60, 2, 367–97.
Schafer, R (1976) The idea of Resistance, *A New Language for Psycho-Analysis*, New Haven, Yale UP, 212–45.
Schauza, M (2000) The concept of substantial equivalence in safety assessment of foods derived from genetically modified organisms, *AgbiotechNet*, 2, April, 1–4.
Scheufele, DA (2000) Agenda-setting, priming and framing revisited: Another look at cognitive effects of political communication, *Mass Communication and Society*, 3, 2–3, 297–316.

Scheufele, DA (2007) Opinion climates, spiral of silence and biotechnology: Public opinion as a heuristic for scientific decision making, in: Brossard, D, J Shanahan and TC Nisbett (eds) *The Public, the Media and Agricultural Biotechnology*, Reading, CAB International, 231–44.

Schleife, K (2007) Regional versus individual aspects of the digital divide in Germany, *ZEW Discussion Papers* no 06–085 (February 2007) [ftp://ftp.zew.de/pub/zew-docs/dp/dp06085.pdf].

Schmidt, B (1985) *Das Widerstandsargument in der Erkenntnistheorie. Ein Angriff auf die Automatisierung des Wissens*. Frankfurt, Suhrkamp.

Schneider, M, A Frogatt and S Thomas (2011) 2010–2011 World nuclear industry status report, *Bulletin of the Atomic Scientists*, 67, 4, 60–77.

Schuman, H and S Presser (1996) *Questions and Answers in Attitude Surveys: Experiments on Question Form, Wording and Context*, Thousand Oaks, Sage Publications.

Schwartz, SI (1998) *The atomic audit: The Costs and Consequences of US Nuclear Weapons since 1940*, Washington, Brookings Institution Press.

Scientific American (1966) *Information: A Comprehensive Review of This New Technology*, San Franscico, WH Freeman.

Scott, JC (1985) *Weapons of the Weak: Everyday Forms of Peasant Resistance*. New Haven, Yale UP.

Scott, JC (1987) Resistance without protest and without organization: Peasant opposition to the Islamic Zakat and the Christian Tithe, *Comparative Studies in Society and History*, 29, 417–53.

Seidmann, P (1974) *Der Mensch im Widerstand. Studien zur anthropologischen Psychologie*, Bern, Francke Verlag.

Seifert, F (2006) Synchronised national publics as functional equivalent of an integrated European public: The case of biotechnology, *European Integration Online Papers*, 10, 8 [http://eiop.or.at/eiop/index.php/eiop/article/view/2006_008a/26].

Sellen, AJ and RH Harper (2003) *The Myth of the Paperless Office*. Cambridge, MA, MIT Press.

Sharkey, N (2008) Ground for discrimination: Autonomous robot weapons, *RUSI Defense Systems*, October, 86–89.

Sharp, G (1973) *The Politics of Non-violent Action: Parts 1–3*, Boston, Porter Sargent Publishers.

Sheingate, AD (2006) Promotion versus precaution: The evolution of biotechnology policy in the US, *British Journal of Political Science*, 36, 2, 243–68.

Sheldrake, PE (1971) Attitudes to the computer and its uses, *Journal of Management Studies*, 8, 39–62.

Shepard, A (2007) *Focus on Digital Age: Use of ICT among Households and Individuals*, London, Office of National Statistics.

Sherif, M (1936) *The Psychology of Social Norms*, New York, Harper.

Shinn, T and R Whitley (1985) (eds) *Expository Science: Forms and Functions of Popularisation*, Dordrecht, D Reidel Publishing Company.

Shneiderman, B (1987) *Designing the User Interface*, Reading MA, Addison-Wesley.

Shotton, M (1989) *Computer Addiction? A Study of Computer Dependency*, London, Taylor and Francis.

Shukla, R and MW Bauer (2012) The Science Culture Index (SCI): Construction and validation; in: Bauer, MW, R Shukla and N Allum (eds) *The Culture of Science: How the Public Relates to Science across the Globe*, New York, Routledge, 179–99.

Siefken, H (1994) What is resistance? in: Siefken, H and H Vieregg (eds) *Resistance to National Socialism*, Nottingham, University of Nottingham, 5–19.

Sieferle, RP (1984) *Fortschrittsfeinde? Opposition gegen Techik und Industrie von der Romantik bis zur Gegenwart*, Munchen, CH Beck.
Sieferle, RP (1986) Menschen gegen neue Technik—Geschichte der Techik, *Bild der Wissenschaft*, April, 77–97.
Simon, B and B Klandermann (2001) Politicized collective identity: A social psychological analysis, *American Psychologist*, April, 319–31.
Simon, HA (1981) *The Science of the Artificial*, $2^{nd}$ edition, Cambridge, MA, MIT Press.
Sjoberg, L (2006) Myths of the psychometric paradigm and how they can misinform risk communication, paper presented to WHO meeting on 'risk perception and communication', Venice, Isola San Servolo, 29–30 May.
Sjoberg, L (2003) Risk perception is not what it seems: the psychometric paradigm revisited (manuscript, Stockholm, 11 March).
Skrentny, JD (1993) Concern for the environment: A cross-national perspective, *IJPOR*, 5, 4, 335–52.
Sloterdijk, P (2005) *Im Weltinnenraum des Kapitals*, Frankfurt, Suhrkamp.
Slovic, P (1987) Perception of risk, *Science*, 1987, 236, 280–85.
Smedley, A and BD Smedley (2005) Race as biology is fiction, racism as social problem is real, *American Psychologist*, 60, 1, 16–28.
Smit, WA (2006) Military technology and politics, in: Goodwin, RE and CH Tilly (eds) *The Oxford Handbook of Contextual Political Analysis*, Oxford, Oxford University Press, 723–44.
Solomon, JJ ([1971] 1984) *Promethee empetre: la resistance au changement technique*, Paris, Pargamon.
Specter, M (1999) Decoding Iceland, *The New Yorker*, 1 January, 40–51.
Specter, M (2000) The pharmageddon riddle. Did Monanto just want more profits, or did it want to save the world? *New Yorker*, 10 April, 58–71.
Sperber, D (1990) The epidemiology of belief, in: Fraser, C and G Gaskell (eds) *The Social Psychology of Widespread Beliefs*, Oxford, Clarendon, 25–43.
Stahel, AA (2006) *Widerstand der Besiegten. Guerillakrieg oder Knechtschaft von der Antike zur Al-Kaida*, Zurich, vdf.
Starker, S (1991) *Evil Influences: Crusades Against the Mass Media*, New Brunswick, Transaction.
Stern Review (2006) On Economics of Climate Change, London, HM Treasury & Cabinet Office, 30 October [http://webarchive.nationalarchives.gov.uk/+/http://www.hm-treasury.gov.uk/sternreview_index.htm; accessed 11 February 2013].
Stevens, MLT (2003) *Bioethics in America: Origins and Cultural Politics*, Baltimore, The Johns Hopkins UP.
Strodthoff, GG, RP Hawkins and AC Schoenfeld (1985) Media roles in a social movement: a model of ideology diffusion, *Journal of Communication*, 35, 2, 134–53.
Struck, F and T Mussweiler (2001) Resisting influence: judgmental corrections and its goals, in: Forgas JP & KD Williams (eds) *Social Influence: direct and indirect processes*, Philadelphia, PA, Psychology Press, , 199–212.
Surowiecki J (2004) *Wisdom of the Crowd. Why the Many are Smarter than the Few*, New York, Anchor Books.
Sutcliffe, A (1988) *Human-Computer Interface Design*, London, Macmillan.
Sugiman, T (2014) Lessons from the 2011 debacle of the Fukushima Nuclear Power plant, *Public Understanding of Science*.23, 254–267
Swade, D (2001) *The Cogwheel Brain: Charles Babbage and the Quest to build the first Computer*, London, Little, Brown.
Swiss Re (1996) *Electrosmog: A Phantom* risk, Zurich, Swiss Reinsurance Company.
Tarde, G ([1890] 2001) *Les lois de l'imitation*, Paris, Edition du Seuil.

Tarde, G ([1901] 2006) *L'opinion et la foule*, Paris, Edition du Sandre.
Taylor, C (2007) *A Secular Age*, Cambridge, MA, The Belknap Press of HUP.
Tennant, C (2012) On the threshold: A social psychological study of different standpoints in the climate change debate. PhD thesis, The London School of Economics and Political Science.
Tenner, E (1997) *Why Things Bite Back: Predicting the Problems of Progress*, London, Forth Estate
Thomas, K (1983) *Man and the Natural World: Changing attitudes in England 1500–1800*, Harmondsworth, Penguin.
Thompson, EP and D Smith (1980) *Protest and Survive*, Harmondsworth, Penguin.
Thompson, J (1985) *Psychological Aspects of Nuclear War*, Chichester, BPS.
Thompson, M (1991) Plural Rationalities: The rudiments of a practical science of the inchoate, in: Hansen, JA (ed) *Environmental Concerns: An Inter-disciplinary Exercise*, London, Elsevier, .
Thorgeirsdottir, S (2004) The controversy on consent in the Icelandic database case and narrow bioethics, in: Arnason, G, S Nordal and V Arnason (eds) *Blood and Data: Ethical, Legal and Social Aspects of Human Genetic Databases*, Reykjavik, University of Iceland Press, 67–77.
Tichenor, PJ, DA Donohue and CN Olien (1980) Community conflict and the press, Berverly, Hills, Sage Publications
Tichenor, PJ, GA Donohue and CN Olien (1970) Mass media flow and differential growth in knowledge, *Public Opinion Quarterly*, 34, 159–70.
Tilly, C (1978) From mobilization to revolution, New York, McGraw-Hill.
Toffler, A (1971) *Future Shock*, New York, Bantam Books.
Torgersen, H (2001) Precautionary openness: Understanding of precaution as an indicator of change in biotechnology policy, *Notizie di Politeia*, 17, 63, 67–79.
Torgersen, H and J Hampel (2012) Calling controversy: Assessing synthetic biology's conflict potential, *Public Understanding of Science*, 21, 2, 149–62.
Touraine, A (1985) An introduction to the study of new social movements, *Social Research*, 52, 749–87.
Touraine, A (1992) *Critique de la Modernite*, Paris, Fayard.
Trebilcock, C (2002) Surfing the wave: The long cycle in the industrial centuries, in: Martland, P (ed) *The Future of the Past: Big Questions in History*, London, Pimlico, 66–88.
Triandafyllidou, A (1995) The Chernobyl accident in the Italian press: a 'media story-line', *Discourse and Society*, 6, 517–36.
Trumbo, C (1996) Constructing climate change: Claims and frames in US news coverage of an environmental issue, *Public Understanding of Science*, 5, 269–84.
Tucker, J (1993) Everyday forms of employee resistance, *Sociological Forum*, 8, 25–45.
Tuckman, BW and MAC Jensen (1977) Stages of small group development revisited, *Group and Organisational Studies*, 2, 419–27.
Turing, A (1950) Computing machinery and intelligence, *Mind*, 59, 433–60.
Turkle S (1984) *The Second Self: Computers and the Human Spirit*, Cambridge MA, MIT Press
Tuomela, R (1995) *The Importance of Us: A Philosophical Study of Basic Social Notions*, Standford, SUP.
Turrow, S (1994) *Power in Movement*, Cambridge, CUP.
Tushman, ML and P Anderson (1986) Technological discontinuities and organizational environments, *Administrative Science Quarterly*, 31, 439–65.
UNDP (2002) The human consequences of Chernobyl nuclear accident: A recovery strategy, A report commissioned by UNDP and UNICEF with support from UN-OCHA and WHO, Oxford, 25 January 2002.

Unmanned Systems (2011) Unmanned Systems Integrated Roadmap, FY2011–2036, US Department of Defence, Reference Number: 11-S-3613; corrected copy: [http://publicintelligence.net/dod-unmanned-systems-integrated-roadmap-fy2011–2036/; accessed 27 Februar 2013]
US Census Bureau (2005) Computer and Internet Use in the US 2003, Washington, issued October 2005.
Useem, B and MN Zald (1987) From pressure group to social movement: Efforts to promote use of nuclear power, in: Zald, MN and JD McCarthy (eds) *Social Movements in an Organizational Society: Collected Essays*, New Brunswick, Transaction Books, 273–92.
Vain, P (2007) Trends in GM crop, food and feed safety literature, *Nature Biotechnology*, 25, 6, 624–26.
Valente, TW and E M Rogers (1995) The origins and development of the diffusion of the innovation paradigm as an example of scientific growth, *Science Communication*, 16, 3, 242–73.
Valiverronen, E (1998) Biodiversity and the power of metaphor in environmental discourse, *Science Studies*, 1, 19–34.
Valsiner, J (2007) Ornamented life [book project presented to LSE seminar, 19 June].
Van der Heijden HA, R Koopmans and MG Giugni (1992) The Western European environmental movement, in: Finger M and L Kriesberg (eds) Research in Social Movements, Conflicts and Change—The Green Movement Worldwide, Supplement 2, pages 1–40.
Van der Pligt, J (1992) *Nuclear Energy and the Public*, Oxford, Blackwell.
Vander Zanden, JW (1958/59) Resistance and social movements, *Social Forces*, 37, 312–15.
Van Duijn, JJ (1983) *The Long Wave in Economic Life*, London, George Allen & Unwin.
Van Reenen J and R Sadun (2005) Information technology and productivity: It ain't what you do, it's the way you do I.T., LSE EDS Innovation Research Programme, Discussion Paper no 02.
Van Tulder, M, A Malmivaara and B Koes (2007) Repetitive strain injury, *The Lancet*, 369, 9575, 1815–22.
Vaughan D (1996) *The Challenger Launch Decision: Risky Technology, Culture and Deviance at NASA*, Chicago, CUP
Veblen, T (1964) *The Instinct of Workmanship: The State of the Industrial Arts*, New York, WW Norton (original publication 1914).
Vettel, EJ (2006) *Biotech: The Countercultural Origins of an Industry*, Philadelphia, PA, U of Pennsylvania Press.
Visschers, VHM and M Siegrist (2012) How a nuclear power plant accident influences acceptance of nuclear power: Results of a longitudinal study before and after the Fukushima disaster, *Risk Analysis*, DOI:10.1111/j.1539-6924.2012.01861.x.
Wadham, M (2008) James Watson's genome sequenced at high speed, *Nature*, 452, 17 April, 788.
Wagner, W and N Hayes (2005) *Everyday Discourse and Common Sense: The Theory of Social Representations*, Basingstoke, Palgrave.
Wagner, W, N Kronberger, G Gaskell, A Allansdottir, N Allum, S deCheveigne, U Dahinden, C Diego, L Montali, AT Mortensen, U Pfenning, T Rusanen and N Seger (2001) Nature in disorder: The troubled public of biotechnology, in: Gaskell, G & MW Bauer (2001) (eds) *Biotechnology, 1996–2000: The Years of Controversy*, London, Science Museum, 80–95.
Wagner, W, N Kronberger, N Allum, S de Cheveigne, C Diego, G Gaskell, M Heinssen, C Midden, M Odegaard, S Ohman, B Rizzo, T Rusanen and A Stathopoulou (2002) Pandora's genes—images of genes and nature, in: Bauer, MW &

G Gaskell (eds) *Biotechnology: The Making of a Global Controversy*, Cambridge, CUP, 244–77.
Waldenfels B (2005) *Phänomenologie der Aufmerksamkeit*, Frankfurt, Suhrkamp.
Waldenfels B (1982) The Despised Doxa—Husserl and the continuing crisis of Western Reason, *Research in Phenomenology*, 12, 1, 21–38; (translated from Japanese SHISO in by J Claude Evans).
Walker, SJ (2004) *Three Mile Island: A Nuclear Crisis in Historical Perspective*, Berkeley, UCP
Wall, PD (1979) On the relation of injury and pain, *Pain*, 6, 253–264.
Wall, PD (1999) *Pain: The Science of Suffering*, London, Weidenfeld & Nicholson.
Wang, Z (1997) Responding to *Silent Spring*: Scientists, popular science communication, and environmental policy in the Kennedy years, *Science Communication*, 19, 2, 141–63.
Warren, S (1995) *From Margin to Mainstream: British Press Coverage of Environmental Issues*, Chichester, Packard Publishers.
Watkins, ES (2001) Radioactive fallout and emerging environmentalism: cold war fears and public health concerns, 1954–1963, in: Allan, GE and RM MacLeod (eds) *Science, History and Social Activism: A tribute to Everett Mendelsohn*, Dortrecht, Kluver Academic Publishers, 291–306.
Watson J (1968) *The Double Helix: a Personal Account of the Discovery of the Structure of DNA*, New York, Scribner.
Watson, J (2004) *DNA: The Secret of Life*, London, Arrow Books.
Weart, SR (1988) *Nuclear fear: A History of Images*, Cambridge, MA, HUP.
Weinberg, A (2004) *Glory and Terror: The Growing Nuclear Danger*, New York, NY Review Books.
Weingart, P (1998) Science and the Media, *Research Policy*, 869–79.
Weingart, P (2012) The lure of the mass media and its repercussions on science, in: Rodder, S, M Franzen and P Weingart (eds) *The Sciences' Media Connection: Public Communication and its Repercussions*, NY, Springer, 17–34.
Weingart, P, A Engels and P Pansegrau (2007) *Von der Hypothese zur Katastrophe. Der anthropogene Klimawandel im Diskurs zwischen Wissenschaft, Politik und Massenmedien*, Opladen, Leske&Budrich (2nd revised edition),
Weingart, P, C Salzmann and S Wormann (2008) The social embedding of biomedicine: An analysis of German media debates 1995–2004, *Public Understanding of Science*, 17, 381–96.
Weiss, L (2003) Did the 50-year old Atoms for Peace program accelerate proliferation? *Bulletin of the Atomic Scientists*, November, 59, 6. 34–44
Weizenbaum, J (1966) ELIZA—a computer program for the study of natural language communication between man and machine, *Communications of the ACM*, 9, 1, 36–45.
Weizenbaum, J (1976) *Computer Power and Human Reason: From Judgement to Calculation*. New York, WH Freeman & Co.
Wejnert, B (2002) Integrating models of diffusion of innovation: A conceptual framework, *Annual Review of Sociology*, 28, 297–326.
Weldon, S and M Levitt (2004) Public databases and privatised property? A UK study of public perceptions of privacy in relation to population based human genetic databases, in: Arnason, G, S Nordal and V Arnason (eds) *Blood and Data: Ethical, Legal and Social Aspects of Human Genetic Databases*, Reykjavik, University of Iceland Press, 181–86.
Wenzel, UJ (2002) Zeitzeichen 'Lebenswissenschaft', *Neue Zuricher Zeitung*, 39, 16/17 February, 33.
Wheelright, S and S Baron-Cohen (2001) The link between autism and skills such as engineering, maths, physics, and computing: A reply to Jarrett and Routh, *Autism*, 5, 223–27.

Whiten, A, V Horner and FBM de Waal (2005) Conformity to cultural norms of tool use in chimpanzees, *Nature*, 437, 29 September, 737–40.
Whitley, E, IR Hosein, IO Angell and S Davies (2006) Reflections on the academic policy analysis process and the UK Identity Cards Scheme, LSE Information Systems Group, working paper no 147; [http://is2.lse.ac.uk/asp/ aspwp/locate.asp; accessed 17 Sep 2007].
Wiener, N (1950) *The Human Use of Human Being: Cybernetics and Society*, Boston, Houghton Mifflin, Riverside.
Wigg, D (2003) Radiation: facts, fallacies and phobias [www.boldenterprise.com.au/bio/radiation/htlm; accessed 25 April 2008].
Wigzell, H (2007) When ministers are well primed, *Nature*, 449, 11 October, 663.
Wildavsky, A and K Drake (1982) Theories of risk perception: who fears what and why? *Daedalus: Journal of the American Academy of Arts and Sciences*.
Wildi, T (2003) *Der Traum vom eigenen Reaktor. Die Schweizerische Atomtechnologieentwicklung 1945–1969*, Zurich, Chronos Verlag.
Williams R (1980) Nuclear power decisions: British policies, 1953–1978, London, Croom Helm.
Williams, R and S Mills (1986) (eds) *Public Acceptance of New Technology: An International Review*, London, Croom Helm.
Wilmut, I, AE Schneike, J McWhir, AJ Kind and K Campbell (1997) Viable offspring derived from fetal and adult mammalian cells, *Nature*, 27 February, 385, 810–13.
Winner, L (1977) *Autonomous Technology: Technics Out-of-Control as a Theme of Political Thought*, Cambridge, MA, MIT Press.
Winner, L (1986) *The Whale and the Reactor*, Cambridge, CUP.
Winston, B (1998) *Media Technology and Society: A history from the Telegraph to the Internet*, London, Routledge.
Witherspoon, S (1994) The greening of Britain: Romance and rationality, in: R Jowell (ed) *British Social Attitudes: The 11th Report*, Dartmouth, Aldershot, 107–39.
Wolsink, M (1994) Entanglement of interests and motives: Assumptions behind the NIMBY-theory on facility siting, *Urban Studies*, 31, 6, 851–66.
Wood, S (1982) *Degrading of Work: Skill, Deskilling and the Labour Process*. London, Hutchinson.
Woolgar, S (1991) Reconfiguring the user: The case of usability trials, in: Law, J (ed) *A Sociology of Monsters*, London, Routledge.
Worcester, R (1993) Public and elite attitudes to environmental issues, *IJPOR*, 5, 4, 315–33.
Wright, S (1986) Molecular biology or molecular politics? The production of scientific consensus on the hazards of recombinant DNA technology, *Social Studies of Science*, 16, 4, 595–96.
Wright, S (1994) *Molecular Politics: Developing American and British Regulatory Policy for Genetic Engineering, 1972–1982*, Chicago, University of Chicago Press.
Wyatt, S (2003) Non-users also matter: The construction of users and non-users of the internet, in: Oudshoorn, N and T Pinch (eds) *How Users Matter: The Co-construction of Technology and Users*, Cambridge, MA, MIT Press, 67–80.
Wynne, B (1982) *Rationality and Ritual: The Windscale Inquiry and Nuclear Decisions in Britain*, Cahlfort St Giles, Bucks, The British Society for the History of Science.
Wynne, B (1992) Sheep farming after Chernobyl: A case study in communicating science information, in: Lewenstein BV (ed) *When Science Meets the Public*, Washington, AAAS.
Wynne, B (1993) Public uptake of science: A case for institutional reflexivity, *Public Understanding of Science*, 2, 4, 321–38.

Wynne, B (1995) Public understanding of science, in: Jasanoff, J, BE Markle, JC Petersen and T Pinch (eds) *Handbook of Science and Technology Studies*, London, SAGE, 361–88.
Yates, FA (1992) *The Art of Memory*, London, Pimlico.
Young R (1971) Darwin's metaphor: does nature select? *The Monist*, 55, 442–503.
Yoxen, E (1983) *The Gene Business: Who Should Control Biotechnology?* London, Pan Books.
Zald, MN (1996) Culture, ideology and strategic framing, in: McAdams, D, JD McCarthy and MN Zald (eds) *Comparative Perspectives on Social Movement: Political Opportunity, Mobilizing Structures and Cultural Framing*, Cambridge, CUP, 261–74.
Zald, MN and B Useem (1987) Movement and countermovement interaction: Mobilization, tactics and state involvement, in: Zald, MN and JD McCarthy (eds) *Social Movement in an Organizational Society: Collected Essays*, New Brunswick, Transaction Books, 247–72.
Zaller JR (1995) *The nature and origins of mass opinion* (Cambridge Studies of Public Opinion and Political Psychology), Cambridge, CUP.
Zeyer, A and WM Roth (2013) Post-ecological discourse in the making, *Public Understanding of Science*, 22, 1, 33–48.

# Author Index

**A**

Adams, J., 165
Ahlemeyer, H.W., 10, 169
Akerman, N., 91, 92, 214, 221
Algom, D., 227
Allansdottir, A., 131, 187, 189, 195, 202
Allerbeck, K.R., 139
Allum, N., 131, 155, 156, 157, 174n2, 174n4, 189, 195
Alston, J.A., 201
Amann, M., 80
Ambrose, S.E., 66n5
Anders, G., 3, 66n10
Anderson, P., 109
Angell, I.O., 142
Arendt, H., 241
Argyris, C., 105
Arnason, G., 180, 207
Arnason, V., 180, 207
Arnold, L., 48
Arundel, A., 178
Asch, S., 238, 240, 243n2

**B**

Baldwin, R., 19
Baltimore, D., 185
Bandura, A., 104, 112n8
Baron-Cohen, S., 241
Barrett, D., 191
Bar-Tal, D., 17
Barthes, R., 19
Bateson, G., 94, 100
Bauer, M.W., 14, 15f, 17, 19, 20, 25, 27, 28, 29n2, 29n5, 43f, 63, 68n28, 74, 83f, 91, 105, 111n1, 111n2, 116f, 119f, 131, 136, 141, 143, 144, 155, 156, 157, 161, 162, 170, 174n2, 174n4, 185, 186f, 187, 189, 190, 191, 193, 194, 195, 197, 203f, 204, 208, 210n4, 212, 215, 221, 232n2, 232n4, 237, 242, 243, 249, 250
Bay, C., 220
Baylis, J., 32
Beck, U., 61, 164
Bednarczyk, B., 171
Behrens, W.W., III, 77
Bell, A., 82
Bell, D., 120
Benford, T.D., 22, 235
Beniger, J.B., 121
Berg, P., 185
Bergman, K., 206
Berthomier, A., 189
Beuzekom, B., 178
Bieri, F., 198
Bigelow, J.H., 109
Bjorn-Anderson, N., 145
Blass, T., 241
Block, C., 62
Blumenburg, L., 11, 23, 24, 173
Bock, C.L., 150 (spelled "Bosk" on p. 150)
Boden, M.A., 125
Boecker, F., 25
Boesch, E.E., 11, 23
Boehme, H., 124
Bonfadelli, H., 21, 28, 132, 197, 243
Borup, M., 172, 177
Botelho, A.J.J., 115
Boulding, K., 225
Bowcock, A.M., 181
Boy, D., 202
Boyer, H.W., 185
Bradford, K.J., 201
Braman, D., 75

Brey, P., 11
Broderick, M., 67n14
Brookes, S.K., 82, 83*f*
Brosnan, M., 141
Brown, 155
Brown, N., 17, 172, 177
Brown, P., 205
Brown, R., 198
Brumfield, G., 21, 172, 177
Bruner, J., 98
Bucchi, M., 20, 62, 160, 161
Buchmann, M., 197
Bud, R., 176, 185
Burger, J.M., 244n3
Burgess, A., 127
Butler, S., 218

## C

Cabral, R., 66n4
Caldwell,
Campbell, K., 188
Camus, A., 96, 219
Cantley, M., 178, 202
Card, S.K., 146
Carlisle, R.P., 165
Caro, P., 161
Carson, R., 76, 82
Cassidy, A., 191, 237
Castells, M., 121, 139
Caswell, M., 201
Ceruzzi, P.E., 115, 117, 120
Cervero, F., 99, 226
Chadarevian, S., 176, 184
Chaiken, S., 55
Chase, A., 130
Chatjouli, A., 189
Chernousenko, W.M., 47
Churcher, J., 52
Cirincione, J., 53, 65, 245*t*
Clark, R.D., 242
Cleary, P.D., 44
Coch, L., 218
Cochlan, E., 82
Cohen, G., 75
Cohen, S.N., 185
Cohn, S.M., 66n11
Collingwood, R.G., 71, 73
Collins, H.H., 126
Combs, B., 166
Conko, G., 210
Converse, J.M., 154
Cook-Deegan, R., 180
Cortada, 116*f*, 119, 135
Cowe, R., 80

Crafts, N., 122
Cranach, M., 4, 73, 222, 223, 225
Crenshaw, M., 221
Crick, F., 184
Crombez, G., 230

## D

Dagwell, R., 146
Dahinden, U., 195
Dalton, R.J., 80
Davies, S., 142
Davis, J., 133, 137*f*
Davis, R.W., 185
Dawson, E., 86
Day, J.C., 133, 137*f*
deCheveigne, S., 189, 195
Degn, L., 62
DeGreen, B.K., 23
Derrida, J., 151
de Semir, W., 161
Devall, B., 81
de Wall, F.B.M., 241
deVries, R., 209
Diego, C., 195
Dierkes, M., 16, 169
Dinello, D., 124
Doise, W., 242
Donohue, D.A., 27
Dorn, H., 178
Douglas, J.D., 120
Douglas, M., 75, 164
Downey, R., 189
Downing, T., 68n23
Downs, A., 24
Drake, K., 167
Dreyfus, H., 125, 126
Dreyfus, H.L.,
Dreyfus, S., 125, 126
Duclos, D., 168
Dunlap, R.E., 45, 46*f*
Durant, J., 43*f*, 83*f*, 116*f*, 119*f*, 162, 185, 187, 190, 191, 194, 195, 210n4, 249, 250
Durrenmatt, F., 66n3
Dutton, W.H., 133, 134
Duveen, G., 113n11

## E

Eagly, A.H., 55
Eccleston, C., 230
Eder, K., 81
Edgerton, D., 5, 32, 37
Edwards, P.N., 43, 52
Ehrlich, P., 77, 82

Einsiedel, E., 82, 158, 189, 202, 205
Engels, A., 84
Entman, R.N., 22
Etzkowitz, H., 12, 170

**F**

Farr, R.M., 154
Farrant, G., 152
Fernandez-Cornejo, J., 201
Ferree, M.M., 23
Festinger, L., 23, 172, 173
Finlay, 140 (there is no reference for this citation)
Fischoff, B., 107, 166, 218
Flam, H., 63
Fogg, B.J., 146
Forsyth, T., 79
Foster, K.R., 127
Frank, D.J., 191
Frankfurt, H.G., 151
Franklin, J., 45, 62, 172, 177
Franzen, M., 172
Franzosi, R., 94, 235
Frayn, M., 65n2
French, J.R.P., 218
Frese, M., 145
Freud, S., 238
Frogat, A., 31f
Fukuyama, F., 209
Fuller, S., 191

**G**

Galanter, E., 125
Galbraith, J.K., 220
Gale, B.G., 215
Galloux, J.C., 178, 209
Galvan, C.G., 37
Gamson, W.A., 22, 41, 191, 235
Gaskell, G., 14, 17, 29n2, 74, 83f, 131, 155, 162, 164, 185, 186f, 187, 191, 194, 195, 210n4, 214
Gastil, J., 75
Gaumont, H., 178, 209
Gehlen, A., 23, 173
Gelperin, D., 145
Geransar, R.M., 205
Gerbner, G., 27
Gergen, K.J., 2, 212
Gervais, M.C., 75
Geyer, M., 93, 96
Giami, A., 155
Gibbon, D., 242
Gibson, W., 124
Gierl, H., 25

Gigerenzer, G., 167, 208
Giles, J., 172
Gingras, Y., 161
Ginneken, J., 151, 238
Giugni, M., 51, 54, 80
Glaser, A., 39
Godin, B., 72, 161
Goepferdt, W., 21, 172, 177
Golan, E., 201
Goldschmidt, G., 32, 33, 60, 245t
Graber, P., 202
Grafee, C., 92, 226
Graff, G.D., 206
Graumann, C.F., 75
Grayson, L., 172
Greely, H.T., 190
Gregory, J., 63, 158, 160, 161
Gross, A.G., 19
Gross, L., 27
Gutteling, J., 21
Gutteling, J.M., 189, 197

**H**

Haber, E., 179
Habermas, J., 20, 108, 149, 151, 154, 189, 240
Hacker, W., 122
Hagerstrand, T., 60
Hampel, J., 202, 214
Hannemyr, G., 136
Hansen, A., 20, 82, 150, 190
Harper, R.H., 138
Harré, R., 4, 25, 91, 111n1, 212, 215, 232n2
Haskar, V., 220
Hawkins, R.P., 25, 102
Hayes, N., 166
Heath, A., 152
Hecht, G., 36, 37, 38, 55, 63
Hechter, M., 79, 236
Hedges, B., 152
Heijs, W.J.M., 195
Heims, S.J., 118
Heinssen, M., 195
Helsper, E.J., 133, 134
Hennen, L., 153f, 154, 195
Hetzel, B., 145
Hewstone, M., 241
Hilgartner, S., 149
Hilty, L.M., 117, 121
Hinde, R.A., 105
Hirschman, A.O., 5, 94, 96, 218, 219
Hirter, H., 198
Hoag, W.J., 139

Hoffmann, U., 16, 169
Hofstadter, D.R., 125]
Hogness, D.S., 185
Holland, J.H., 223
Holyoak, K.J., 223
Horner, V., 241
Hosein, I.R., 142
Hood, C., 19, 164
Houts, T.S., 44
Howard, G.S., 145
Howe, J.F., 228
Hu, T-W., 44
Huber, M., 164
Hunger, F., 115

I

Iacono, R., 136
Iggo, A., 99, 226
Inglehart, R., 79, 80
Inglis, D.R., 31
Isaacs, J., 68n23
Iversen, L.L., 99, 226
Iyengar, S., 22

J

Jackson, J., 213
Jackson, T., 87
Jacques, J., 161
Jaensch, G., 154
James, C., 202, 203f
Jamison, A., 76
Janus, A., 133, 137f
Jasanoff, S., 17, 62, 108, 158, 169, 178, 201
Jasper, J.M., 51
Jaufmann, D., 154
Jelsoe, E., 202
Jensen, C., 25, 91, 111n1, 212, 215, 232n2, 237
Jensen, M.A.C., 13
Jensen, P., 161
Joffe, H., 29n1
Johnson-Laird, P., 125
Jonas, H., 208
Jordan, A.G., 82, 83f
Jordan, T., 129
Jost, G., 105, 221
Jovchelovitch, S., 74, 113n11
Jowell, R., 152
Juma, C., 200
Jungk, R., 38

K

Kahan, D.M., 75

Kahnemann, D., 218
Kai Ho, M., 22, 235
Kalantzandonakes, N., 201
Kaldor, M., 36
Kamara, M.W., 202
Kantor, E.J., 171
Karasek, R.A., 122
Katz, D., 152
Katz, J.E., 129
Kendall, P., 155
Kepplinger, H.M., 27, 44, 45, 47, 154
Kevles, D., 176, 184, 191
Key, V.O., 149
Kielbowicz, R.B., 19
Kiesler, S., 127
Kimber, R.H., 82, 83f
Kind, A.J., 188
Kinder, D., 22
Kipphardt, H., 66n3
Kirby, D.A., 39, 126
Kish, L., 152
Kistler, E., 154
Klein, H.K., 11
Kleinmann, D.L., 11
Kling, R., 136
Knight, D., 12
Knowles, E.S., 218
Koes, B., 127
Köhler, A., 117
Kohring, M.,
Konrad, K., 172, 177
Koopmans, R., 51, 54, 80
Kolata, G., 189
Koput, K.W., 179
Koulaidis, V., 161
Kraut, R., 127
Kriese, H.P., 79
Krige, J., 24
Krimsky, S., 12, 170, 178
Kristensen, H.M., 31f
Kroll, G., 76
Kronberger, N., 195
Kruglanski, A., 105, 221
Kruse, L., 75
Kuchler, F., 201
Kueppers, G., 62
Kuhn, T.S., 112n11

L

Lage, E., 242
Lahteenmaki, R., 180
Langer, M., 32, 39
Lassen, J., 187
Latour, B., 10, 14, 23, 36, 152, 243n1

Lawrence, P.L., 3, 101, 111n2, 218
Lawrence, S., 178, 180
Lear, L.J., 76
LeBon, G., 149
Lee, J., 82
Lee, S., 120, 131
Leiserowitz, A.A., 86
Lenhart, A., 133, 134
Leonarz, M., 189
Levidow, L., 192, 202
Levitt, M., 207
Levy, S., 129
Lewenstein, B., 213
Lewin, K., 217, 218
Lewis, C.S., 227
Leydesdorff, L., 12, 170
Liakopoulos, M., 162, 187
Lichtenstein, S., 166
Lieven, E., 52
Linder, W., 198
Lindsey, N., 187, 202
Linke, S., 184, 191
Linn, J.A., 218
Litmanen, T., 48, 60
Liu, J.T., 58
Livingstone, S., 127
Lodge, D., 69
Lorenz, K., 222
Lovins, 50, 57f
Luebbe, H., 62
Luhmann, N., 20, 73, 111n2, 149, 150, 164, 223, 225
Lundmark, V., 127
Lynn, P., 152
Lyon, D., 122, 142

**M**

Maass, A., 242
MacCulloch, D., 79
Machlup, F., 120
Macleod, R., 58
Maddison, A., 23
Maguire, P., 141
Maibach, E.W., 86
Malmivaara, A., 127
Mandelbaum, M., 32, 33, 36, 42, 52, 53, 56, 245t
Mangano, J., 44
Mannheim, K., 81
Manzoli, F., 189
Markus, M.L., 145
Marlier, E., 195
Martin, R., 241
Martin, S., 195

Martin, W.K., 141, 142
Martineau, B., 179, 187
Marz, L., 16, 169
Mathesonm, J.C., 205
Matusow, H., 131
Mayer, R.N., 88
Mazur, A., 21, 67n15, 82, 103
McCarthy, J.D, 10, 135, 169
M'charek, A., 207
McCombs, M., 21
McWhir, J., 188
Meadows, D.H., 77
Meadows, D.L., 77
Mejlgaard, N., 62
Melber, B.D., 42, 45, 46f, 55
Mellor, F., 172
Melzack, R., 92, 226, 227
Mensch, G., 23
Merton, R.M., 155, 170, 218
Meyer, M., 19
Michael, M., 17
Midden, C., 162
Midden, C.J.H., 195
Milgram, S., 241
Miller, G.A., 125
Miller, H.I., 210
Miller, J.G., 225
Miller, S., 155, 156, 161, 174n2
Mills, S., 132f, 140
Mirovski, P., 12, 170
Modigliani, A., 22, 41, 191, 235
Mitchell, L., 201
Monbiot, G., 86
Montali, L., 195
Moran, T.P., 146
Morgan, M., 27
Morris, D.B., 226, 227, 229
Mortenson, A.T., 187, 195
Moscovici, S., 29n3, 71, 155, 242
Mugny, M., 242, 243
Mukopadhyay, T., 127
Myers, D.J., 135
Myers, T.,

**N**

Nader, R., 88
Naess, A., 81
Naffrechoux, M., 242
Nash, H.T., 33
Nathans, D., 185
Nealey, S., 42, 45, 46f, 55
Neidhardt, F., 17, 20, 26, 150, 233, 237
Neisser, U., 125
Nelkin, D., 118

Newell, A., 146
Newig, J., 24, 234
Nielsen, M.W., 62
Nisbett, R.E., 223
Noelle-Neumann, E., 26, 149
Nordal, S., 180, 207
Norman, D.A., 99, 146
Norris, P., 80
Norris, R.S., 31f, 34
Novas, C., 182
Nowotny, H., 62, 215
Nussbaum, M.C., 189

O

Ochsenbein, G., 222, 225
O'Connor, C., 29n1
Odegaard, M., 195
Ohman, S., 195
Olien, C.A., 127
Olafsson, A., 187, 189
Osgood, K., 39, 232n4
Otway, H., 166

P

Pakulski, J., 80
Pansegrau, P., 84
Papert, S., 138
Patterson, M., 127
Percovich, G., 32
Perez, J.A., 242
Perrow, C., 44, 168
Peters, H.P., 26
Peters, T., 116
Pfenning, U., 195
Phillips, D.P., 171
Piaget, J., 112n9
Plein, L.C., 191
Polletta, F., 22, 235
Post, J.M., 221
Powell, W.W., 179
Presser, S., 152
Pribram, K.H., 125
Priest, S., 26
Przestalski, A., 189
Pylyshyn, Z.W., 125

R

Rabinovici, D., 96
Radkau, J., 35, 36, 38, 44, 52, 76, 191, 195
Ragnarsdottir, A., 43f, 116f, 119f, 249, 250
Raichvarg, D., 161
Ramana, M.V., 50

Ramberg, B., 47
Randers, J., 77
Rankin, W.L., 42, 45, 46f, 55
Raschke, J., 79
Ravn, T., 62
Rawls, J., 96, 220
Read, S., 166
Rees, J., 19
Reicher, S., 238, 239
Reiss, M.J., 194
Renn, O., 47
Rhinow, A.R., 96, 219, 220
Rhodes, R., 32, 66n3
Ribeiro, D., 23, 37
Richardson, J.J., 82, 83f
Riecken, H.W., 172, 173
Rifkin, J., 178, 187
Rizzo, B., 195
Roblin, R., 185
Roedder, S., 172
Rogers, E.M., 24, 25, 94, 107, 132, 136, 243
Rollin, B., 226, 228
Rosa, E.A., 45, 46f
Rose, H., 191
Rose, S., 191
Rosen, I.D., 141
Roser-Renouf, C., 86
Roth, W.M., 87
Rothman, S., 26
Rothschild, E., 32
Rothstein, H., 19, 164
Rubin, B.P., 206
Rubinstein, M.I., 66n8
Rucht, D., 17, 51, 59, 63, 68n25, 95, 237
Ruddy, T., 117
Rudenko, L., 205
Rudolfsdottir, A., 43f, 116f, 119f, 249, 250
Ruedig, W., 30, 47, 51, 59f
Rusanin, T., 189, 195
Rusenik, B.A., 38

S

Sadun, R., 122
Sakelaris, G., 202
Salzman, C., 193, 205
Sammut, G., 15f, 113n11, 237 ("and" should replace ampersand on p. 113, l. 8)
Sandbach, F., 82
Saner, H., 219
Schaefer, M.S., 19

Schacter, S., 172, 173
Schauza, M., 201
Scherer, C., 19
Scherlis, W., 127
Scheufele, D.A., 22, 26
Schleife, K., 133
Schmidt, B., 110
Schneider, M., 31*f*, 34
Schneike, A.E., 188
Schoenfeld, A.C., 25, 102
Schön, D.A., 105
Shultz, 155
Schumann, H., 152
Schwartz, S.I., 32
Scott, J.C., 95
Seger, N., 195
Seidman, P., 111n2
Seifert, F., 189
Sellen, A.J., 138
Sent, E.M., 12, 170
Shanahan, J., 27
Sharkey, N., 126
Sharp, 94, 219
Sheingate, A.D., 201, 205
Shepard, A., 137*f*
Sheldrake, P.E., 131
Sherif, M., 239
Shinn, T., 161
Shneiderman, B., 146
Shotton, M., 141
Shukla, R., 157, 162, 174n4
Sieferle, R.P., 2, 81
Siefert, F., 187
Siegrist, M., 47
Sjoberg, L., 167
Skrentny, J.D., 80
Sloterdijk, P., 4, 173, 220
Slovic, P., 75, 166, 218
Smedley, A., 207
Smedley, B.D., 207
Smith, K., 58
Smith, N., 86
Smith-Doerr, L., 179
Snow, D.A., 22, 235
Solomon, J.J., 5
Spackman, J., 19
Specter, 179, 201
Sperber, D., 239
Stahel, A.A., 94
Stathopoulou, A., 189, 195
Staveloz, W., 161
Stevens, M.L.T., 208, 209
Stevenson-Hinde, J., 105
Stevers, E., 178, 209

Stoeckle, T., 195
Straughan, R., 194
Strodthoff, G.G., 25, 102
Sugiman, T., 49
Sundlof, S.F., 205
Sunstein, C.R., 189
Surowiecki, J., 241
Sutcliff, A., 146
Swade, D., 114

T
Tait, J., 195
Tarde, G., 21, 149, 152, 238, 239
Tassone, J.R., 129
Tastad, P.L., 171
Taylor, C., 3, 38, 74, 149, 238
Taylor, P., 129
TenEyck, T., 213
Tennant, C., 83*f*
Thagaard, P.R., 223
Thomas, S., 31*f*
Thompson, E.P., 52
Thompson, M.,
Thorgeirsdottir, S., 180
Tichenor, P.J., 27
Tilly, C., 10, 169
Toffler, A., 120, 121
Torgersen, H., 202, 214
Touraine, A., 79, 219, 220
Trebilcock, C., 23, 109
Trumbo, C., 82
Tuckman, B.W., 13
Tuomela, R., 13
Turing, A., 125, 137
Turkle, S., 124
Tushman, M.L., 109
Tversky, A., 218

U
Useem, B., 10, 12, 90, 169, 170

V
Vain, P., 190
Valach, V., 225
Valente, T.W., 24
Valsiner, J., 100
Van der Heiden, H-A., 51, 54, 80
Van der Pligt, J., 47, 55, 62
Vander Zanden, J.W., 90, 234
Van Dujin, J.J., 23
van Lente, H., 172, 177
Van Reenen, J., 122
Van Scheéle, F., 117
Van Tulder, M.,

Vargas, R., 161
Vaughan, D., 39, 167
Veblen, T., 42, 62
Veltri, G., 213
Vettel, E.J., 178
Vicari, S., 235
Victoroff, J., 221
Visschers, V.H.M., 47
von Hippel, F.N., 39

**W**

Wadham, M., 181
Wagner, W., 166, 189, 195
Waldenfels, B., 101, 150
Walker, C.C., 220
Walker, S.J., 44
Wall, P.D., 92, 226
Wang, Z., 76
Warren, S., 82, 83*f*
Waterman, R.H., 116
Watkins, E.S., 76
Watson, J., 180, 184
Watson, J.D., 184, 185
Weart, S.R., 33, 38, 41, 44, 68n28, 242, 245*t*
Weber, R., 146
Weinberg, A., 65
Weiner, N., 118
Weingart, P., 84, 162, 171, 172, 173, 193, 205
Weiss, L., 39
Weissman, S., 185
Weizenbaum, J., 124
Weldon, S., 207

Wenzel, U.J., 200
Wheelright, S., 241
Whiten, A., 241
Whitley, E., 142
Whitley, R., 161
Wigg, D., 51
Wigzell, H., 205
Wildavsky, A.B., 75, 167
Wildi, T., 36
Williams, R., 33, 35, 36, 43, 44, 67n17, 132*f*, 140
Williams, S., 80
Wilmut, I., 188
Winner, L., 11
Winston, B., 116*f*, 121, 135, 140, 141
Witherspoon, S., 80
Wolsink, M., 45, 53
Wood, S., 122
Woolgar, S., 140
Wormann, S., 193, 205
Wyatt, S., 134
Wynne, B., 48, 155, 166

**Y**

Young, R., 215
Yoxen, E., 185
Yates, F.A., 114

**Z**

Zah, R., 117
Zald, M.N., 10, 12, 90, 169, 170
Zaller, J.R., 26
Zeyer, A., 87
Zinder, N.D., 185

# Subject Index

## A
AAAS (American Association for the Advancement of Science), 11, 159
abductive logic, 224
abortion debate, 23, 94, 183, 190, 206, 235
absence of resistance, 3, 118, 135, 146
acceptance, 46f, 107, 145, 209–210: of nuclear weapons, 39; problem, 139–140, research, 214
accommodation, 14–16, 26–29, 58, 105, 106f, 112n9, 112n11, 213, 242; efforts, 109–110
accountability, 62; public, 40, 189–192, 250
acid rain, 24, 82
action repertoire, 10, 80, 94, 169–170
active resistance, 94–95, 130
activist groups, 10, 22–23, 183–192; education of, 79; mobilising, 14–16, 134–135; and nuclear protest, 63; and project formation, 12–13
acute pain, 223–227
adult nucleic transfer, 171, 188
age of disinhibition, 4, 173
age of inhibition, 220
agenda setting, 21–22
agricultural biotechnology, 88, 108, 183, 188–190
air pollution, 71, 74–75, 82
allergenic GM, 188–190
analogy between pain and resistance, 4, 91, 221–225, 230
analysis of resistance, 3, 6t, 97, 230–231
animal cloning, 195, 205
anthropogenic climate change, 43–44, 73, 82–83, 86, 126

anti-ballistic missile treaties, 52
Apple, 117, 120, 139–140, 246–247t
ARPANET, 117–118
artificial intelligence (AI), 12–13, 125–126
artillery guidance system, 155, 244n4
Asian countries, 57–58, 136
Asilomar conference, 184–185
assimilation and accommodation, 14–16, 29, 105, 112n9
atomic age, 9, 43, 168
atomic bombs: driving force of 35–36; production of nuclear power, 30–31; and physics, 31–32
atomic energy, 52–57, 191
Atomic Energy Act, 35
atomic society, 37–38
Atoms for Peace, 30, 34–35, 39, 43, 52–53
attention cycle, 83–84, 136
attention span: and pain, 228, public, 24, 150
attention to an issue, 20, 26–27, 187
attitude change, 84–86; and news cycles, 42–44, 156t
attitudes to nature, 70–71
attitudes to science: German, 153f, items measuring, 155, 158–161, 212
Austria: GM crop debate in, 188, 195 levels of resistance in, 50–51; and nuclear ambitions, 58, 196; levels of resistance in, 50–51; and nuclear ambitions, 58; nuclear reactors, 68n27; and radioactive dust from Chernobyl, 46–47;
Australia: awareness of climate change in, 83; green commitments in, 80;

## 292  Subject Index

and identity card scheme, 142;
nuclear ambitions in, 58–59;
nuclear reactors, 68n27
Autopoiesis, 73–74, 225
authority of science, 16, 169–172
automation debate, 119–122. *See also*
office automation
avoidance learning, 107–108, 112n8,
229
awareness of climate change, 41, 83
axial transition of society, 139

## B

ballistic missile, 52, 115, 120, 126
big science, 31–32, 180
bioethics, 188–190; boom, 180; loss of
authority of, 208–210
biological warfare, 191
biomedical biotechnology, 108, 160,
192–198
biotech sector, 177–178
biotechnology: and genetic engineering, 192, 196; movement,
175–179, 198–201; traditional,
177
blame avoidance, 164
blame shifting, 94, 110, 213
body image, 102, 228–230
Bove, José, 183
BRAER oil spill, 75, 78
Brazilian soya miracle, 203–204
breast cancer genes, 185, 189
BRCA genes, 189, 201n3
Brent Spar, 78
Britain: and club of nuclear powers,
33; committee for nuclear disarmament, 52; identity cards, 129;
nuclear electricity, 56; and cost
of nuclear power, 53; sustainability in, 85f; white heat of
technology speech, 43
BSA (British Science Association), 11,
159
Bt corn, 188, 200
Bulgaria: alteration of nuclear projects,
58; clean air act, 82; long-term
trends, 160; and nuclear power,
47, 50

## C

Caesium-137, 46, 49
Calgene, 179, 199
Cambodia: households with Internet,
133

Canada: atomic network, 33, 36; attitude towards computers, 131;
design decisions, 67n17; level of
resistance in, 51, 58; support of
nuclear power, 48
carbon footprint, 41, 86–88
catastrophising, 228
causal attribution, 93–95
CCTV cameras in public places,123,
128, 137
Celera, 181
CERN, 32, 35, 117–118
challenge of resistance, 29, 107, 111
change agent, 216–218
CHAOS computer club, 129–130, 141
Chernobyl: and climate change, 164;
disaster, 3, 12; global consequences of, 61; impact of
accident on public opinion,
45–46
China Syndrome, 40, 67n20,
chronic pain, 111, 226–230
civil defence, 52
civil disobedience, 169, 219–221, 236
civil nuclear power, 39, 43, 53, 64–65
civil society, 181–184
climate change, 41, 43, 64, 69–71,
83–87, 126, 164, 219
climate of opinion, 26, 41, 152
clustering, 152–153, 193; of installations in nuclear parks, 60
CND, 34, 52
codex alimentarius, 199, 201
cognitive deficit, 96, 146, 156, 158
collateral damage, 5, 164, 13
collective action, 9–12, 80, 128, 130
223, 243
collective attention, 20, 152, 239
collective learning, 16, 29, 63, 105–
106, 181
command and control, 73–74, 96
common good, 42, 173
common interest, 79, 149
common reference, 10, 169, 216, 239
common sense, 2, 12, 17, 27, 34, 62,
65, 69; new, 87–89, 99, 124,
166, 174, 210, 242
common understanding, 108, 149, 151,
168, 239
communicative action, 151, 239
community building, 10, 15, 19–20
complex system, 44, 168, 195
compliance, 16; costs, 201, 240; public, 240–241

computer: addiction, 141, 146; culture, 123–126; intelligence, 117; literacy, 138; phobia, 141
conformity pressure, 182–183, 190, 233, 238, 240–241
consensus conference, 108, 158
consumer movement, 69, 87–88
consumer organisations, 40, 87–88, 128, 135, 160, 203–204
court actions, 88, 203
critical mass, 25, 29n4, 118; of Unabomber, 130, 134–135, 141
crowd psychology, 238
cybernetics, 72, 118–120
Cyborg, 1, 41, 126
cycles of public attention, 36–37, 86
Czech Republic and Slovakia, 68n26; favoring nuclear power, 47

### D

data protection, 130, 142
dealing with newcomers, 25–28
deCode (Icelandic company), 180, 208
deficit concept, 91–94
developed/developing world, 57–58, 127
digital divide, 132–136, 140
dignity of resistance, 219–220
disarmament, 33, 52–54, 68n23
dissociation, 227–228
doctrine of suggestion, 238–239
Dolly the sheep, 28, 171, 175, 187–189, 195–198, 205, 249t
Dr. Strangelove, 38, 40
Dutch: atomic network, 33, 35; opinion polls, 154; sustainable lifestyle, 87
duty to resist, 220–221

### E

Eastern Europe: atomic network, 33; nuclear ambitions
ecological awareness, 7, 76, 81
economic prospect, 189–191
economics of nuclear power, 36, 64
effects of resistance, 55–64
electrosmog, 127–128, 144
elite consensus, 59–60, 63
embryonic stem cell research, 104, 205–206, 249t
emergent technologies, 147, 274, 214
emotion and pain, 4, 224
endogenous pain control (modulation), 229–230

energy too cheap to meter, 35, 43, 107. 168
enriched uranium, 35–36
environmental awareness, 81, 89
environmental impacts, 78, 109
environmental movement, 30, 63, 79, 130
environmental risk assessment, 78
epidemiological studies, 88, 127, 144
error, 99–101
ethos of resistance, 16
eugenics, 12, 176, 184, 191
EURATOM, 38
Europe: anti-nuclear mobilisation in, 51, 104; awareness in, 83; biotechnology across, 21, 27–28, 178–179; digital divide in, 135–136; effects of Chernobyl on, 46–47; encouraging disarmament in, 56; green issues in, 76, 89; households with Internet, 133; interest in science and technology in, 157–162; Peace Movement in, 52; post-material values in, 80; public opinion in, 2, 131; TMI, 45–47
Europe vs. Facebook, 129–130
European Federation of Biotechnology (EFB), 176
evolutionary psychology, 184, 191, 237
expert consensus, 25–26
external attribution of resistance, 100
expectations: hyping, 171–172; modulating, 56–57; sociology of, 177

### F

Facebook: privacy issues, 122–123, 129, 151
factors of resistance, 216–217, 231
fait accompli, 3, 33, 36, 168, 191–192
false consciousness, 5–6, 121, 233
familiarity, 164–165, 168
field theory, 218
Flavr Savr tomato, 179, 187
focus of attention, 101–102, 145, 187
food: safety, 182, 190, 199, 201, 205; security, 179, 182, 190, 199
force field, 217–218
form of life, 6–7
fossil fuels, 34–35, 44, 53, 56
frame of mind, 242
frame of reference, 28, 228, 239
framing of resistance, 98, 102

## Subject Index

France: atomic network, 33, 35; nuclear project, 37–38
freedom of speech, 80, 128, 149
Friends of the Earth, 63, 77, 182
fuel cycle, 35–38, 53–54, 61
functional analogy, 4, 223–232
functionality of resistance, 91, 110, 216
functionally equivalent, 59, 155, 165, 233
future orientation, 72–73
Fukushina, 170
futility, 96, 121–122, 218–219

## G

gate control, 227–228
gene therapy, 177, 189, 190–191
Genentech, 112n10, 178–179, 185
general reisistology, 2, 92, 111n1, 214, 216, 221, 231
genetic engineering, 6–8, 10–11, 16, 19, 23, 27, 29, 72, 107, 110–111, 191–194, 208, 248t; history of 175–177; mobilising for and against, 177–184
genetic enhancement, 183–192
genetic modification, 177, 183–184, 194–195, 198–199, 211n8
genetic testing, 182, 190, 195
genome project, 180–182, 210n2, 214
Gestalt theory, 108, 112n11
global warming, 21–22, 26–27, 71, 82–84
GM crops, 108–110, 166–167, 182–184, 187–188, 190–205, 210n6, 211n8, 218
GM food debate, 187–188, 195, 202
GM soya, 187, 189, 200–204
GM wheat, 108, 198
Gnosis, 22, 124
Goettinger manifest, 32, 36
golden rice, 179, 190
Google: file sharing, 118; privacy issues, 122–123
gradient of resistance, 17, 29n3
grand narratives, 23, 94
graphite moderation, 36
green biotechnology, 108, 110, 197, 200, 203
green consumer, 85–89
green investors, 86, 88
green movement, 52, 76–77, 81, 87, 170
green parties, 81, 86, 130

greenhouse effect, 82, 84f
grey goo, 13, 213
group dynamics, 13, 29n5,
group formation, 9, 239
growing pains, 222–223
gurt technology, 108, 198, 204–205
gut feelings, 167, 208

## H

hacker: culture, 117; movement, 129–134
hardware and software, 139–144
health hazard, 76–78
heavy water, 36, 67n17
herbicide tolerant, 202
Hiroshima and Nagasaki, 31–34, 46, 66n3
Holy Spirit, 23
homo faber, 163, 168
human activity, 11, 173, 224
human brain project, 9–11, 28, 213
human cloning, 24, 188–189, 209
human dignity, 22
human genome project, 180–184
human rights, 23, 209
hunger strikes, 236
hyperbole, 64–65, 118, 144, 177

## I

IBM, 116–120, 131, 135, 139, 143–144
Iceland: entrepreneurialism in, 179–180; internet access in, 133, 136; projects in 207–208
ID cards, 135, 142
identity theft, 128, 142
image of science, 162
impact of resistance, 29, 63, 140
India: correlation between knowledge and attitude, 162; farming in, 189; nuclear movement in, 17, 32–33, 50, 58; software development outsourcing in, 116
information society, 119, 168–169
insight learning, 108, 229
institutional learning, 62, 65, 135, 144–147
intelligence gathering, 122
intensity of resistance, 58–59, 63
interest groups, 24–25, 237
internal attributions, 93, 213
internet access, 132–136, 140
Iron Curtain, 56, 66n10
issue cycles, 24–25, 81–84, 118–121
issue entrepreneurs, 95, 213

issue salience, 81–84, 250
Israel: Google Earth Street View challenge, 123; nuclear reactor in, 68n27; policy of opacity in, 33; sponsors of protest in, 234
Italy: effects of Chernobyl in, 46–50; professional activism, 78; project alterations in, 58–59; public opinion in, 45; science news, 160

## J

Japan: anti-nuclear movement in, 56; Fukushima nuclear reactor, 49–50; Google Earth Street View challenge, 123 levels of resistance, 50–51; nuclear attack, 31–32; privacy concerns, 129, 134; projects in, 14, 58–59; public opinion in, 47–48
jeopardy, 96, 219
joint intentionality, 13–16, 28

## K

knowledge gap hypothesis, 132, 197
knowledge society, 180, 220
Kyoto protocol, 83, 86

## L

Latin America, 57, 189, 200
law of kinship, 96, 220
law of the suppression, 121, 135
learned societies, 11, 159, 169, 213
learning difficulties, 65, 74, 81, 91–92, 95, 100, 105, 110, 173–174, 230f
levels of analysis, 224–225
Libya and Iraq, 33, 58
life science vision, 179, 198–201
line extension, 107, 112n10
linear model, 24, 176
Lithuania: closure of power plant in, 51; electricity production in, 35; nuclear electricity in, 56; opinion polls in, 47–48
local resistance, 100, 143
loss of control, 20, 122
Lotus, 142
Luddites, 130, 146

## M

Mac(Intosh), 117, 139–140, 146
MAD strategy, 33, 68n23
mainframe computers, 115–117, 123, 136, 146

majority view, 26, 240m, 242
managing change, 111n2, 223
Manhattan project, 31–32, 66n3, 180
mass communication, 26, 227
mass media, 18–28, 243, 244n4; British, 81–82; database 249–250
mechanistic worldview, 71–72
media events, 82, 161
medicalisation of science, 190–191
mentality of resistance, 99–100
Microsoft, 116–117, 134, 139–140
Milgram experiments, 241, 244n3
mindset, 154, 168, 2212, 228–231, 232n3, 239, 242
minority influence, 242–243
missile defence systems, 52
mobile phones, 127, 144
mobilisation process, 213, 224
mode I to mode II, 105–106, 110, 111
modern biotechnology, 177–183, 193
Monarch butterfly, 188
Monsanto, 106–107, 179–180
moral panics, 126–128
moratorium: on GM, 198–199; on nuclear power, 55–57
motor learning, 105, 145–146
mutually assured destruction, 33

## N

Naming and framing the issues, 37, 235
nanotechnology, 9, 13, 15, 23, 63, 87, 131, 213–214. *See also* synthetic biology
narratives, 23–24, 63, 93–94, 98–99
natural law, 220–221
Nature (as in natural world), 69–70
Nature/nurture frame, 184, 191–192, 250
nautical ecstasy, 4–5, 173
New Zealand and Australia, 58, 59
news routines, 235–236
NIABY attitudes, 6t, 45
NIMBY attitude, 6t 45–53, 67n19, 79, 107
nocipation, 99, 228
nonliner, 133–134
non-users, 132–134
Novartis, 179, 198, 200, 210n5, 210n7
nuclear accidents, 78–80; impact on public opinion, 44–50
nuclear age, 37, 42
nuclear ambitions, 58–60, 65, 68n25
nuclear capability, 53–54

## Subject Index

nuclear complex, 32, 39
nuclear industry, 32, 39–41, 61; British, 61
nuclear parks, 60
nuclear phobia, 62, 68n28, 216
nuclear priesthood, 32, 62
nuclear renaissance, 42, 50
nuclear safety, 61, 64
nuclear society, 28, 37
nuclear warfare, 33, 52
nuclear waste disposal, 166, 218
nuclear winter, 34, 43, 52, 82
nuclear phobia, 62, 68n28, 216

## O

obedience to authority, 238 241
object of desire, 137–139, 143
objective reality, 19–20, 240–241
objective risks, 165, 218
office automation, 2, 137–140, 145, 147n6. *See also* automation debate
oil crises, 40–44
oil spill, 74–75, 78
OncoMouse, 192, 195
open access, 129–130, 141, 207
opinion measurement, 148–155
opinion polling, 152–158
opportunity costs, 3, 64, 97
opposition, 45–48, 54–55, 64, 95, 170–171
organic farmers, 183
overcome resistance, 107, 217–218

## P

pain: affects, 103; analogy, 3, 105, 111, 213, 223–231; avoidance, 107, 228; chronic, 113, 113n2; experience, 222, 225–229; model, 69, 93; is modulated, 92; relief, 112n8; resistance, 3–4, 91–92, 97; response, 228; threshold, 228
Pandora's Box frame, 190–195, 250
paperless office, 107, 138, 145
paradox of pain, 227
participatory design, 142–147
patents, 178–179, 189, 198, 206, 237
patient groups, 181–182
pathos and ethos, 19
Peace Movement, 52
peer pressure, 240–241
personal computer, 120, 137
perspectives on resistance, 216–222

persuasive design, 146
perversity, 96, 121, 218–219
phantom pain, 226–227
phantom risk, 144–147, 167
pharmaceutical industry, 107, 182
phobia. *See* computer phobia *and* nuclear phobia
pirate party, 130
policy makers, 15, 138, 142
political elites, 63–64
politics by other means, 166, 233
popular culture, 36–37, 45
popular science, 118–119, 138, 161
positive attitudes, 158, 161–162
post-industrial society, 120, 220
precautionary principle, 189–190, 202
prejudice, 72, 100, 111n1, 156, 167, 215, 243n2
pressurised water, 44, 67n17
pretend play, 109, 112n9
primary learning, 105
Prince Charles, 82, 187–188, 213–214
privacy actions, 147n2
privacy protection, 123, 129
private patronage, 12, 170, 178
probabilistic risk assessment, 53, 107
product safety, 69, 87–88
professionalisation of science communication, 170, 172
progress frame, 172, 192
proliferation anxieties, 37, 58–60
prophecy fails, 84, 172
protection of privacy, 128, 134–135
psychology of things, 3, 66n10
public accountability, 40, 189–192, 250
public attention, 10, 14, 16–17, 20, 24–25, 28, 80, 84–86, 214, 234–235
public attitudes, 221, 27, 81, 84
public concerns, 60, 144, 158
public consultation, 60–61, 68n29, 108
public controversy, 2, 14, 19, 27, 113n11,
public debates, 68n28, 107, 179–180, 184–185, 201
public deliberation, 15, 62, 65, 213
public discourse, 19, 23, 42, 92–93, 103–107, 183, 192–197
public engagement, 29–30, 108, 155–159
public interest, 40, 162, 190
public opinion measurement, 151–154
public opinion process, 150–152

## Subject Index 297

public perception, 194–198
public relations, 14–15, 39, 151, 158, 161, 172
public sentiment, 21, 26, 40, 45, 49, 58, 60, 204, 212
public sphere, 28–29, 60, 148–155, 168, 243
public understanding of science, 156–161, 174n2

## Q

quality of coverage, 82
quality of working life, 122, 134, 141

## R

radiation poisoning, 32, 65n1
radioactive fall-out, 32, 45–50, 82
radioactive waste, 35, 53, 60
rain forests, 24, 204
raise attention, 13, 230, 236
raise awareness, 128, 130
raising expectations, 136, 173
rational choice, 99, 235, 240
rebound effect, 45–47, 121
recalcitrant public opinion, 2, 91, 146–148, 155, 168, 209, 212
recipation, 99–101
recombinant DNA, 176–177, 185
recursive effects, 92–96
red and green biotechnology, 197
reduce complexity, 20–22
refuseniks, 132–134
regulatory capture, 49, 61, 209
religious groups, 183
renewable energy, 29, 77, 206
repertoire of actions, 12, 109, 112n9
repetitive strain injury (RSI), 127, 144, 221
representation: of the future, 139–140; of resistance, 100
reproductive cloning, 209
resistance as pain, 3–4, 213
resistance is functional, 170–174, 213
resistance is registered, 98–100
resistance to change, 5–6, 19, 107
resistology, 2, 111n1, 214–216, 221, 231
resource mobilisation, 177–178, 213, 233–235
response to resistance, 90–91, 231
response to the challenge, 107, 111
rhetoric of reaction, 121, 218–219
rhetoric of resistance, 96
Rio summits, 78–79, 82, 86

rise of environmentalism, 76–81
risk assessment, 107, 163–168, 208, 213
risk aversion, 5, 106, 202
risk communication, 52, 166
risk compensation, 165
risk management, 163–165
risk perception, 30, 62, 108, 148, 163–168
risk society, 61, 65, 164–165
robots, 115, 125, 137
Rockefeller Foundation, 199, 205
root metaphors, 71, 138, 215, 221
roundup ready, 179
Royal Society of London, 11, 159, 169, 174n4, 211n8
Royal Society for the Protection of Birds (RSPB), 76
runaway frame, 189

## S

safety culture, 53
safety regimes, 42, 65, 78
salience figures, 81–85
science as a social movement, 12, 169
science centres, 159, 214
science fiction, 159
science journalism, 62–63, 159, 172
science literacy, 108, 155–161
science news, 27, 47, 159–160, 190, 206
science writers, 45, 65, 66n11
scientific authority, 169–174
scientific literacy, 108, 156–163
scientific management, 119, 138
scientific reputation, 171
scientific worldview, 74
secondary learning, 105
self-active systems, 73–75
self-monitoring, 223–230
self-observation, 223–225
sense making, 91, 104–105, 108, 121
sequencing and mapping, 181, 189, 207
Sierra Club, 63, 76–77
*Silent Spring*, 76–77
Silicon Valley, 115, 179
similarity between pain and resistance, 221–222
singularity, 126
sleeper effect, 97, 242
social conformity, 26, 241
social engineering, 5, 107, 210
social identities, 9, 235
social influence, 237–243
social intelligence, 239–241

social milieus, 17, 236–237
social mobilisation, 168–174, 213, 221
social movement organisations (SMOs), 10, 14, 16, 19–24, 169, 213, 233
social networks, 129–130
social norms, 240–241, 243n2
social representation, 17, 29n3, 36, 93, 113n11, 155, 235
social values, 9, 79
Social psychological liquidation of public opinion, 154–155
sociological imagination, 3–5
sociology of expectations, 172, 177
soft path, 40–42, 54
software design, 145–146
software development, 116
Solow's productivity paradox, 143–144
somatic stem cells, 190, 198, 205–206
sound science, 20, 157, 182, 189
South Africa: nuclear projects, 33, 58, 65
South America: digital divide in, 133; seed operations in, 199
South Korea: nuclear electricity in, 56; project alterations in; 58, 59f
soya miracle, 203–204
space race, 33, 119–120
spiral of silence, 25–26, 28
Star Wars, 33, 52
stem cell cloning, 189, 195–197, 205
stem cell debate, 188–193
stem cell research, 191–193, 198, 204, 249t; adult, 206; embryonic, 104, 190, 209
strategic adaptation, 15, 29, 110, 198–200
strategic bomber command, 3, 38, 41, 67n13
structural learning, 108–110, 213
Supercollider project, 14, 170
suppression of radical potential, 121, 135
surveillance society, 122–123
Swiss referendum, 197–198
synthetic biology, 9, 13–14, 168, 213–214. See also nanotechnology
system failure, 49, 64, 143
system thinking, 73

T

tactical retreats, 97. 107, 110
tampering with nature, 167, 195
target intelligence, 155, 244n4
technical fix, 36–39
technocracy, 41–42, 62, 65, 123

technocratic elite, 37, 62, 120
technological determinism, 11, 36, 140
technologies of humility, 62, 108, 158, 166
techno-science, 1, 2, 5, 7, 16–17, 20–21, 27–28, 42, 63, 90, 93, 98–99, 103, 107, 109, 168–170, 174, 213, 220, 225, 237
Tepco and the regulatory authority, 49–50
terminator technology, 204–205
tertium comparison, 223–225
Test Ban Treaty, 33, 51
theory of resistance, 215–217, 226, 231
TMI and Chernobyl, 47–48, 54–55, 58
totemic resistance, 96
trade unions, 37, 88, 134–135, 141, 220
transatlantic puzzle, 183, 195
transformative potential, 213, 222
Turing Test, 125, 137–138
Two cultures, 161–163
types of learning, 104–105

U

UK government, 142, 187
unabomber, 95, 130
Urban information systems, 137
USSR: awareness of IT in, 135; military threat, 66n5; nuclear electricity, 56; nuclear projects, 33–35

V

VDT units, 143–144
venture capital, 15, 179, 206
virtual reality, 124–125
vita contemplative, 70, 163

W

watershed years, 191–193
Waldsterben, 74–75, 82, 166
weapons capability, 56, 68n27
Wellcome Trust, 181, 207
Windscale Fast-Breeder Project, 44, 47–48, 64
word processing, 2, 138, 143, 145
work and leisure, 139
work and privacy, 121, 134

X

Xerox Parc, 142, 145

Y

Yuck-factor, 167, 188, 208, 209